Leech Biology and Behaviour

LEECH BIOLOGY AND BEHAVIOUR

Leech Biology and Behaviour

Volume III   Bibliography

Roy T. Sawyer

BIOPHARM (UK) LIMITED
2/8 Morfa Road, Swansea SA1 2HT, UK

CLARENDON PRESS · OXFORD
1986

Oxford University Press, Walton Street, Oxford OX2 6DP

Oxford New York Toronto
Delhi Bombay Calcutta Madras Karachi
Petaling Jaya Singapore Hong Kong Tokyo
Nairobi Dar es Salaam Cape Town
Melbourne Auckland

and associated companies in
Beirut Berlin Ibadan Nicosia

Published in the United States by
Oxford University Press, New York

British Library Cataloguing in Publication Data

Sawyer, Roy T.
Leech biology and behaviour.
Vol. 3: Bibliography
1. Leeches
I. Title
595.1'45   QL391.A6
ISBN 0-19-857623-4

Library of Congress Cataloging in Publication Data

Sawyer, Roy T.
Leech biology and behaviour.
Includes bibliographies and indexes.
Contents: v. 1. Anatomy, physiology, and behaviour—
v. 2 Feeding biology, ecology, and systematics—
v. 3. Bibliography.
1. Leeches—Collected works.   I. Title.
QL391.A6S28   1986   595.1'45   84-25532
ISBN 0-19-857377-4 (v. 1)
ISBN 0-19-857622-6 (v. 2)
ISBN 0-19-857623-4 (v. 3)

Printed in Great Britain
by St Edmundsbury Press,
Bury St Edmunds, Suffolk

This bibliography includes those papers concerned with some aspect of the biology of leeches published since the last comprehensive bibliography by Autrum (1939c). In addition to items not included in Autrum, this bibliography includes from Autrum a few selected items of especial importance. Titles are in their original language except for Slavonic and non-European languages. Items are not generally annotated except for species clarifications and a few other instances. To assist workers without access to a large library, selected items include the Biological Abstract reference number (e.g 1976BA:34567 indicates 1976 Biological Abstract number 34567).

The bibliography is comprised of two parts, those dealing with (i) Branchiobdellida and (ii) Euhirudinea and Acanthobdellida. A comprehensive bibliography of papers dealing with the biology of branchiobdellids has never been published heretofore and merits separate compilation.

'Better know nothing than half-know many things! Better be a fool on one's own account, than a sage on other people's approbation! I . . . go to the basis: . . . What matter if it be great or small? If it be called swamp or sky? A hand-breadth of basis is enough for me, if it be actually basis and ground! . . . A handbreadth of basis: thereon can one stand. In the true knowing-knowledge there is nothing great and nothing small.'

'Then thou art perhaps an expert on the leech?' asked Zarathustra; 'and thou investigatest the leech to its ultimate basis, thou conscientious one?' 'O Zarathustra', answered the trodden one, 'that would be something immense; how could I presume to do so!

That, however, of which I am master and knower, is the *brain* of the leech: that is *my* world!'

F. Nietzsche, *Thus Spake Zarathustra*,
c. 1883

Contents

VOLUME III BIBLIOGRAPHY

Branchiobdellida

Anderson, D.T. (1971). Embryology. In *Aquatic oligochaeta of the world* (eds. Brinkhurst, R.D. and Jamieson, B.G.M.) pp. 73–103. Oliver and Boyd, Edinburgh.

André, J. (1962). Contribution à la connaissance du chondriome. Étude de ses modifications ultrastructurales pendant la spermatogenèse. *J. ultrastruct. Res. Suppl.* **3**, 1–185. [*Branchiobdella pentodonta.*]

Berry, J.W. and Holt, P.C. (1959). Reactions of two species of Branchiobdellidae (Oligochaeta) to high temperatures and low oxygen tensions. *Va. Agr. exp. Stat., Virginia Polytechnic Instit. Tech. Bull.* **141**, 1–11.

Bilek, F. (1908). Über den feineren Bau des Gefässsystems von *Branchiobdella*. *Zool. Anz.* **33**, 466–73.

Bishop, J.E. (1968). An ecological study of the branchiobdellid commensals (Annelida: Branchiobdellidae) of some mid-western Ontario crayfishes. *Can. J. Zool.* **46**, 835–43.

Blackford, S. (1966). A study of certain aspects of the ecology and morphology of branchiobdellid annelids epizooic on *Callinectes sapidus*. Unpublished Senior Thesis, Newcomb College, Tulane University. [Intolerant of salt water.]

Blanchard, E. (1849). [In Gay, C. (1849).] Historia fisica y politica de Chile. *Zoologia* **3**, 51–2. (Page 51, *Temnocephala* gen. of (?) Branchiobdellida, body oblong; cephalic divided; eyes 2; only and dubious record of branchiobdellid in southern Hemisphere.)

Bogojawlensky, N. (1922). Zum Studium der Spermatogenesis bei *Branchiobdella*. *Rev. Zool. Russe* **3**, 57–70. [Russian.] [Pages 70–3 German summary.]

Bolsius, H.S.F. (1894a). Communication préliminaire sur certains détails de l'anatomie d'*Astacobdella branchialis*. *Extr. Ann. Soc. Sci. Bruxelles* **18** (part 1), 27–32.

—— (1894b). Sur l'anatomie de la *Branchiobdella parasita* et de la *Mesobdella gemmata*. *Ann. Soc. Sci. Bruxelles* **18**, 57–61.

Bondi, C. (1962a). Osservazioni sull'apparato maschile in *Branchiobdella pentodonta* Whitman. *Revista Biol.* **55**, 7–36.

—— (1962b). Indagini morfologiche sugli spermi di *Branchiobdella pentodonta* Whitman, eseguite mediante microscopio elettronico. *Rivista Biol.* **55**, 177–86.

—— (1962c). Ulteriori ricerche sulla morfologia del nemasperma di *Branchiobdella pentodonta* Whitman eseguite mediante microscopio elettronico. *Boll. Zool.* **29**, 433–52.

—— (1963a). Osservazioni sulla morfologia e sulla fine struttura del nemasperma di *Branchiobdella pentodonta* Whitman esaminate al microscopio elettronico. *Rivista Biol.* **56**, 21–48.

—— (1963b). Note sulla spermiogenesi in *Branchiobdella pentodonta* Whitman. *Rivista Biol.* **56**, 129–60.

—— (1963c). Sulla fine struttura del nemasperma di *Branchiobdella pentodonta* Whitman indagata mediante microscopio elettronico. *Boll. Soc. ital. Biol. sper.* **34**, 576–8.

—— (1963d). Indagine al microscopio elettronico sull'origine e sviluppo della struttura nastriforme della coda dello spermio di *Branchiobdella pentodonta* Whitman. *Arch. Zool. Ital.* **48**, 1–12.

—— and Facchini, L. (1972). Observations on the oocyte ultrastructure and vitellogenesis of *Branchiobdella pentodonta* Whitman. *Acta embryol. exp.* **2**, 225–41.

—— and Farnesi, R.M. (1976). Electron microscope studies of spermatogenesis in *Branchiobdella pentodonta* Whitman (Annelida, Oligochaeta). *J. Morph.* **148** (1), 65–88. [1976BA:8435.]

Braun, I.F.P. (1805). *Systematische Beschreibung einiger Egelarten*. Berlin.

Brown, G.G. (1961). Some ecological studies of the Branchiobdellidae found in Sinking Creek, Giles County, Virginia. Virginia Polytechnic Institute. Unpublished Master's Thesis. (Conclusions summarized in Hobbs *et al.* (1967).)

—— (1962). Some ecological studies of the Branchiobdellidae. *Virginia J. Sci.* **13**(4), 241–42 [Abstr.]

Canegallo, M.A. (1928). Una nuova specie di Branchiobdella — *Branchiobdella italica. Atti Soc. ital. Sci. Nat. Museo Civ. Milano* **67**, 214–24.

Causey, D. (1955). Branchiobdellidae in Arkansas. *Proc. Arkansas Acad. Sci.* **7**, 43–6.

Dahm, A.G. (1959). Kräftigeln *Branchiobdella* — en parasitisk oligochaet i den svenska faunan. *Fauna och Flora* **54**, 60–8.

D'Angelo, L. (1963). Osservazioni sui coelomociti in *Branchiobdella pentodonta* Whit. *Arch. Zool. Ital.* **48**, 329–40.

—— (1965a). Osservazioni sull'apparato riproduttore femminile in *Branchiobdella pentodonta* Whitman. *Arch. Zool. Ital.* **50**, 29–40.

—— (1965b). Osservazioni al microscopio elettronico sulla spermateca in *Branchiobdella pentodonta* Whitman. *Arch. Zool. Ital.* **50**, 325–8.

Del Roscio, D. (1962). Osservazioni e ricerche sulla *Branchiobdella pentodonta* Whit. parassita del gambero d'acqua dolce *Potamobius pallipes* Lereb. *Rivista Biol.* **55**, 73–88.

Dorner, H. (1865). Ueber die Gattung *Branchiobdella*. *Z. wiss. Zool.* **15**, 464–94.

Ellis, M.M. (1912). A new discodrilid worm from Colorado. *Proc. US nat. Mus.* **42**, 481–6.

—— (1918). Branchiobdellid worms (Annelida) from Michigan crawfishes. *Trans. Am. microsc. Soc.* **37**, 49–51.

—— (1920 (for 1919)). The branchiobdellid worms in the collections of the United States National Museum with descriptions of new genera and new species. *Proc. US nat Mus.* **55**, 241–65.

Evans, C.R. (1939). The Branchiobdellidae (Annelida) on crayfishes of Champaign County, Illinois. *J. Parasitol.* **25**, 448.

Facchini, L. (1970). Prime osservazioni al M.E. sulla vitellogenesi in *Branchiobdella pentodonta* Whit. (oligochete). *Boll. Soc. ital. Biol. sper.* **47**(13), 381–2.

Farnesi, R.M. (1973). Ultrastructural examination of the cuticle in *Branchiobdella pentodonta* Whit. *Boll. Zool.* **40**(3–4), 371–73. [Italian summary.]

—— and Tei, S. (1975). Indagini istochimiche ed ultrastrutturali sul bozzolo di *Branchiobdella pentodonta* Whit. (Anellide, Oligochete). *Boll. Soc. ital. Biol. sper.* **51**(18), 1184–89. [In Italian.] [Cocoons.]

—— and Vagnetti, D. (1972). Morphological investigations on the parietal musculature in *Branchiobdella pentodonta* Whitman at the optical and ultrastructural levels. *Boll. Zool.* **39**, 343–50.

—— —— (1973). Il clitello in *Branchiobdella pentodonta* Whit.: Idagini istochimiche ed ultrastrutturali. *Riv. Idrobiol.* **12**(1), 21–32. [In Italian; English summary.] [1976BA:37682.]

—— —— (1975). The fine structure of the myoneural junctions in the body wall muscles in *Branchiobdella pentodonta* Whit. (Annelida, Oligochaeta). *Anat. Rec.* **182**(1), 91–102.

—— Marinelli, M., Tei, S. and Vagnetti, D. (1981). Morphological and ultrastructural aspects of *Branchiobdella pentodonta* Whit. (Annelida, Oligochaeta) suckers. *J. Morph.* **170**, 195–205.

—— (1982*a*). The body surface of *Branchiobdella pentodonta* Whit. (Annelida, Oligochaeta) examined by scanning microscopy. *J. Morph.* **174**, 17–24.

—— (1982*b*). Ultrastructural aspects of mechano- and chemoreceptors in *Branchiobdella pentodonta* (Annelida, Oligochaeta). *J. Morph.* **173**(2), 237–45. [1983 BA:49675.]

Ferraguti, M. and Lanzavecchia, G. (1977). Comparative electron microscopic studies of muscle and sperm cells in *Branchiobdella pentodonta* Whitman and *Bythonomus lemani* Grube (Annelida, Clitellata). *Zoomorphologie* **88**, 19–36.

Franzèn, A. (1962). Notes on the morphology and histology of *Xironogiton instabilia* (Moore, 1893) (Fam. Branchiobdellidae) with special reference to the muscle cells. *Zoologiska Bidrag Fran, Uppsala* **35**, 369–83. Published 31.12.62.

Freeman, C.J. (1963). An anatomical study of the posterior nephridial systems in four species of the Branchiobdellidae. Unpublished Master's thesis, Virginia Polytechnic Institute.

Gelder, S.R. (1978). Observations on selected free-living and symbiotic polychaetes and oligochaetes (Annelida). Ph.D. thesis, University of Leeds, England.

Georgévitch, J. (1955). Sur les Branchiobdellides des écrevisses du Lac Dojran. *Acta Mus. Maced. Sci. Nat.* Skopje **2**(10/21), 199–221.

—— (1957). Les Branchiobdellides de Yougoslavie. *Bull. Acad. Serbe Sci. Cl. Sci. Math. Nat., N.S. 18, Sci. Nat.* **5**, 5–25.

Goodnight, C.J. (1939). Geographical distribution of North American branchiobdellids. *J. Parasitol.* Suppl. **25**, 11. (Abstr.)

—— (1940a). New records of branchiobdellids (Oligochaeta) and their crayfish hosts. *Report Reelfoot Lake Biol. Station* **4**, 170–1.

—— (1940b). The Branchiobdellida (Oligochaeta) of North American crayfish. *Ill. Biol. Monog.* **17**(3), 1–75.

—— (1941). The Branchiobdellidae (Oligochaeta) of Florida. *Trans. Am. microsc. Soc.* **60**(1), 69–74.

—— (1942). A new species of branchiobdellid from Kentucky. *Trans. Am. microsc. Soc.* **61**(3), 272–3.

—— (1943). Report on a collection of branchiobdellids. *J. Parasitol.* **29**(2), 100–2.

Grubda, E. and Wiezbecka, J. (1969). The problem of parasitism of the species of the genus *Branchiobdella* Odier 1823. *Pol. Arch. Hydrobiol.* **16**, 93–104.

Gruber, A. (1883). Bemerkungen über die Gattung *Branchiobdella*. *Zool. Anz.* **6**, 243–8.

Hall, M.C. (1914). Descriptions of a new genus and species of the discodrilid worms. *Proc. US nat. Mus.* **48**, 187–93.

Henle, J.F.G. (1835). Über die Gattung *Branchiobdella*. *Arch. Anat. Physiol. wiss. Medizin* 574–608.

Hobbs, H.H. Jr, Holt, P.C. and Walton, M. (1967). The crayfishes and their epizootic ostracod and branchiobdellid associates of the Mountain Lake, Virginia, region. *Proc. US nat. Mus.* **123**, 1–84. [Branchiobdellids:47 ff.]

—— and Villalobos, A. (1958). The exoskeleton of a freshwater crab as a microhabitat for several invertebrates. *Virginia J. Sci.* **9**(4), 395–6. (Abstr.) [Branchiobdellid on freshwater crab.]

Hoffman, R.L. (1963). A revision of the North American annelid worms of the genus *Cambarincola* (Oligochaeta: Branchiobdellidae). *Proc. US nat. Mus.* **114**, 271–371.

Holt, P.C. (1949). A comparative study of the reproductive systems of *Xironogiton instabilius instabilius* (Moore) and *Cambarincola philadelphica* (Leidy). (Annelida, Oligochaeta, Branchiobdellidae.) *J. Morph.* **84**(3), 535–72.

—— (1951). The genera *Xironodrilus* and *Pterodrilus* in North America, with notes on other North American genera of the family Branchiobdellidae. Unpublished dissertation, University of Virginia.

—— (1953). Characters of systematic importance in the family Branchiobdellidae (Oligochaeta). *Virginia J. Sci.* **4**(2), 57–61.

—— (1954). A new branchiobdellid of the genus *Cambarincola* (Oligochaeta, Branchiobdellidae) from Virginia. *Virginia J. Sci.* **5**(3), 168–72.

—— (1955). A new branchiobdellid of the genus *Cambarincola* Ellis, 1912 (Oligochaeta, Branchiobdellidae) from Kentucky. *J. Tenn. Acad. Sci.* **30**(1), 27–31.

—— (1960*a*). The genus *Ceratodrilus* Hall (Branchiobdellidae, Oligochaeta) with the description of a new species. *Virginia J. Sci.* **11**(2), 53–77.

—— (1960*b*). On a new genus of the family Branchiobdellidae (Oligochaeta). *Am. Midl. Nat.* **64**(1), 169–76.

—— (1963). A new branchiobdellid (Branchiobdellidae: *Cambarincola*). *J. Tenn. Acad. Sci.* **38**(3), 97–100.

—— (1964). A new branchiobdellid (Annelida) from Costa Rica. *Tulane Stud. Zool.* **12**(1), 1–4. [On freshwater pseudothelphusid crab.]

—— (1965*a*). On *Ankyrodrilus*, a new genus of branchiobdellid worms (Annelida). *Virginia J. Sci.* **16**(1), 9–21.

—— (1965*b*). The systematic position of the Branchiobdellidae (Annelida: Clitellata). *Syst. Zool.* **14**(1), 25–32.

—— (1967*a*). *Oedipodrilus oedipus*, n.g., n. sp. (Annelida, Clitellata: Branchiobdellida). *Trans. Am. microsc. Soc.* **86**(1), 58–60.

—— (1967*b*). Status of the genera *Branchiobdella* and *Stephanodrilus* in North America with description of a new genus (Clitellata: Branchiobdellida). *Proc. US nat. Mus.* **124**, 1–10.

—— (1968*a*). The Branchiobdellida: epizoötic annelids. *Biologist* **50**(3–4), 79–94.

—— (1968*b*). New genera and species of branchiobdellid worms (Annelida: Clitellata). *Proc. biol. Soc., Wash.* **81**, 291–318.

—— (1968*c*). The genus *Pterodrilus* (Annelida: Branchiobdellida). *Proc. US nat. Mus.* **125**, 1–44.

—— (1969). The relationships of the branchiobdellid fauna of the southern Appalachians. *The distributional history of the biota of the southern Appalachians*. Part I: *Invertebrates* (ed. P.C. Holt) Vol. 1, pp. 191–219. *Research Division Monograph, Virginia Polytechnic Institute*.

—— (1973*a*). Epigean branchiobdellids (Annelida: Clitellata) from Florida. *Proc. biol. Soc., Wash.* **86**(7), 79–104.

—— (1973*b*). A summary of the branchiobdellid (Annelida: Clitellata) fauna of Mesoamerica. *Smithson. Contr. Zool.* **142**, i–iii, 1–40.

—— (1973c). Branchiobdellids (Annelida: Clitellata) from some eastern North American caves, with descriptions of new species of the genus *Cambarincola*. *Int. J. Speleol.* **5**, 219–56. [On cave isopods; good paper.]

—— (1973d). A free-living branchiobdellid (Annelida: Clitellata)? *Trans. Am. microsc. Soc.* **92**(1), 152–3.

—— (1973e). An emended description of *Cambarincola meyeri* Goodnight (Clitellata: Branchiobdellida). *Trans. Am. microsc. Soc.* **92**(4), 677–82.

—— (1974a). An emendation of the genus *Triannulata* Goodnight, 1940, with the assignment of *Triannulata montana* to *Cambarincola* Ellis 1912 (Clitellata: Branchiobdellida). *Proc. biol. Soc., Wash.* **87**(8), 57–72.

—— (1974b). The genus *Xironogiton* Ellis, 1919 (Clitellata: Branchiobdellida). *Virginia J. Sci.* **25**(1), 5–19.

—— (1976). The branchiobdellid (Annelida: Clitellata) associates of astacoidean crawfishes. In: *Freshwater Crayfish* (ed. J.W. Avault, Jr.), pp. 337–96. Proc. Internat. Symp. on Crayfish, Louisiana. State University Press, Baton Rouge.

—— (1977a). An emendation of the genus *Sathodrilus* Holt 1968 (Annelida: Branchiobdellida), with the description of four new species from the Pacific drainage of North America. *Proc. biol. Soc., Wash.* **90**(1), 116–31. [1978BA:3232.]

—— (1977b). A gill-inhabiting new genus and species of the Branchiobdellida (Annelida: Clitellata). *Proc. biol. Soc., Wash.* **90**(3), 726–34.

—— (1978). The reassignment of *Cambarincola elevatus* Goodnight, 1940 (Clitellata: Branchiobdellida) to the genus *Sathodrilus* Holt, 1968. *Proc. biol. Soc., Wash.* **91**(2), 472–83.

—— (In press). A resumé of the members of the genus *Cambarincola* (Annelida: Branchiobdellida) from the Pacific drainage of the United States. *Proc. biol. Soc., Wash.*

—— and Hoffman, R.L. (1959). An emended description of *Cambarincola macrodonta* Ellis, with remarks on the diagnostic characters of the genus (Oligochaeta: Branchiobdellidae). *J. Tenn. Acad. Sci.* **34**(2), 97–104.

—— and Weigl, A.M. (1979). A new species of *Xironodrilus* Ellis 1918 from North Carolina (Clitellata: Branchiobdellida). *Brimleyana* **1**, 23–9. [1980BA:65164.]

Janda, V. (1928). Über die Regenerationsfähigkeit von *Branchiobdella parasita* Henle (Ein Beitrag zur Kenntnis der regenerativen Potenzen der Parasiten). *Roux' Arch. Entwicklungsmechanik* **113**, 530–5.

Järvekülg, A. (1958). *Jõevähk Eestis*. 186 pp., Tartu. [*Branchiobdella parasita*, *B. pentodonta* and *B. astaci* in southern part of Estonia.]

Jennings, J.B., and Gelder, S.R. (1979). Gut structure, feeding and digestion in the branchiobdellid oligochaeta *Cambarincola macrodonta*

Ellis 1912, an ectosymbiote of the freshwater crayfish *Procambarus clarkii. Biol. Bull.* **156**(3), 300–14. [1980BA:3230.]

Karaman, S.M. (1970). Beitrag zur Kenntnis der europäischen Branchiobdelliden (Clitellata, Branchiobdelloidea). *Int. Revue ges. Hydrobiol.* **55**, 325–33. [Key to *pentodonta* group of spp.]

Keferstein, W. (1863). Anatomische Bemerkungen über *Branchiobdella parasita* (Braun) Odier. *Arch. Anat. Physiol.* 509–20.

Kozarov, G., Michailova, P., and Subchev, M. (1972). [Studies on the Branchiobdellidae (Oligochaeta, Annelida) of Bulgaria.] *God. So. Univ. (Biol. Fak.)* **64**(1), 77–89.

Lankester, E. Ray (1878). Vascular system of *Branchiobdella J. Anat. Physiol.* **12**, 591–2.

Lanzavecchia, G. and Ferraguti, M, (1976). Indagini sullo spermatozoo degli Anellidi: *Branchiobdella pentodonta. Boll. Zool. Atti XLIV Convegno UZI.*

Leidy, J. (1851*a*). Contributions to helminthology. *Proc. Acad. nat. Sci. Phila.* **5**, 205–9.

—— (1851*b*). A description of *Astacobdella philadelphica. Proc. Phila. Acad. nat. Sci.* **5**, 209.

Lemoine, V. (1880). Recherches sur l'organisation des Branchiobdelles. *Ass. franc. Adv. Sci. Reims* **9**, 745–74.

Liang, Yan-Lin (1963). Studies on the aquatic Oligochaeta of China. I. Descriptions of new naids and branchiobdellids. *Acta zool. sinica.* **15**(4), 560–70. [On freshwater shrimp.]

Lui, Sz-Cheng and Chen-An Chang (1964). [On four new species of Branchiobdellidae from crayfish, *Cambaroides dauricus* (Pallas).] *J. Zool.* **1**(16), 37–8.

Luther, A. (1911). [*Branchiobdella parasita* Braun (s. lat.).] *Medd. Soc. Fauna Flora Fenn.* **37**, 113–14.

McManus, L.R. (1960). Some ecological studies of the Branchiobdellidae (Oligochaeta). *Trans. Am. microsc. Soc.* **79**(4), 420–8.

Mazhilis, A.A. [Mažylis, A.] (1973). [On the infection of *Astacus astacus* with Branchiobdellae and their control.] *Lietuvos TSR Mokslu Akademijos Darbai. Ser. C* **3**(63), 107–13. [In Russian; English and Lithuanian summaries.]

Mazzarelli, G. (1906). Su di alcune malattie di pesci e gamberi osservate in Lombardia. *Atti del 3° Congresso Nazionale della Pesca.* **23**, 259–87.

Moore, J.P. (1894). On some leech-like parasites of American crayfishes. *Proc. Acad. nat. Sci. Phila.* (for 1893), 419–28.

—— (1895*a*). *Pterodrilus*, a remarkable discodrilid. *Proc. Acad. nat. Sci. Phila* (for 1894, part 3), 449–54.

—— (1895*b*). The anatomy of *Bdellodrilus illuminatus*, an American discodrilid. *J. Morph.* **10**(2), 497–540.

—— (1897). On the structure of the discodrilid nephridium. *J. Morph.* **22**(1), 327–80.

Moszyński, A. (1938). Kilka uwag o przedstawicielach rodziny Branchiobdellidae w Europie. *Ann. Mus. Zool. Pol.* **13**(9), 89–103. [In Polish.]

Nurminen, M. (1966). Rapulaiset (Branchiobdellida), ravun salamatkustajat. *Luonnon Tutk.* **70**, 36–40. [In Finnish.] [See also *Ann. Zool. Fenn.* **3**, 70–2.]

Odier, A. (1823). Mémoire sur le Branchiobdelle nouveau genre d'Annèlides de la famille des Hirudines. *Mem. Soc. Hist. nat. Paris* **1**, 69–78.

Ostroumoff, A. (1883). Ueber die Art der Gattung *Branchiobdella* Odier auf der Kiemer der Flusskrebs. (*Astacus leptodactylus* Eschh.). *Zool. Anz.* **6**, 76–8.

Overstreet, R.M. (1979). Crustacean health research at the Gulf Coast Research Laboratory (Mississippi). *Proceedings of the Second Biennial Crustacean Health Workshop*, Texas A and M University, College Station, Texas, July 1979. TAMU – SG-79, **114**, 300–14. [Page 303 branchiobdellid *Cambrincola mesochoreus* on blue crabs in very slightly brackish water.]

Penn, G.H. (1959). Survival of branchiobdellid annelids without a crayfish host. *Ecology* **40**(3), 514–15.

Pierantoni, U. (1905). *Cirrodrilus cirratus* n. g. n. sp. parassita dell' *Astacus japonicis. Annuario Mus. R. Zool. Un. Napoli* (new series) **1**(3), 1–3.

—— (1906*a*). Osservazioni sul *Branchiobdella* Odier. *Annuario Mus. R. Zool. Un. Napoli* (new series) **2**, 1–10; *Riv. Mensile di Pesca, Milano* **8**, 1–11.

—— (1906*b*). Nuovi discodrilidi del Giappone e della California. *Annuario Mus. R. Zool. Un. Napoli* (new series) **2**(11), 1–9.

—— (1912). Monografia dei Discodrilidae. *Annuario Mus. Zool. Un. Napoli* (new series) **3**(24), 1–28.

Pop, V. (1965). Systematische Revision der europäischen Branchiobdelliden (Oligochaeta). *Zool. Jahrb.* (*Abt. Syst., Ökol., Geog. der Tiere*) **92**, 219–38.

Rappini, M. (1926–27). Sulla *Branchiobdella pentodonta* Whit. *Atti. Soc. nat. Mat. Modena* (6) **5/6**, 1–9.

Richards, B.O. (1971). A comparative study of the digestive system of *Cambarincola philadelphica* (Leidy, 1851) and *Xironodrilus formosus* (Ellis, 1918). (Annelida, Oligochaeta, Branchiobdellidae). Ph.D. dissertation, Ohio State University, Columbus, Ohio.

Rioja, E. (1940). Estudios hidrobiologicos II. Datos sobre los Branchiobdellidae de Xochimilco Zempoala y Texcoco. *Sobretiro Anales Inst. Biol.* **11**(1), 249–53.

—— (1943). Estudios hidrobiologicos. IX. Anotationes sobre branqui-obdelidos Mexicanos. *Sobretiro Anales Inst. Biol.* **14**(2), 541–6.

Robinson, D.A. (1954). *Cambarincola gracilis*, sp. nov., a branchiobdellid oligochaete commensal on western crayfishes. *J. Parasitol.* **40**, 466–9.

Rohde, E. (1885). Die Muskulatur der Chaetopoden. *Zool. Beitr.* **1**, 164–205.

Rosalba, M.F., Marinelli, M., Tei, S. and Vagnetti, D. (1982). The body surface of *Branchiobdella pentodonta* Whit. (Annelida, Oligochaeta) examined by scanning microscopy. *J. Morph.* **174**(1), 17–24.

Rösel von Rosenhof, A.I. (1755). Monatliche herausgegebene Insekten-belustigungen. Nürnburg, Teil 3, p. 120.

Salensky, W. (1887). Études sur le développement des annélides. II. Développement de *Branchiobdella*. *Arch. Biol.* **6**, 1–64.

Schmidt, F. (1902). Die Körpermuskulatur von *Branchiobdella parasita*. *Nachr. Kgl. Gesellsch. Wiss. Göttingen.* (*Mathem.-Physik. Klasse.*) No. 5, 284–90.

—— (1903). Die Muskulatur von *Branchiobdella parasita*. *Z. wiss. Zool.* **75**, 596–705.

—— (1905). Zur Anatomie und Topographie des Centralnervensystems von *Branchiobdella parasita*. *Z. wiss. Zool.* **82**, 664–92.

—— (1907). Über die Verbreitung des Flusskrebses sowie der sogenannten Krebsegel in der Umgegend von Osnabrück. Nebst Bemerkungen über die einzelnen Krebsegelarten selbst. *Jber. naterw. Ver. Osnabrück* **16**(1903–1906), Anhg., 1–37.

Smallwood, W.M. (1906). Notes on *Branchiobdella*. *Biol. Bull.* **11**, 100–111.

Stjerna-Pooth, I. (1958). *Branchiobdella*. En för Sverige ny kräftparasit. *Svensk Fisk. Tidskr.* **67**, 122–3.

Subchev, M. (1978). A new branchiobdellid — *Branchiobdella kozarovi* sp. n. (Oligochaeta, Branchiobdellidae) from Bulgaria. *Acad. Sci. Bulg. Acta zool. bulg.* **9**, 78–80. [Bulgarian summary.] [1979BA:61185.]

Sz-Cheng Liu see Liu, Sz-Cheng.

Tannreuther, G.W. (1915). The embryology of *Bdellodrilus philadelphicus*. *J. Morph.* **26**, 143–215.

Tichomirnowa, O. (1928). [Neurosomite of *Branchiobdella parasita*.] *Proc. 3rd Congr. Russian Zool., Leningrad.* [According to Livanow (1913) *B. parasita* has 18 neurosomites.]

Vagnetti, D. and Farnesi, R.M. (1978). Morphological and ultrastructural study of the ventral nerve cord in *Branchiobdella pentodonta* Whitman (Annelida, Oligochaeta). *J. comp. Neurol.* **178**(2), 365–82. [1978BA:2810.]

Vallot, — (1840). Mémoire sur l'*Astacobdelle* branciale. *Act. Acad. Roy. Sci.*, Bordeaux 2, 3. trim., 483; *Mem. Acad. Dijon* 1843/44.

—— (1841). Mémoire sur l'*Astacobdelle*. *C. r. hebd. Séanc. Acad. Sci.*, *Paris* **13**, 941.

Vejdovsky, F. (1884). *System und Morphologie der Oligochaeten*. F. Rivnác, Prague.

Voigt, W. (1883). Die Varietäten der *Branchiobdella astaci* Odier. *Zool. Anz.* **6**, 121–5.

—— (1885*a*). Untersuchungen über die Varietätenbildung bei *Branchiobdella varians*. *Arb. Zool.-Zootom. Inst. Würzburg* **7**, 41–94.

—— (1885*b*). Ueber Ei-und Samenbildung bei *Branchiobdella*. *Arb. Zool. Zootom. Inst. Würzburg.* **7**, 300–68.

—— (1886). Beiträge zur feinen Anatomie und Histologie von *Branchiobdella varians*. *Arb. Zool. -Zootom. Inst. Würzburg.* **8**, 102–28.

Voinov, D.O. (1986). Sur les néphridies des *Branchiobdella varians* (V. *astaci*.). *Mem. Soc. Zool. France* **9**, 363–94; preliminary communications: *C. hebd. Séanc. Acad. Sci., Paris* **122**, 1066–71; *Ann. nat. Hist.* **18**, 199–200.

Weismann. A. (1862). Über die zwei Typen kontraktilen Gewebes. *Z. ration. Med.* III. Reihe. **15**, 84.

Whitman, C.O. (1882). A new species of *Branchiobdella*. *Zool. Anz.* **5**, 636–7.

Wojtas, F. (1964). [Notes on the family Branchiobdellidae (Oligochaeta) of the river Grabia.] *Prz. Zool.* **8**(2), 149–51. [In Polish; English summary.]

Yamaguchi, H. (1932*a*). On the genus *Cirrodrilus* Pierantoni, 1905, with a description of a new branchiobdellid from Japan. *Annationes Zool. Jap.* **13**(4), 361–7.

—— (1932*b*). Description of a new branchiobdellid, *Carcinodrilus nipponicus*, n. g. et n. sp. *J. Fac. Sci. Hokkaido Imp. Un. Ser. VI* (*Zool.*) **2**(1), 61–7.

—— (1932*c*). A new species of *Cambarincola*, with remarks on spermatic vesicles of some branchiobdellid worms. *Proc. Imp. Acad., Japan* **8**(9), 454–6.

—— (1933). Description of a new branchiobdellid, *Cambarincola okadai*, n. sp., parasitic on American crayfish transferred into a Japanese lake. *Proc. Imp. Acad., Japan* **9**(4), 191–3.

—— (1934). Studies on Japanese Branchiobdellidae with some revisions on the classification. *J. Fac. Sci. Hokkaido Imp. Un. Ser. VI.* (*Zool.*) (3), 177–219.

Yan-Lin Liang, see: Liang, Yan-Lin.

Young, W. (1966). Ecological studies of the Branchiobdellidae (Oligochaeta). *Ecology* **47**(4), 571–8.

Euhirudinea and Acanthobdellida

Anonymous (1822). On the use of lavemens and leeches. *Med.-Chir. Rev.* and *J. med. Sci.* **2**, 357–61.

Anonymous (1829*b*). The conservation and reproduction of the medicinal leech. *Boston Med. Surg. J.* **1**, 522–4.

Anonymous (1831). Rapport du conseil de santé de la Guyane française sur la reproduction des sangsues. *Ann. mar. colonial.* t. 45, 2ᵉpart, t. 1, pp. 304–8. [Published by M. Bajot, Paris in 8.]

Anonymous (1854). [Medicinal leech.] [Great Soviet Encyclopaedia izd. 2-e, T.26.] [In Russian.]

Anonymous (1898). Piyavka Meditsinskaya. *Entsikloshed. Slovar', izd. Brokgauz-Kphrov* **23A**, 784–5. [In Russian.]

Anonymous (1931). *Haementeria costata* Fr. Müll. in der Grenzmark Posen-Westpreussen. *Abh. Ber. Grenzmärk. Ges. Naturwiss. Abt. Schneide-mühl.* **6**, 155.

Anonymous (1952). [*Haemopis sanguisuga* in Ireland.] *Ir. nat. J.* **10**, 271.

Anonymous (1955). Hirudinea. *Illustrated encyclopaedia of the fauna of Japan* (*exclusive of insects*), pp. 1379–89. Tokyo.

Anonymous (1959). [Leech jar of white earthenware, about 16 inches high.] [From collection of Philip Harris, Ltd., 144 Edmund St., Birmingham 3.] *Chem. Drugg.* **171**, 54, illustrated.

Anonymous (1965*a*). When leeches were used for black eyes. Fourth leader in *The Times*, Saturday, 13 April 1965.

Anonymous (1965*b*). [Leeches at the breakfast club.] *Svenks. Farm. T.* **69** (20 Jan.), 60–1. [In Swedish.]

Anonymous (1968*a*). Grooming leeches for a star turn. *New Scient.* **39**, 530–1.

Anonymous (1968*b*). Leeches to be put into orbit. *Sci. J.*

Anonymous (1969*a*). Survey of Brookham common. Twenty-eighth year. Progress report for 1969. *London Nat.* **49**, 93–104.

Anonymous (1969*b*). Catalogi del Museo istituto di zoologia sistematica dell'universita di Torino (Italia), Vol. 6. Formicidae: Isoptera: Acari: Pseudo-scorpiones: Onycophora: Hirudinea: Myriapoda. Universita, Museo Zoologica, Turin.

Anonymous (1972). Reflections of a lexicographer. (Stethoscope. Leech.) *Lancet* **i**, 312–13.

Anonymous (1973*a*). Australian leeches. *Aust. Mus. Leaflet.* No. 64, 1–10.

Anonymous (1973*b*). Leeches, p. 1. In *Preliminary list of endangered plant and animal species in North Carolina.* North Carolina Department of Natural and Economic Resources.

Anonymous. 'Leeches.' (Bibliography on therapeutic and biological aspects.) *Index Catalog of the Surgeon General's Library.* Washington, DC. (date undetermined)

Anonymous. *Hirudo medicinalis.* (Bibliography on therapeutic and biological aspects.) *Waring's Bibliotheca Therapeutica,* Vol. 2, p. 460. (date undetermined)

Anonymous. Leeches. *Clin. Exc.* 155–8, illustrated. (In WIMH Libr.) (date undetermined)

Anonymous. Hansjochem Autrum. *J. comp. Physiol. A. Sens. Neurol. behav. Physiol.* **120**(1), 101–7. [1978BA:14598.] (date undetermined)

Aalam, D. *et al.* (1968). [On 42 cases of foreign bodies (leeches) at the level of the tracheo-oesophageal junction.] *Ann. Otolaryngol., Paris* **85**, 705–6. [In French.]

Abdallah, A. and Tawfik, J. (1972). *Helobdella punctato-lineata* as a potential predator of *Biomphlaria alexandrina* snail populations under laboratory conditions. *J. Egypt. publ. Hlth Ass.* **47**(1), 43–52.

Abeloos, M, (1924). See Abeloos, M. (1924*a*) in Autrum (1939*c*).

—— (1925). Recherches histologiques et histophysiologiques sur le parenchyme et les néphridies des Hirudinées Rhynchobdelles. *Bull. biol. France Belg.* **59**, 436–97.

Able, K.W. (1976). Cleaning behaviour in the cyprinodont fishes: *Fundulus majalis, Cyprinodon variegatus* and *Lucania parva. Chesapeake Sci.* **17**(1), 35–9. [1976BA:23590.]

Abolarin, M.O. (1970). A note on the trypanosomes from the African freshwater fish and some comments on the possible relationship between taxonomy and pathology in trypanosomes. *Bull. epizoot. Dis. Afr.* **18**(3), 221–8.

Abraham, A. (1958). [Innervation of the intestinal tract of the medicinal leech, *Hirudo medicinalis* L.] *Bull. Biol. Sect. Hung. Acad. Sci.* **2**, 139–55. [In Hungarian.]

—— and Minker, E. (1958). Über die Innervation des Darmkanales des medizinischen Blutegels (*Hirudo medicinalis* L.). *Z. Zellforsch.* **47**, 367–91.

Achard, J.B. (1823). Notice sur la sanguse officinale, sa reproduction aux Antilles. *St. Pierre Turban.* in 8.

Acharya, L.N., Acharya, B.N., and Patnaik, M.M. (1974). Leech infestation of wild animals. *Indian veter. J.* **51**, 574. [*Hirudinaria granulosa* in *Melursus ursinus* and *Elephas maximus.*]

Adamczyk, P. (1979). Messung der Spektralen Empfindlichkeit einzelner Sehzellen in den Ocellen des Blutegels, *Hirudo medicinalis* L. Thesis, Neurobiologie, University of Ulm, Donau.

Adamic, S. (1974). Accumulation of choline by the segmental ganglia of the leech. *Biochem. Pharmacol.* **23**(18), 2595–602. [1975BA:25535.]

—— (1975). [³H]-Choline entry and [³H]-acetylcholine formation in leech segmental ganglia. *Biochem. Pharmacol.* **24**(19), 1763–6. [1976BA:20033.]

Agapova, A.I. (1956). Parazit' p'b vodoemov Zapadnogo Kazahstana. *Tr. Inst. Zool. A.N. Kazahsk. SSR* **5**, 5–60. [In Russian.]

—— (1966). *Parazit' P'b Vodoemov Kazahstana,* Alma-Ata, 1–343. [In Russian.]

Agapow, L. (1968). [Leeches (Hirudinea) of the peat waters on lower Noteć River.] *Prezegl. Zool.* **12**(4), 398–401. [In Polish; English summary.]

—— (1973). [The occurrence of *Hirudo medicinalis* L. in the Voivodeships Szczecin and Zielona Gora.] *Przegl. Zool.* **17**(4), 436–9. [In Polish; English summary.] [First report of *Batracobdella paludosa* in Poland.] [1974BA:31817.]

—— (1975). [Leeches (Hirudinea) of middle and lower Odra and some of its tributaries.] *Bodania Fizjogr. Pol. Zachod. (Zool.)* **28**C, 79–100. [In Polish; English summary.] [Effect of industrial pollution.]

—— (1976). [The leeches (Hirudinea) of the lower course of the Obra River.] *Acta un. Lodziensis Nauki Mat. — Przyr.,* Ser. II (9), 43–8. [In Polish; English summary.]

—— (1977*a*). [*Haementeria costata* (Fr. Müller) in the water reservoirs at Stupsk.] *Przegl. Zool.* **21**(3), 214–15. [In Polish; English summary.] [1979BA:48151.]

—— (1977*b*). [New stations of three rare species of leeches in vicinity of Gorzów Wielkopolski.] *Badania Fizjograficzne Nad Polska Zachodnia* **30** (series C – Zoologia), 187–9. [In Polish; English summary.]

—— and Klarman, A. (1983). [*Boreobdella verrucata,* new record (Hirudinea) in the river Rega on the Pomeranian lake district (Poland).] *Przegl. Zool.* **27**(1), 63–6. [Polish; English summary.] [1984BA:28069.]

Aisenstadt, T.B. (1964). [Cytological study of ovogenesis. I. Morphology of the gonad of *Glossiphonia complanata* L. based on optical and electron microscopy.] *Tsitologiya* **6**, 19–21. [Russian.]

—— (1965). [Certain peculiarities of the oocyte ultrastructure in relation to yolk synthesis.] *Zh. Obshch. Biol.* **26**(2), 230–6. [In Russian; English summary.] [*Glossiphonia complanata.*]

—— and Brodskii, V. Ya. (1963). The fine structure of the ovular membranes of the leech. *Dokl. Akad. Nauk SSSR* (English transl.) **148**, 70–3. [Original Russian article **148**, 728–30.] [*Glossiphonia complanata.*]

—— —— and Gazaryan, K.G. (1967). [Autoradiographic study of the RNA and protein synthesis in gonads of animals with different types of oogenesis.] *Tsitologiya* **9**(4), 397–406. [In Russian; English summary.] [*Glossiphonia complanata.*]

—— Brodskii, V. Ya. and Ivanova, S.N. (1964). [Cytological studies on ovogenesis. II. Cytochemical studies on the growth of ovocytes of *Glossiphonia complanata*, based on ultraviolet cytophotometry and interference microscopy.] *Tsitologiya* **6**, 77–81. [In Russian.]

Aitchinson, J.E.T. (1889). The zoology of the Afghan Delimitation Commission. *Trans. Linn. Soc. Lond, Zool.* (2), V, part 3, pp. 53, 105.

Akhmedov, K.R. (1973). [Data on the reproductive biology of the blackbird *Turdulus merula intermedia* Richm. on the southern slope of the Gissar Ranges (Tadzhik SSR).] *Izv. Akad. Nauk Tadzh. SSR. Otd. Biol. Nauk.* **2**, 77–81. [Food of nestlings excl. insects, partially leeches and larvae of Odonata.] [1974BA:59380.]

Akhmerov, A.Kh. (1941). K izucheniu parazitophaun' p'b ozera Balhash. *Uch. Zap. Leningr. Gos. Un. Ser. Biol. Nauk.* **18**, 37–51. [In Russian.]

—— (1954). [Parasitic fauna of fishes in Kamchatka river.] *Trans. Spec. Probl. Confer.* (7th conf. on problems of parasitology), Moscow, Vol. 4, pp. 89–98. [In Russian.]

—— (1955). Parazitophauna p'b R. Kamchatki. *Izv. Tihookeansk. N.-Issl. Inst. R'bn. Khoz. i Okeanogr.* **43**, 99–137. [*Acanthobdella livanowi* freeliving from Kamchatka.]

Akhrorov, F. (1967). [On the bottom fauna of the Lakes in Pamir.] *Izv. Akad. Nauk Tadzhik. SSSR (Biol.)* **2**(27), 64–75. [In Russian; English summary.]

Akoev, G.N. (1975). [An electrophysiologic and electron-microscopic study of the electrical connection between the giant nerve cells of the leech *Hirudo medicinalis.*] *Nerv. Sist.* **15**, 92–6.

—— and Krasnikova, T.L. (1967). [Dynamics of changes in membrane potential of giant neurons of *Hirudo medicinalis* in calcium-free solutions.] *Dokl. Acad. Sci. USSR (Biol. Sci. Sect.)* **177**(4), 967–9. [English transl. **177**(4), 815–17.]

—— and Sizaya, N.A. (1969). The effect produced by lithium upon the electric activity of the nerve cells in *Hirudo medicinalis*. *Dokl. Akad. Nauk SSSR* **184**, 1447–8.

—— —— (1971). Influence of lithium ions on the electrical activity of nerve cells of the leech. *Neurophysiol. Transl.* **2**, 484–9. [Translated from *Neurofiziologiya* **2**, 636–72 (1970); in Russian.]

—— *et al.* (1971). [Fluorescence of leech Retzius' cells vitally stained with acridine orange.] *Dokl. Akad. Nauk SSSR* **197**, 1448–51.

Aksenova, O.P., Antonov, V.F., and Tereshkov, O.D. (1973). [Effect of dinitrophenol on the paired giant cells in the central nervous system of leeches.] *Fiziol. Zh. SSSR IM I M Sechenova* **59**(3), 407–13. [In Russian; English summary.]

Alavi, K. (1969). Expistaxis and hemoptysis due to *Hirudo medicinalis* (medical leech). *Arch. Otolaryngol.* **90**(2), 178–9.

Albach, E. (1939). [Frankfurt am Main.] Beobachtungen bei der therapeutischen Blutegelverwendung mit besonderer Berücksichtigung anaphylaktischer Erscheinungen. 30 pp. [Inaugural Dissertation zur Erlangung de Doctorwarde in der gesamten Medizin.] Gelnhausen. Kalbfleisch.

Alekseev, V.A. and Uspenskaya, N.E. (1974). [Toxicological characteristics of acute phenol intoxication in some freshwater worms.] *Gidrobiol. Zh.* **10**(4), 48–55. [In Russian; English summary.] [1976BA:69737.]

Alexeieff, A. (1911, 1914). See Autrum (1939c).

Ali, D.S. *Hirudo medicinalis* (medical leech) in larynx. *J. Laryngol. Otol.* **62**, 752–3.

Aliff, J.V., Smith, D., and Lucas, H. (1976). Some metazoan parasites from middle Georgia USA, fishes and frogs. *Bull. Ga. Acad. Sci.* **34**(2), 53–4.

—— (1977). Some metazoan parasites from fishes of middle Georgia. *Trans. Am. micros. Soc.* **96**(1), 145–8.

Aligadzhiev, A.D. (1969). [Effect of long term draining of a body of water (Lake Mekteb, USSR) on the parasites of the fish in it (Cnidosporida, Sporozoa, Mollusks, Trematodes, Leeches).] *Parazitologiya* (Leningrad) **3**(2), 144–8. [In Russian.]

Allen, D.M. and Allen, W.B. (1981). Seasonal dynamics of a leech–mysid shrimp interaction in a temperate salt marsh. *Bio. Bull.* **160**(1), 1–10.

Allison, L.N. (1950). *Common diseases of fish in Michigan.* Mich. Dept. of Conservation, Ann Arbor. Misc. Publ. No. 5.

Almallah, Z. (1968). Internal hirudiniasis in man with *Limnatis nilotica* in Iraq. *J. Parasitol.* **54**(3), 637–8.

Altman, L.K. (1981). Leeches still have their medical uses. *The New York Times* 17 February 1981, p.C.2.

Alvarez, J. (1971). La sistemática de los anélidos y la posición de los oligoquetos. *Bol. R. Soc. Esp. Hist. nat.* (*Biol.*) **69**, 265–72.

—— and Selga, D. (1968). Observaciones sobre invertebrados dulceanicolas de las alrededores de Madrid. *Bol. R. Soc. Esp. Hist. nat.* (*Biol.*) **65**, 171–97. [English summary.]

Amin, O.M. (1969). Helminth fauna of suckers (Catostomidae) of the Gila River system, Arizona. II. Five parasites from *Catostomus* spp. *Am. Midl. Nat.* **82**(2), 429–43.

—— (1977). Helminth parasites of some southwestern Lake Michigan fishes. *Proc. Helminthol. Soc. Wash.* **44**(2), 210–17. [*Placobdella montifera*, nec *P. parasitica.*] [1978BA:3632.]

—— (1978). Notes on *Dina lineata* (O.F. Müller) (Hirudinea: Erpobdellidae) from the gut of some Nile fishes in Egypt. *Proc. helm. Soc. Wash.* **45**(2), 272–5. [Mainly in eel *Anguilla vulgaris.*]

—— (1981). Leeches (Hirudinea) from Wisconsin, and a description of the spermatophore of *Placobdella ornata. Trans. Am. microsc. Soc.* **100**(1), 42–51.

Amos, W.H. (1970). Teeming life of a pond. *Natn. geogr. Mag.* **138**, 274–98. [Illustrated article on pond life, including leeches.] [See also *The life of the pond.* McGraw-Hill.]

Ananichev, A.V. (1961). [Comparative biochemical characteristics of some freshwater invertebrates and fish.] *Biokhimiia* **26**, 18–29. [In Russian; *Erpobdella nigricollis* cuticle.]

Andersen, C. (1962). [Some observations on the biology of *Acanthobdella peledina* in summer 1962.] *Fauna (Blindern)* **15**(3), 177–81. [In Norwegian.]

Anderson, D.T. (1966). The comparative early embryology of the Oligochaeta, Hirudinea and Onychophora. *Proc. Linn. Soc. New S. Wales* **91**(1), 10–43.

—— (1971). Embryology. In: *Aquatic oligochaeta of the world.* (eds. Brinkhurst, R.O. and Jamieson, B.G.M.) pp. 73–103. Oliver and Boyd, Edinburgh.

—— (1973). Oligochaetes and leeches. In *Embryology and phylogeny in annelids and arthropods*, pp.51–92. Pergamon Press, New York.

Anderson, W.A. and Personne, P. (1970). The localization of glycogen in the spermatozoa of various invertebrate and vertebrate species. *J. cell Biol.* **44**, 29–51.

Andersson, E. (1965). Ecological notes on *Acanthobdella peledina* Grube found on grayling and brown trout. *Rep. Inst. freshwater Res. Drottningholm* **46**, 185–99.

Andonian, M.R., Barret, A.S., and Vinogradov, S.N. (1975). Physical properties and subunits of *Haemopis grandis* erythrocruorin. *Biochim. biophys. Acta* **412**(2), 202–13. [1976BA:37680.]

—— and Vinogradov, S.N. (1975). Physical properties and subunits of *Dina dubia* erythrocruorin. *Biochim. et Biophys. Acta* **400**, 244–54.

André, E. (1925). See Autrum (1939*c*).

—— (1966). Catalogue des invertebres de la Suisse FASC. 16. Hirudinées, Branchiobdelles et Polychetes. *Pub. du Museum D'Histoire Naturelle De Geneve.* (Revue Suisse de Zoologie.)

André, P. (1934). A propos des sangues des poissons. *Bull. franc. Piscicult.* No. 68.

Andrew, W. (1965). *Comparative hematology.* Grune and Stratton, New York.

Andrikovics, S. (1973). [Hydroecological and zoological examinations in pondweed fields of Lake Fertö.] *Különlenyomat az Allattani Közlemenyck. Számaból.* **60**(1–4), 39–50. [In Hungarian; English summary.] [1975BA:24562.]

—— (1976). Contribution to the knowledge of the invertebrate macrofauna living in the pondweed fields of Lake Fertö, Hungary. *Opusc. Zool., Budapest* **16**(1–2), 59–66.

Angelov, A. (1963). Nyakoi novi danni b'erhu morphologiyata i ekologiyata na *Dina lineata* var. *arndti* Augener, 1925 (Hirudinea, Herpobdellidae). *Izv. Zool. Inst. Muz. (B'lg. Akad. Nauk) Kn* **14**, 245–9. [In Russian.]

Anglas, J. (1933). See Anglas, J. (1933*b*) in Autrum (1939*c*).

Annandale, N., Prashad, B., and Amin-ud-Din (1921). The aquatic amphibious molluscs of Manipur. *Records Indian Mus.* **22** Part 4(28), 553. [Almost every specimen of the snail *Vivipara oxytropis* was infested with *Alboglossiphonia weberi*; more than 30 in branchial chamber of one specimen.]

Annenkova, N.P. (1930). Piyavki (Hirudinea) v sborah iss'kkul'skoi ekspeditsii 1928 g. Mat. Komis. Eksped. Issl., II. Iss'kkul'skaya Ekspeditsiya 1928 g., **1**, 49–50. [In Russianl.]

Apathy, St V. (1888*a*). Analyse der äusseren Körperform der Hirudineen. *Mittheil. Zool. Stat. Neapel.* **8**, 153–232.

—— (1888*b*). see Autrum (1939*c*).

—— (1888*c*). Systematische Streiflichter I. Marine Hirudineen. *Arch. Naturgesch.* **54**(1), 43–61.

—— (1897). Das leitende Element des Nervensystems und seine topographischen Beziehungen zu den Zellen. *Mittheil. zool. Stat. Neapel.* **12**, 495–748.

—— (1902). See Autrum (1939*c*).

—— (1907). [An article criticizing Cajal and others who support the neuron hypothesis; Apathy believing in the continuity, i.e. syncytial, hypothesis.] *Anat. Anz.* **31**(18–20).

Appy, R.G., and Cone, D.K. (1982). Attachment of *Myzobdella lugubris* (Hirudinea: Pisciolidae) to logperch, *Percina caprodes*, and brown bullhead, *Ictalurus nebulosus. Trans. Am. microsc. Soc.* **101**(2), 135–41.

—— and Dadswell, M.J. (1981). Marine and estuarine pisciolid leeches (Hirudinea) of the Bay of Fundy and adjacent waters with a key to species. *Can. J. Zool.* **59**(2), 183–92. [French summary.]

Arbas, E.A. (1981). Rate modification in the leech heartbeat central pattern generator. *Neurosci. Abstr.* **7**, 137.

Archer, T.C.R. (1957). Medical problems of the operational infantry soldier in Malaya. *Jl R. Army Corps.* **104**, 1–13.

Aricidiacona, G. (1979). Differentiation of the *Batracobdella paludosa* sperm cell. *Acta émbryol. éxp.* **2**, 209–28. [1980:BA2925.]

Arndt, W. (1940). Als Heilmittel gebrauchte Stoffe. *Q. Blutegel*, pp. 524–73. *Die Rohstoffe des Tierreiches Q.* Berlin.

—— (1943). Beiträge zur Kenntnis der Susswasserfauna Bulgarie. *Inz. tsarsk. prirodonauch. Inst. Sof. (Mitt. K. naturw. Inst. Sofia)* **16**, 188–206.

Arnesen, E. (1904). Über den feineren Bau der Blutgefässe der Rhynchobdelliden mit besonderer Berücksichtigung des Rückengefässes und der Klappen. Jena. *Z. Naturwiss.* **38**, 771–806.

Arnold. S.J. (1981). Behavioural variation in natural populations. 1. Phenotypic, genetic and environmental correlations between chemoreceptive responses to prey in the garter snake *Thamnophis elegans*. *Evolution* **35**(3), 489–509.

Arthur, J.R., Margolis, L., and Arai, H.P. (1976). Parasites of fishes of Aishihik and Stevens Lakes, Yukon Territory, and potential consequences of their interlake transfer through a proposed water diversion for hydroelectrical purposes. *J. fish. Res. Bd Can.* **33**(11), 2489–99.

Arts, L. Des. (1909). See Autrum (1939*c*).

Ascoli, G. (1911). See Ascoli, G. (1911*a*) in Autrum (1939*c*).

Aston, R.J., and Brown, D.J.A. (1975). Local and seasonal variations in populations of the leech *Erpobdella octoculata* (L) in a polluted river warmed by condenser effluents. *Hydrobiologia* **47**, 347–66.

Audit, C., Viala, B., and Robin, Y. (1967). Biogenèse des derivés diguanidiques chez la sangsue, *Hirudo medicinalis* L. I. Origine des groupements guanidiques et de la chaîne carbonée. *Comp. Biochem. Physiol.* **22**(3), 775–85.

Audy, J.R. (1952). Leeches. In *Federation of Malaya: Annual Report of the Institute for Medical Research*, pp. 65–70. London.

—— and Harrison, J.L. (1952). Field trials of repellants and poisons against aquatic and terrestrial leeches in British North Borneo, 1952. (A report of research investigations supported by funds supplied by the Office of the Surgeon General, US Army.) Colonial Office Research Unit. Typescript report.

Augener, H. (1925, 1926, 1930, 1931*a,c*). See Autrum (1939*c*).

—— (1936). Hirudineen aus Deutsch-Südwestafrika. *SB. Ges. naturf. Fr. Berlin* 381–97.

—— (1954). Field tests of repellent M-1960 against leeches. *Med. J. Malaya* **8**(3), 240–50.

—— and Autrum, H. (1939). Hirudinea, Egel. In *Tierwelt Mitteleuropas*, hrsg. v. Brohmer, Ehrmann, Ulrich, Wächtler, 1; XIIa, pp. 1–26.

Austin, O.L. (1961). *Birds of the world.* Golden Press, London.

Austin, W.C., Preker, M.V., Bergey, M.A., and Leader, S. (1982). Species list for invertebrates recorded from the Barkley Sound Region. Bamfield Marine Station Report. [MS]. 48 pp. [Marine leeches, Pacific.]

Autrum, H. (1936). Hirudineen. Systematik. In Bronn's *Klassen und Ordnungen des Tierreichs,* Bd. 4, Abt. III, Buch 4, T.1: pp. 1–96.

—— (1939*a*). Hirudineen. Geographische Verbreitung. In Bronn's *Klassen und Ordnungen des Tierreichs,* Bd. 4, Abt. III, Buch 4, T.2: pp. 497–520.

—— (1939*b*). Die Stellung der Hirudineen in Systems. In Bronn's *Klassen und Ordnungen des Tierreichs,* Bd. 4, Abt. III, Buch 4, T.2: pp. 521–37.

—— (1939*c*). Literatur über Hirudineen bis zum Jahre 1938. In Bronn's *Klassen und Ordnungen des Tierreichs*, Bd. 4, Abt. III, Buch 4, T.2: pp. 539–642.

—— (1958). Hirudinea. Egel. In *Die Tierwelt Mitteleuropas* (ed. P. Brohmer, P. Ehrmann, and G. Ulmer) Vol. 1(7b), pp. 1–30. Leipzig.

—— and Graetz, E. (1934). Vergleichende Untersuchungen zur Verdauungsphysiologie der Egel. I. Die lipatischen Fermente von *Hirudo* und *Haemopis*. *Z. vergl. Physiol.* **21**, 429–39.

Avdeeva, N.I., Bazanova, I.S., Mashanskii, V.F., and Merkulova, O.S. (1976). [The strength of synaptic stimulation in the morphofunctional readjustment of the neuron.] *Fiziol. Zh. SSSR IM.I. M. Sechenova* **62**(9), 1292–9. [In Russian.] [1977BA:57173.]

—— —— (1974). [Changes of the ultrastructure and bioelectrical activity of medical leech Retzius cells upon synaptic and direct stimulation.] *Fiziol. Zh. SSSR IM.I. M. Sechenova.* **60**(10), 1526–30. [In Russian.] [1975BA:8769).]

Avel, M. (1959). Classe des annélides oligochètes. In: *Traité de zoologie.* Grassé, Vol. **V**, pp. 224–470, 1066–7.

Awachie, J.B.E. (1971). Unpublished report on preliminary observations on the fish and fish parasites of Nike Lake, an oligotrophic lake. [leeches destroyed large numbers of fish fry during flood seasons in Nigeria.] [see Mbahinzireki 1980.]

Babaskin, A.W. (1931). Über das Bindegewebe der Hirudineen. *Zool. Jahrb. Anat.* **53**, 1–102.

Babu, S.J. (1967). Two new fish-leeches from Pulicat Lake. *Curr. Sci.* **36**(20), 548–9.

Bacq, A.M., and Coppée, G. (1937). Action de l'esérine sur la préparation neuromusculaire du siponcle et de la sangsue. *C. r. Soc. Biol.* **124**, 1244–7.

Badham, C. (1916). On an Ichthyobdellid parasitic on the Australian Sand Whiting (*Sillago ciliata*). *Q. Jl microsc. Sci.* (new series) **62**, 1–41.

—— (1923). On *Centropygus joseensis*, a leech from Brazil. *Q. Jl microsc. Sci.* (new series) **67**, 243–56.

Baer, J.G. (1951) Annelida. In *Ecology of animal parasites*, pp. 41–2. University of Illinois Press, Urbana.

Bagdasarova, A.V. (1969). [Effect of hirudin therapy on some functions and the hydrodynamics of the eye of glaucoma patients.] *Vestnik Oftalmologii* **2**, 43–5. [In Russian.]

—— and Medvedev, A.N. (1969). [The evaluation of phlebotomy in some eye diseases.] *Med. Zh. Uzb.* **10**, 37–9.

Bagdy, D., and Török, S. (1962). [Extraction of hirudin — whole leeches.] Hung. Pat. No. 150 600, 8 March 1962.

—— Barabas, E., and Graf, L. (1973). Large scale preparation of hirudin. *Thromb. Res.* **2**(3), 229–38. [1974BA:27462.]

—— Bihari, I., and Török, S. (1962). [Extraction of hirudin — whole leeches.] Hung. Pat. No. 150833, 14 December 1962.

—— Barabas, E. Jecsay Gy., and Kazi, E.F. (1962). [Chromatography of crude hirudin on Ecteola Cellulose.] Hung. Pat. No. 152836.

—— —— Graf, L., Petersen, T.E., and Magnusson, S. (1976). Hirudin. *Meth. Enzymol.* **45**, 669–78.

Bagnoli, P., and Magni, F. (1975). Synaptic inputs to Retzius' cells in the leech. *Brain Res.* **96**(1), 147–52.

—— Brunelli, M., and Magni, F. (1969). Una via ad elevata velocità di conduzione nel sistema nervosa centrale della sanguisuga *Hirudo medicinalis. Boll. Soc. ital. Biol. sper.* **45**, fasc. 20 bis, n. 103.

—— —— (1970). Analisi delle risposte registrate nel S.N.C. di '*Hirudo medicinalis*' alla stimolazione fotica. *Arch. Fisiol.* **68**, 36–7. [Responses to light.]

—— —— (1972). A fast-conducting pathway in the central nervous system of the leech *Hirudo medicinalis. Arch. Ital. Biol.* **110**(1), 32–51.

—— —— (1973). Afferent connections to the fast-conducting pathway in the central nervous system of the leech *Hirudo medicinalis. Arch. Ital. Biol.* **111**(1), 58–75.

—— —— —— Pellegrino, M. (1974). Suprasegmental inputs to the fast-conducting system in the central nervous system of *Hirudo medicinalis. Arch. Ital. Biol.* **112**(4), 307–29. [1975BA:26181.]

—— —— —— (1975). The neurons of the fast conducting system in *Hirudo medicinalis:* identification and synaptic connections with primary afferent neurons. *Arch. Ital. Biol.* **113**(1), 21–43.

Bai, Thai Tran (1966). See Thai Tran Bai.

Baird, W. (1869). Descriptions of some new suctorial Annelides in the collection of the British Museum. *Proc. zool. Soc., Lond.* 310–18.

Baker, F.C. (1924). The fauna of the Lake Winnebago region. *Trans. Wisconsin Acad. Sci.* **21**, 109–46.

Baker, J.C. and Crites, J.L. (1976). Parasites of channel catfish, *Ictalurus punctatus* Rafinesque, from the island region of western Lake Erie. *Proc. helm. Soc., Wash.* **43**(1), 37–9. [1976BA:9754.]

Baker, J.R. (1974). The evolutionary origin and speciation of the genus *Trypanosoma. Symp. Soc. gen. Microbiol.* XXIV. Evolution in the microbial world, pp. 343–66. [See also: *Exp. Parasitol.* **13**, 219 (1963); *3rd symp. Br. Soc. Parasitol.* Blackwell (1965).]

Ball, G.H. (1958). A hemogregarine from a water snake, *Natrix piscator*, taken in the vicinity of Bombay, India. *J. Protozool.* **5**(4), 274–81. [Developmental stages of hemogregarines from the snake in "*Hirudinaria granulosa*".]

Bando, S. (1924). See Autrum (1939c).

Bangham, R.V. (1941). Parasites of fish of Algonquin Park Lakes. *Trans. Am. fisheries Soc.* **70**, 161–71.

—— and Hunter, G.W. (1939). Studies on fish parasites of Lake Erie. Distribution studies. *Zoologica, NY* **24**, 385–448.

—— and Vernard, C.E. (1942). Studies on parasites of Reelfort Lake fish. IV. Distribution studies and check list. *Tenn. Acad. Sci.* **17**, 22–38.

—— (1946). Parasites of fish of Algonquin Park Lakes. *Un. Toronto Stud., Biol. Ser.* **65**, 33–46.

Banko, W.E. (1960), The Trumpeter Swan. *N. Am. Fauna* No. 63. [Leeches in nose and on body.]

Baradany, N. (1961). Organisation ultramicroscopique de la membrane nucléaire de la spermatide de la sangsue (*Hirudo medicinalis*). *C. r. hebd. Séanc. Acad. Sci., Paris* **253**, 1125–9.

Barboza, A. (1948). Novo metodo ed coloracao e montagem de hirudineos, helmintos e pequenos artropodos. *An. Acad. Bras. cienc. Rio.* **20**(3), 277–80.

Baring, C. (1915). *Ir. Nat.* **24**. 68–71. [*Haemopis sanguisuga* in Ireland.]

Barnes, R.D. (1981). *Invertebrate zoology*. 4th ed., Holt-Saunders, Philadelphia. 1089 pp.
   Georgetown University. 1977. pp. 316. (*Dissertation Abstr.* B 38:4004.)

—— and Bick, K.L. (1977). Histochemical characterization of the hemocoelomic tissue of the leech *Haemopis* sp. *J. Histochem. Cytochem.* **25**(3), 235. [1977BA:91569.]

Barnes, R.D. (1981). *Invertebrate zoology*. 4th ed., Holt-Saunders, Philadelphia. 1089 pp.

Barnhart, C.S. (1964). Leech repellent development and testing. USA Limited War Laboratory Technical Report TR-3. 1-20.

Barrois, Th. (1896). Recherches sur la faune des eaux douces des Açores. *Lille* 121–2.

Barrow, J.H. (1953). The biology of *Trypanosoma diemyctyli* (Tobey). I. *Trypanosoma diemyctyli* in the leech *Batrachobdella picta* (Verrill). *Trans. Am. microsc. Soc.* **72**, 197–216.

—— (1954). Observations of some host specificity and immunological reactions of trypanosome infections in some freshwater fish of Europe. *Anat. Rec.* **120**, 110.

—— (1958). The biology of *Trypanosma diemyctyli* (Tobey). III. Factors influencing the cycle of *Trypanosoma diemyctyli* in the vertebrate host *Triturus viridescens viridescens*. *J. Protozool.* **5**, 161–70.

Barrows, C. (1893). [Brief leech review.] *Q. Bull. Un. Minn.* **1**(3), 87–8.

Barthelmes, D. (1970). Die Wirkung der Trockenlegung auf das Makrozoobenthos des Karpfenleiches. *Z. Fisch.* **18**, 55–80 [English and Russian summaries.] [Quantitative changes of leeches in partially and completely drained carp-ponds.]

Bartholomeus, A. (1485). Van den proprieteyten der dinghen, vert. door Mr. Jb. Beyaert, XVIIIe boeck, Cap. LXXXVII, Sign. Q.III recto.

Barton, D.R. and Hynes, H.B.N. (1978). Wave-zone macrobenthos of the exposed Canadian shores of the St. Lawrence Great Lakes, *J. Great Lakes Res.* **4**, 27–45.

Bartonek, J.C. (1972). Summer foods of American Widgeon, Mallards, and a Green-winged Teal near Great Slave Lake, Northwest Territories. *Can. Field-Nat.* **86**, 373–6. [Leeches as food; around oesophagus.]

—— and Hickey, J.J. (1969*a*). Selective feeding by juvenile diving ducks in summer. *The Auk* **86**, 457–93. [Leeches in gut contents.]

—— (1969*b*). Food habits of canvasbacks, redheads and lesser scaup in Montana. *Condor* **71**, 280–90. [Leeches in gut contents.]

—— and Murdy, H.W. (1970). Summer foods of Lesser Scaup in Subarctic taiga. *Arctic* **23**, 35–44. [Leeches as food: around oesophagus.]

—— and Trauger, D.L. (1975). Leech (Hirudinea) infestations among waterfowl near Yellowknife, Northwest Territories. *Can. Field-Nat.* **89**(3), 234–43.

Basche, R.E., Pecor, C.H., Waybrant, R.C., and Denaga, D.E. (1980). Limnology of Michigan's nearshore waters of Lakes Superior and Huron. EPA-600/3-80-059. US Environmental Protection Agency, Duluth, Minnesota. 189 pp.

Baskova, I.P. and Cherkesova, D.U. (1980). [Comparative properties of hirudin from whole leeches and from leech head and bodies.] *Biokhimiya* **45**(2), 266–72. [In Russian; English summary.] [English translation: *Biochemistry* **45** (2 Part 1), 195–9.]

—— —— and Mosolov, V.V. (1976). [Purification of hirudin by the method of isoelectric focusing.] *Biokhimya* **41**(5), 939–41. [In Russian; English abstract.] [1977BA:26678.] [English translation: *Biochemistry* **41**(5), 774–5.]

—— —— —— Malova, E.L. and Belyanova, L.A. (1980). Comparative study of hirudin and pseudohirudin. *Biokhimiya* **45**(3), 463–7. [In Russian; English translation, *Biochemistry* **45**(3 Part 1), 348–51.]

—— Memon, M.S., and Cherkesova, D.U. (1977). Use of *N*-benzoyl-phenylalanylvalyl arginine, P-nitro-anilide and hirudin for the detection of enzymatic activity in prothrombin acylation by citraconic anhydride. *Biokhimiya* **42**(8), 1487–90. [In Russian; English translation, *Biochemistry* **42**(8), 1161–4.] [1978BA:43988.]

Bassewitz, E. von (1920). A sanguessuga, *Haementeria officinalis*, transmissora da pyroplasmose equina Sul-America, 'Mal de Cadeiras'. *Brazil-Med.* **24**(18), 122–4.

Baster, J. (1760). De bloedzuiger der Vissen. *Natuurkundige uitspanningen* **12**, pp. 94–5. Haarlem.

Batchelor, A.C.G., Davidson, P., and Lully, L. (1984). The salvage of congested skin flaps by the application of leeches. *Br. J. Plastic Surg.* **37**, 358–60.

Batchelor, M.D. (1960). Leeches and amphibian blood. *Malayan nat. J.* **15**, 71.

Bauer, B.H. (1976), Notes on the leeches found parasitizing some perciformes fishes in Tennessee. *J. Tenn. Acad. Sci.* **51**(1), 9–10. [1976BA:60949.]

—— and Branson, B.A. (1975). The leech *Piscicolaria reducta* parasitizing some percid fishes. *Trans. Ky. Acad. Sci.* **36**(1–2), 18–19. [1976BA:4293.]

Bauer, O.N. (1942). [New localities and new hosts of *Acanthobdella peledina* Grube.] *Zool. Zhurn.* **21**(6), 282–3. [In Russian, English summary.]

—— (1948). Parazit' p'b reki Eniseya. *Izv. Vsesouzn. N.-Issl. Inst. Ozern. i Rechn. R'bn. Khoz.* **27**, 157–74. [In Russian.]

—— (1953). Sanitarno-ozdorovitel'n'e meropriyatiya v prudov'kh Khozyaistvakh pri parazitarn'kh zabolevniyakh p'b, pp. 1–36. Moscow.

—— (1958). Parazitarn'e zabolevaniya p'b v prudov'kh, nerestova-b'rastn'kh Khozyaistvakh i r'bopitomnikakh i mer' bor'b' s nimi. *Osnovn. Prodl. Parazitol. P'b*, pp. 267–300. Leningrad.

—— (1961). Parasitic diseases of cultured fishes and methods of their prevention and treatment. In *Parasitology of fishes* (ed. V.A. Dogiel, G.K. Petrushevski, and Y.I. Polyanski) pp. 265–98. Oliver and Boyd, Edinburgh.

—— and Greeze, V.N. (1948). Parazit' p'b ozera Taim'r. *Izv. Vsesouzn. N-Issl. Inst. Ozern. i Rechn. r'bn. Khoz.* **27**, 186–94. [In Russian.] [Observation on *Acanthobdella*.]

—— and Nikol'skaya, N.P. (1948). K poznaniu parazitov r'b reki Anad'r'. *Izv. Vsesouzh. N-Issl. Inst. Ozern. i Rechn. r'bn. Khoz.* **27**, 175–6.

Baugh, S.C. (1960*a*). Studies on Indian Rhynchobdellid Leeches. I. *Parisitology* **50**, 287–301.

—— (1960*b*). Studies on Indian Rhynchobdellid Leeches, II. *Zool. Anz.* **165**, 468–77.

Bay, E.C. (1974). Predator–prey relationships among aquatic insects. *Ann. Rev. Entomol* **19**, 441–53.

Bayanov, M.G. (1980). [Circulation of helminths in aquatic paludial birds in water bodies of different trophic level and mineral content in the southern Urals, USSR.] *Ekologiya* **5**, 56—62. [In Russian.]

—— and Kussaya, N.A. (1972). [Leeches in the Bashkir ASSR as intermediate hosts of bird helminths.] *Trudy Bashkirskogo Sel'skok-hovaistvennogo Instituta* (Mery bor'by s invazionnymi boleznyami zhirotnykh) **17**, 33–45. [In Russian.]

Bayer, E. (1898). Hypodermis und neue Hautsinnesorgane der Rhynchobdelliden. *Z. wiss. Zool.* **64**, 648–96.

Bayliss, H.A. (1922). See Baylis, H.A. (1922) in Autrum (1939*c*).

Baylor, D.A. and Nichols, J.G. (1968). Receptive fields, synaptic connections and regeneration, patterns of sensory neurons in the C.N.S. of the leech. In *Physiological and biological aspects of nervous integration* (ed. F.D. Carlson) pp. 3–16. Prentice-Hall, Englewood Cliffs, N.J.

—— (1969*a*). Changes in extracellular potassium concentration produced by neuronal activity in the central nervous system of the leech. *J. Physiol.* **203**, 555–69.

—— (1969*b*). After-effects of nerve impulses on signalling in the central nervous system of the leech. *J. Physiol.* **203**(3), 571–89.

—— (1969*c*). Chemical and electrical synaptic connexions between cutaneous mechano-receptor neurones in the central nervous system of the leech. *J. Physiol.* **203**(3), 591–609.

—— (1971). Patterns of regeneration between individual nerve cells in the central nervous system of the leech. *Nature, Lond.* **232**, 268–9.

—— —— and Stuart, A.E. (1967). Synaptic connexions of paired homologous neurons in the central nervous system of the leech. *J. Physiol.* **189**, 10–11P.

Baźal, K., Lucký, Z., and Dyk, V. (1969). Localization of fish-lice and leeches on carp during the autumn fishing. *Acta vet. brno* **38**(4), 533–44. [*Piscicola geometra.*]

Bazan, H. (1962). Nowe obserwacje nad wymoczkiem *Opercularia clepsinis* Popow (Peritricha) *Fragm. Faun., Warsaw* **9**(17), 247–54.

Bazanova, I.S., Kazanskii, V.S., and Sergeev, V.V. (1976). [Membrane resistance of medicinal leech Retzius cells during prolonged influences.] *Fiziol. Zh. SSSR IM.I.M. Sechenova* **62**(5), 789–92 [In Russian.] [1977BA:2385.]

—— Kusainova, N.I., Mashanskii, V.F., and Merkulova, O.S. (1979). [The bioelectrical activity and the ultrastructure of the Retzius neuron of the medicinal leech during long-lasting direct and synaptic activation.] *Fiziol. Zh. SSSR IM. I. M. Sechenova* **65**(4), 575–81. [In Russian; English summary.] [1979BA:67766.]

Bazyluk, W. (1951). [Contribution to the study of the leeches (Hirudinea) in the Podlasia region.] *Fragm. Faun. Mus. Zool. Polon.* **6**(6), 129–34. [In Polish.]

—— (1971). *Trocheta bykowskii* Gedroyć i niektóre inne pijawki (Hirudinea) z Bieszczadów Zachodnich. *Fragmenta Faun.* **17**(4), 49–52.

Beauchamp, P. de. (1904, 1905). See de Beauchamp, P.

Beaufort, L.F. de (1937). Aantekeningen over de Fauna van de omgeving van Buzenol. (prov. Luxemburg). *Biol. Jaarb.* **4**, 213–22. [*Glossiphonia complanata.*]

Beccari, O. (1904). See Autrum (1939*c*).

Beck, D.E. (1954). Ecological and distribution notes on some Utah Hirudinea. *Utah Acad. Proc.* **31**, 73–8.

Becker, C.D. (1964). The parasite–vector–host relationship of the hemoflagellate, *Cryptobia salmositica* Katz, the leech *Piscicola salmositica* Meyer and certain freshwater teleosts. Ph.D. thesis, University of Washington.

—— (1970). Haematozoa of fishes, with emphasis on North American records. In: *A Symposium on diseases of fishes and shellfishes* (ed. S.F. Snieszko). *American Fisheries Society Special Publication* **5**, 82–100.

—— (1977). Flagellate parasites of fish. In: *Parasitic protozoa. Vol. I. Taxonomy, kinetoplastids and flagellates of fish* (ed. J.P. Kreier), **1**, 357–416. Academic Press: New York.

—— (1980). Hematozoa from resident and anadromous fishes of the central Columbia River, Washington, USA: a survey. *Can. J. Zool.* **58**(3), 356–62. [1980BA:23968.]

—— and Cloutman, D.G. (1975). Parasites of selected game fishes of Lake Fort Smith, Arkansas. *Ark. Acad. Sci. Proc.* **29**, 12–18.

—— and Dauble, D.D. (1979). Records of piscivorous leeches (Hirudinea) from the central Columbia River, Washington Stage. *US Natl. mar. Fish. Serv. Fishery Bull.* **76**(4), 926–31. [1980BA:25711.]

—— and Holloway, H.L. (1968). A survey of haematozoa in Antarctic vertebrates. *Trans. Am. micros. Soc.* **87**, 354–60.

—— and Katz, M. (1965*a*). Distribution, ecology, and biology of the salmonid leech, *Piscicola salmositica* (Rhynchobdellae: Piscicolidae). *J. fish. res. Bd. Can.* **22**(5), 1175–95.

—— (1965*b*). Infections of the hemoflagellate, *Cryptobia salmositica* Katz, 1951, in freshwater teleosts of the Pacific Coast. *Trans. Am. fish. Soc.* **94**(4), 327–33.

—— (1965*c*). Transmission of the hemoflagellate, *Cryptobia salmositica* Katz, 1951, by a rhynchobdellid vector. *J. Parasitol.* **51**, 95–9.

—— (1966). Host relationships of *Cryptobia salmositica* (Protozoa: Mastigophora) in a western Washington hatchery stream. *Trans. Am. fish. Soc.* **95**(2), 196–202.

—— and Overstreet, R.M. (1979). Haematozoa of marine fishes from the northern Gulf of Mexico. *J. fish Dis.* **2**, 469–79. [Piscicolids implicated as potential vectors.]

Becker, D.A., Heard, R.G., and Holmes, P.D. (1966). A preimpoundment survey of the helminth and copepod parasites of *Micropterus* spp. of Beaver Reservoir in northwest Arkansas. *Trans. Am. fish. Soc.* **95**(1), 23–34.

—— and Cloutman, D.G. (1975). Parasites of selected game fishes of Lake Fort Smith, Arkansas. *Ark. Acad. Sci. Proc.* **29**, 12–18.

Beckerdite, F.W. and Corkum, K.C. (1973*a*). Observations on the life history of the leech *Macrobdella ditetra* Moore, 1953. *Proc. La. Acad. Sci.* **36**, 61–3.

—— (1973*b*). Studies on the life cycle of a species of the trematode genus *Alloglossidium* (Trematoda: Macroderoididae) parasitic in *Macrobella ditetra* (Hirudinea: Hirudinidae). *Bull. Southeast. biol. Assoc.* **20**(2), 29.

—— (1974). *Alloglossidium macrobdellensis* sp. n. (Trematoda: Macroderoididae) from the leech, *Macrobdella ditetra* Moore, 1953. *J. Parasitol.* **60**(3), 434–6.

Beer, G.R. de (1958). See de Beer, G.R. (1958).

Behning, A.L. (1953) [Uber die qualitative und quantitative Zusammensetzung der Bodenfaunoe einiger Seen Jakutiens.] *Explor. Lacs. URSS, Leningrad* **8**, 125–40. [In Russian; German summary.]

Belar, K. (1921). See Autrum (1939c).

Belardetti, F., Biondi, C., Colombaioni, L., Brunelli, M., and Trevisani, A. (1982). Role of serotonin and cyclic AMP on facilitation of the fast conducting system activity in the leech *Hirudo medicinalis*. *Brain Res.* **246**(1), 89–104. [1983BA:49678.]

—— Brunelli, F., and Simoni, A. (1978). Visually driven neurons in the central nervous system of the leech. *Neurosci. Lett.* 1978 (Suppl. 1) S368. [1979BA:35915.]

Beleslin. B.B. (1968a). Studija spontanih praznjenja potencijala siljaka Retzius-ovih ganglijskih ćelija pijavica pomoću mikroelektroda. M.Sc. thesis. *Inst. za patofiziologiju. Med. fak., Beograd.* 1–69.

—— (1968b). Leech ganglion cells: is metabolism involved into the process of intercellular ion communication? *Proc. XXIV Inter. Cong. Physiol. Sci., Washington,* Vol. 7, p. 36.

—— (1971). Effects of different external media on the leech ganglion cells interaction. *Period. Biol.* **73**(2), 63–7. [Serbo-Croatian summary.]

—— (1974). O prirodi dejstva nekih katjona na gigantske nerve ćelije pijavice. *Saopstenja 50 god. Farmakol. Inst. Med. fak. Beograd* 40–1.

—— (1975a). Lack of electrotonic transmission between Retzius nerve cells in Vth and VIth free ganglion of leech. *Proc. Int. Biophys. Cong.* (Copenhagen), P-513.

—— (1975b). Basic electrophysiological properties of Retzius nerve cells in fifth and sixth free ganglion of horse leech *Haemopis sanguisuga*. *Period. Biol.* **77**, 71.

—— (1976). [Effect of lithium on leech ganglion cells.] *Period. Biol.* **78**(2), 154. [1978BA:5392.]

—— (1977a). Lack of electrotonic transmission between Retzius nerve cells in the fifth and sixth free ganglion of horse leech *Haemopis sanguisuga*. *Comp. Biochem. Physiol. A Comp. Physiol.* **56**(4), 509–2. [1977BA:20997.]

—— (1977b). The influence of external lithium on some passive electrophysiological properties of Retzius nerve cells of horse leech *Haemopis sanguisuga*. *Comp. Biochem. Physiol. A Comp. Physiol.* **56**(4), 513–18. [1977BA:20998.]

—— (1979). Voltage clamp studies of Retzius nerve cells of *Haemopis sanguisuga*. *Iugoslav. Physiol. Pharmacol. Acta* **15**, 237–9.

—— (1982). Membrane physiology of excitable cells in annelids. In: *Membrane physiology of invertebrates.* (ed. R.B. Podesta) pp. 199–260. Marcel Dekker, New York and Basel.

—— and Dekleva, N.D. (1981). Outward currents in the Retzius nerve cell of the horse leech. *Periodicum Biologorum* **83**(1), 113—14.

—— —— and Osmanović, S.Ś. (1982). Frequency dependent decay of the outward current in leech ganglion cell. *Periodicum Biologorum* **84**(2), 115–16.

—— —— and Pešić, B. (1977). The effects of K-free Ringer on Retzius nerve cells of leech. *Proc. Int. Union physiol. Sci.* **13**, 63.

—— Katz, G., and Dekleva, N. (1980). Voltage clamp experiments in leech ganglion cells. *Proc. Int. Union physiol. Sci.* **14**, 864.

—— and Majic V.V. (1974). The effects of pulsed magnetic field on the nerve cells of leeches. *Period. Biol.* **76**(1), 39.

—— and Mihailović, Lj. T. (1967). Effect of tetrodotoxin on the spontaneous discharge of spike potentials of the nerve cells of Retzius in a segmental ganglion of leech. *Iugoslav. physiol. pharmacol. acta* **3**, 85–6.

—— —— (1969). Leech ganglion cells: effect of 2:4-dinitrophenol on junctional membrane coupling. *Acta Biol. Iugoslav. Ser. C Iugoslav. Physiol. Pharmacol. Acta* **5**(3), 295–300.

Beneden, P.J. Van, and Hesse, C.E. (1863). Recherches sur les Bdellodes (Hirudinées) et les Trématodes marins. *Mém. Acad. Sci. Belg.* **34**, 1–59.

Benham, W.B. (1904). On a new species of leech (*Hirudo antipodum*) recently discovered in New Zealand. *Trans. Proc. N. Zealand Inst.* **36**, 185–92.

—— (1907). Two new species of leech in New Zealand. *Trans. Proc. N. Zealand Inst.* **39**, 180–92.

—— (1909). See Autrum (1939*c*).

Bennet, J.A., and Van Oliver, G. (1826). Naamlijst van Wormen in Nederland aanwezig. *Natuurk. Verh. Holl. Mij. d. Wetenschappen* **XV** 2, 63–7.

Bennike, S.A.B. (1939). The Greenland variety of *Theromyzon garjaewi* (Livanow). *Medd. Gronland.* **125**(2), 1–8.

—— (1940). On some Iranian freshwater Hirudinea. *Dan. scient. Inves. Iran* Part II, 1–10.

—— (1943). Contributions to the ecology and biology of the Danish freshwater leeches. *Folia Limnol. Scand.* **2**, 1–109.

—— and Bruun, F.R. (1939). *Pterobdellina jenseni*, n. subgen. n. sp., a new Ichthyobdellid from the North Atlantic. *Vidensk. Medd. dansk. naturk. foren. Kjobenhavn.* **103**, 517–21.

Bere, R. (1929). Reports of the Jasper Park Lakes Investigations, 1925–26. III. The leeches. *Contrib. Can. Biol. Fish.* **4**(14), 177–83.

—— (1931) Leeches from the lakes of northeastern Wisconsin. *Trans. Wisc. Acad. Sci.* **26**, 437–40.

Berg, K. (1938). See Autrum (1939*c*).

Bergh, R.S. (1885). Die Metamorphose von *Aulastoma gulo*. *Arb. zool. Inst. Würzburg* **7**, 231–91.

Berglund, B.E., and Digerfeldt, G.A. (1970). A palaeoecological study of the late-glacial lake at Torreberga, Scania, South Sweden. *Oikos* 21, 98–128. [*Piscicola geometra.*]

Beritov, I.S. (1950). [Spontaneous activity.] *Zh. Obshch. Biol.* 11, 31. [In Russian.]

—— and Gorgava, M.V. (1944). [Spontaneous activity.] *Soobshch. AN Gruzinsk. SSR* 9, 927. [In Russian.]

—— —— (1949). [On the spontaneous activity in the ganglia of the central nervous system of leech.] *J. gen. Biol., Moscow* 10, 421–31. [See also: *Trans. physiol. Inst., Acad. Sci., Gruzinsk SSR* 8, 17. Tbilisi (1950).]

Berland, B. (1973). [On parasites of fish.] *Fiskets Gang.* No. 26, 486–93. [In Norwegian; English summary.] Leeches of marine fish briefly reviewed.]

Bern, H.A. (1962). The properties of neurosecretory cells. *Gen. comp. Endocr.* Suppl. 1, 117–32.

—— and Hagadorn, I.R. (1965). Neurosecretion. In *The structure and function of the nervous system of invertebrates* (ed. T.H. Bullock and G.A. Horridge) Vol. I, pp. 353–429. Freeman, San Francisco.

—— Nishioka, R.S., and Hagadorn, I.R. (1960). Association of elementary neurosecretory granules with the Golgi complex. *J. ultrastruct. Res.* 5, 311–20.

—— —— (1962). Neurosecretory granules and the organelles of neurosecretory cells. *Proc. 3rd Int. Conf. Neurosecretion,* Memoir No. 12, Soc. Endocr., pp. 21–34.

Bernstein, J. (1952). The lowly leech, bedside companion of the surgeon in the good old days. *Frontiers* 17(1), 12–13.

Beron, P., and Gueorguiev, V. (1967). Essai sur la faune cavernicole de Bulgarie. 2. Résultats des recherches biospéologiques de 1961. *Izv. zool. Inst. Sof.* 24, 151–212.

Bertelli, D. (1888). Glandule salivari della *Hirudo medicinalis. Atti Soc. Toscana Sci. nat. Processi verbali, Pisa.*

Besch, W., Hofmann, W., and Ellenberger, W. (1967). Das Macrobenthos auf Polyäthylensubstraten in Fleissgewassern. 1. Die Kinzig ein Fluss der Unteren Salmoniden und Oberen Barbenzone. *Annls Limnol.* 3(2), 331–68. [English summary.] [Leeches on polyethylene plates in running water of a stream, Germany.]

Best, C.H. (1961). A short account on the history of anticoagulants. In *Anticoagulants and fibrinolysin* (ed. McMillan and Mustard). Lea and Febiger, Philadelphia. (incomplete reference)

Bettmann, O.L. (1956). *A pictorial history of medicine.* Thomas, Springfield, Ill.

Bhatia, M.L. (1928a). Hermaphrodite organs of Indian leech. *Proc. Indian Sci. Congr., Calcutta* 15, 201.

—— (1928b). Copulation in the common Indian leech. *Proc. Indian Sci. Congr., Calcutta* 15, 201.

—— (1928*c*). Cocoon formation of *Hirudinaria granulosa*. *Proc. Indian Sci. Congr., Calcutta* **15**, 201.

—— (1930*a*). On the anatomical details of *Placobdella emydae* Harding. A leech parasitic on Indian turtles. *Zool. Anz.* **91**, 225–43.

—— (1930*b*). Sur une nouvelle hirudinée rhynchobdelle, *Glossiphonia cruciata* n. sp., provenant du vivier à tuitre d'Achha Bal, Kashmir. *Ann. Parasitol. hum. comp.* **8**, 344–8.

—— (1931). On a new rhynchobdellid leech, from the trout hatchery, Achha Bal. *Proc. Indian Sci. Congr., Calcutta* **18**, 222.

—— (1932). Some abnormalities in the Indian leech *Hirudinaria granulosa*. *Proc. Indian Sci. Congr., Calcutta* **19**, 261–2.

—— (1934). Nouvelle sangsue rhynchobdellide *Glossiphonia lobata* n. sp. de l'éstablissement de pisciculture d'Achha Bal (Kashmir). *Ann. Parasitol. hum. comp.* **12**, 121–9.

—— (1936). The excretory system of the leech *Hirudinaria*. *Proc. Indian Sci. Congr., Calcutta* **23**, 349.

—— (1938*a*). On structural variations in the Indian leech *Hirudinaria granulosa*. *Curr. Sci.* **6**, 439–42.

—— (1938*b*). On the structure of the nephridia and 'funnels' of the Indian leech *Hirudinaria*, with remarks on these organs in *Hirudo*. *Q. microsc. Sci.* **81**(part 1), 27–80.

—— (1939*a*). The prostomial glands of the Indian leech, *Hirudinaria granulosa*. *J. Morph.* **64**, 37–46.

—— (1939*b*). Fauna of the Dal Lake, Kashmir. I. On some leeches from the Dal Lake, Kashmir. *Bull. Dept. Zool. Punjab. Univ. Lahore* **2**, 1–16.

—— (1940). On *Haemopis indicus* n. sp., a new arhynchobdellid carnivorous leech from Kashmir. *Proc. natn. Acad. Sci.* **10**(4), 133–48.

—— (1941). *Hirudinaria* (The Indian cattle leech.) *Ind. zool. Mem. Lucknow* **8**, 1–85.

—— (1945). On the nephridial system of the Indian carnivorous leech *Haemopis indicus* Bhatia. *Proc. natn. Inst. Sci.* **11**(1), 7–13.

—— (1947). *Limnatis nilotica*, a leech causing laryngo-pharyngitis in man. *Proc. Indian Sci. Congr., Bangalore* **33**(3), 121–2. (Abstr.)

—— (1956). Extra ocular photoreceptors in the land leech, *Haemadipsa zeylanica agilis* (Moore) from Nainital, Almora (India). *Nature, Lond.* **176**, 420–1.

—— (1970). The segmentation of the Gnathobdellid leeches with special reference to the Indian leech *Hirudinaria* and the medicinal leech *Hirudo*. *J. Morph.* **132**(3), 361–76.

—— (1975). Land leeches, their adaptation, and responses to external stimuli. *Zool. Pol.* **25**(2–3), 31–53.

—— (1977). *Hirudinaria*. (The Indian cattle leech) with appendix on *Hirudo medicinalis* (the medicinal leech) and *Haemopis sanguisuga* (the horse leech). *Memoirs on Indian Animal Types*. Emkay Publications, Delhi.

—— and Bora, S.S. (1973). Bionomics and distribution of the land leeches of Kumaon Hills, U.P. *J. Bombay nat. Hist. Soc.* **70**(1), 36–56.

Bhatnagar, K.P., and Shrivastava, A.K. (1966). On the abnormal genital system in the leech *Hirudinaria grandulosa* (Savigny). *Curr. Sci.* **35**(4), 97–8.

Bhatt, B.D. (1959*a*). Excretion of some terrestrial and freshwater leeches. *Proc. Indian Sci. Congr.* **46**, 374–5.

—— (1959*b*). Observations on the behaviour of some land and aquatic leeches. *Proc. Indian Sci. Congr.* **46**, 406.

—— (1960*a*). Observations on the behaviour of some land and aquatic leeches. *Proc. Indian Sci. Congr.* **47**, 444–5.

—— (1960*b*). On some fresh-water leeches from Kumaon (U.P.). *Proc. Indian Sci. Congr.* **47**, 445.

—— (1961). On some water leeches from Naini Tal district (U.P.) *Proc. natn. Acad. Sci. India* **31**(3), 311–14.

—— (1963). On the excretion of some terrestrial and fresh-water leeches. *Zool. Beitr.* n. F. **8**(2), 167–72.

—— and Bhatia, M.L. (1958). Investigations on the control of land leeches in Kumaon. *Proc. Indian Sci. Congr.* **45**, 344.

Bhattacharya, B.K., and Feldberg, W. (1958). Comparison of the effects of eserine and neostigmine on the leech muscle preparation. *Br. J. Pharmacol. Chemother.* **13**, 151–5.

Bianchi, S. (1962). Ricerche istochimiche e fluoromicroscopiche su neuroni cromaffini degli Irudinei. *Arch. Zool. Ital.* **47**, 339–51.

—— (1963). Sul comportamento del nucleolo dei neurociti dei gangli di *Hirudo medicinalis* L. di fronte al nitrato de agrento ammoniacale. *Atti Soc. Peloritana Sci. Fis. Mat. Nat.* **9**(3/4), 207–19.

—— (1964*a*). Neurosecrezione e sostanze fenoliche nei gangli degli Anellidi: 1.— Ricerche sugli Irudinei (*Hirudo medicinalis* L.). *Atti Soc. Peloritana Sci. Fis. Mat. Nat.* **10**, 287–300.

—— (1964*b*). Istochimica del neurosecreto delle cellule nervose di *Hirudo medicinalis* L. *Atti Soc. Peloritana Sci. Fis. Mat. Nat.* **10**(3), 319–25.

—— (1967*a*). On the different types of fluorescent neurons in the leech (*Hirudo medicinalis*). *Atti Soc. Peloritana Sci. Fis. Mat. Nat.* **13**, 39–47.

—— (1967*b*). Sulla secrezione monoaminica dei gangli nervosi di *Hirudo medicinalis*. *Boll. Soc. Ital. Biol. Sper.* **43**, 1136–8.

—— (1969). Sur la signification des neurons fluorescents chez la sangsue *Hirudo medicinalis*. *Ann. Endocr.* **30**, 545–8.

—— (1974). The histochemistry of the biogenic monoamines in the central nervous system of *Hirudo medicinalis*. *Gen. comp. Endocr.* **22**(2), 245–9. [1974BA: 61149.]

Biedermann, W. (1910). Die Aufnahme, Verarbeitung und Assimilation der Nahrung. IV. Die Anneliden, A. Hirudineen. In *H. Winterstein's Handbuch der vergleichenden Physiologie*, Vol. 2(1), pp. 540–51.

Biedl, A. (1910). See Autrum (1939*c*).

Bielecki, A. (1976). (Fauna of leeches (Hirudinea) of the fish ponds in Lagow near Zgorzelec). *Przegl. Zool.* **20**(3), 328–30. [1978BA:65208.]

—— (1976/77). [The leeches of the fishes living in the rivers and streams of the Dale of Klodzko.] *Przegl. Zool.* **21**(2), 141–3. [In Polish; English summary.] [*Cystobranchus fasciatus* and *C. respirans.*] [1979BA:35163.]

—— (1979). [*Trocheta bykowskii,* new record Hirudinea in western Sudety Mountains, Poland.] *Przegl. Zool.* **23**(1), 31–3. [In Polish.] [1980BA:25946.]

—— and Tarnawski, D. (1980). [*Haementeria costata* (Hirudinea, Glossiphoniidae) in Bulgaria and its host.] *Przegl. Zool.* **24**(4), 457–62. [In Polish.]

Biesiadka, E., Kasprzak, K., and Kolasa, J. (1978). Effects of artificial rise of temperature on stagnant waters and their biocoenoses. *Int. Rev. Gesamten Hydrobiol.* **63**(1), 41–56. [1979BA:27187.]

Bilinski, W. (1961). Fauna pijawek powiatow koninskiego i kolskiego. Zakland Zoologii Ogolnej Uniwersytetu Lodzkiego, Lodz. MS. (see Kasprzak, 1977).

Billet, A. (1904). See Autrum (1939*c*).

Biondi, C., Belardetti, F., Brunelli, M., Portolan. A., and Trevisani, A. (1982). Increased synthesis of cyclic AMP and short-term plastic changes in the segmental ganglia of the leech *Hirudo medicinalis. Cell. mol. Neurobiol.* **2**(2), 81–92. [1983BA:65146.]

—— Belardetti, F., Brunelli, M., and Trevisani, A., (1982). Modulation of cyclic AMP levels by neurotransmitters in excitable tissues of the leech *Hirudo medicinalis. Comp. Biochem. Physiol. C Comp. Pharmacol.* **72**(1), 33–8. [1983BA:11038.]

Birshtein, Ya.A. (1970). Caractéristique zoogéographique de la faune souterraine de l'Union Soviétique. In *Livre du centenaire Émile G. Racovitza* 1868–1968 (ed. T. Orghidan and M. Dumitresco) pp. 211–21. Editions de L'Académie de la Réplic Socialiste de Roumanie, Bucarest. [*Dina absoloni ratchaensis,* sub-terranean habits, Crimea and central Asia.]

—— *et al.* (1968). [The atlas of invertebrates of the Caspian Sea.] *Pischevaya Promyshlennost Moscow* 51–115. [In Russian.][Leeches: *Archaeobdella esmonti,* fig. *Caspiobdella tuberculata,* fig. *"Piscicola" caspica,* fig.]

Bishop, A. (1932, 1935, 1937). See Autrum (1939*c*).

Biswas, T., and Mukherjee, A.K. (1978). Isolation of a collagen fraction from the body-wall glycoproteins of the leech (*Hirudo medicinalis*), and characterization of its carbohydrate-amino acid portion. *Carbohydrate Res.* **63**, 173–81.

Blackshaw, S.E. (1981*a*). Morphology and distribution of touch cell terminals in the skin of the leech. *J. Physiol.* **320**, 219–28.

—— (1981*b*). Sensory cells and motor neurons. In: *The neurobiology of the leech.* (eds. K.J. Muller, J.G. Nicholls and G.S. Stent) Chap. 5, pp. 51–78. Cold Spring Harbor, New York.

—— (1981*c*). Morphology of nociceptive terminals in the body wall of the leech. *J. Physiol.* **317**, 81–2.

—— Mackay, D.A. and Thompson, S.W.N. (1984). The fine structure of a leech stretch receptor neurone and its efferent input. *Proc. Physiol. Soc.* 1984 (January), page 79P.

—— and Nichols, J.G. (1979). Distribution and morphology of touch cell endings in leech skin. *J. Physiol.* **292**, 26P–28P. [See also: *Neurosci. Abstr.* **5**.]

—— —— and Parnas I. (1982*a*). Physiological responses, receptive fields and terminal arborizations of nociceptive cells in the leech. *J. Physiol.* **326**, 251–60. [1983BA:34288.]

—— —— —— (1982*b*). Expanded receptive fields of cutaneous mechanoreceptor cells following deletion of single neurons in the CNS of the leech. *J. Physiol.* **326**, 261–8. [1983BA:34289.] [see also: *Neurosci. Abstr.* **7**.]

—— Parnas, I., and Thompson, S.W.N. (1984). Changes in the central arborization of primary afferent neurones during development of the leech nervous system. *Proc. Physiol. Soc.* 1984 (March): page 70P.

Blagovidova, L.A. (1973*a*). [Effect of environmental factors on lake zoobenthos in southern part of western Siberia.] *Gidrobiol. Zh.* **9**(1), 55–61. [In Russian; English summary.]

—— 1973*b*. Effect of environmental factors on the zoobenthos of lakes in southwestern Siberia. *Hydrobiol. J.* [English transl. of *Gidrobiol. Zh.*] **9**(1), 22–6.

Blair, S.S. (1982). Interactions between mesoderm and ectoderm in segment formation in the embryo of a glossiphoniid leech. *Devel. Biol.* **89**, 389–96.

—— (1983). Blastomere ablation and the developmental origin of identification monoamine-containing neurons in the leech (*Helobdella triserialis*). *Devel. Biol.* **95**(1), 65–72. [see also *Neurosci. Abstr.* 1982.] [1983BA:49783.]

—— and Weisblat, D.A. (1982). Ectodermal interactions during neurogenesis in the glossiphoniid leech *Helobdella stagnalis*. *Devel. Biol.* **91**(1), 64–72. [1983BA:11036.]

—— —— (1984). Cell interactions in the developing epidermis of the leech *Helobdella triserialis*. *Devel. Biol.* **101**, 318–25.

Blair, W.N. (1928). Notes on *Hirudo medicinalis*, the medicinal leech as a British species. *Proc. zool. Soc. Lond.* 1927, 999–1002. [Entered as Blaire, W.N. (1928) in Autrum (1939*c*).]

Blake, I.H. (1945). An ecological reconnaissance in the Medicine Bow Mountains. *Ecol. Monogr.* **15**, 207–42.

Blanchard, E. (1849). Annelides. In *Gay's Historia fisca y politica de Chile. Zoologia, Paris* **3**, 43–50.

Blanchard, R. (1888). See Autrum (1939*c*).

—— (1891). Courtes notices sur les Hirudinées. I. Sur la sangsue du cheval du nord de l'Afrique (*Limnatis nilotica* Sav. 1820). *Bull. Soc. zool. France* **16**, 218–21. (and *C. r. Soc. Biol. Paris* **3**, 693–6.) [See also: Brumpt, É. (1919).]

—— (1892*a*). Présence de *Glossiphonia tesselata* au Chili. *Acta Soc. Sci. Chili* **2**, 177–87.

—— (1892*b*). Courtes notices sur les Hirudinées. II. Sur la *Typhlobdella* Kovátsi Diesing. *Bull. Soc. zool. France* **17**, 35–9

—— (1892*c*). Courtes notices sur les Hirudinées. III. Description de la *Nephelis atomaria* Carena. *Bull. Soc. zool. France* **17**, 165–72.

—— (1892*d, e, f*). Courtes notices sur les Hirudinées. IV. Description de la *Glossiphonia marginata* (O.F. Müller). *Bull. Soc. zool. France* **17**, 173–8. V. Description de la *Glossiphonia sexoculata* Bergmann. *Bull. Soc. zool. France* **17**, 178–82. VI. Sur le *Branchellion punctatum* Baird 1869. *Bull. Soc. zool. France* **17**, 222–3.

—— (1892*g*). Description de la *Glossiphonia tessellata*. *Mém. Soc. zool. France* **5**, 56–68.

—— (1892*h*). Description de la *Xerobdella lecomtei*. *Mém. Soc. zool. France* **5**, 539–53.

—— (1892*i*). Sur la présence de la *Trocheta subviridis* en Ligurie et description de cette Hirudinée. *Atti Soc. Lingust. Sci. nat.* **3**(4), 1–31.

—— (1893*a–h*). Courtes notices sur les Hirudinées. *Bull. Soc. zool. France* **18**, (*a*) VII. Sur le *Theromyzon pallens* Philippi, 1867, 14–16 (and *Acta. Soc. Sci. Chili* 3₅, 25–7); (*b*) VIII. Sur *l'Hirudo brevis* Grube, 1871, 26–29; (*c*) IX. Variations de la constitution du somite, 30–5; (*d*) X. Hirudinées de l'Europe boréale, 92–8; (*e*) XI. Description de la *Placobdella catenigera* M.-Tn. 1846, 98–104; (*f*) XII. Description de la *Placobdella carinata* Diesing, 1850, 104–8; (*g*) XIII. Sur les *Hirudo cylindrica* et *H. gemmata* Blanch., 1849, 110–11; (*h*) XIV. Sur la *Blennobdella depressa* Em. Blanchard, 1849, 111–13.

—— (1893*i*). *Torix mirus* (n. g., n. sp.). *Bull. Soc. zool. France* **18**, 185–6.

—— (1893*j–m*). Courtes notices sur les Hirudinées. *Bull. Soc. zool. France* **18**. (*j*) XV. Sur la *Nephelis sexoculata* Schneider, 1883, 194–5; (*k*) XVI. Sur la *Nephelis scripturata* Schneider, 1885, 195–6; (*l*) XVII. Sur la *Nephelis crassipunctata* Schneider, 197; (*m*) XVIII. Encore la *Glossiphonia tessellata*, 197–8.

—— (1893*n*). Sur une Sangsue terrestre du Chili. *C.r hebd. Séanc. Acad. Sci. Paris* **116**, 446–7.

—— (1893*o*). Sur une Sangsue terrestre du Chili. *Ann. nat. Hist.* **12**, 75–6.

—— (1893*p*). Révision de Hirudinées du Musée de Turin. *Boll. Mus. Zool. Anat. Torino* **8**, No. 145

—— (1893*q*). Sur quelques Hirudinées du Piémont. *Boll. Mus. Zool. Anat. Torino* **8**, No. 146.

—— (1893*r*). Viaggio del Dr. E. Festa in Palestina, ne Libano e regioni vicine. 3. Hirudinées. *Boll. Mus. Zool. Anat. Torino* **8**, No. 161.

—— (1893*s*). Voyage du docteur Th. Barrois aux Açores. Hirudinées. *Rev. biol. Nord France* **6**, 40.

—— (1893*t*). Voyage du docteur Th. Barrois en Syrie. Hirudinées. *Rev. biol. Nord France* **6**, 41–6.

—— (1893*u*). Verzeichnis der im Gr. Plöner See gesammelten Hirudineen. *Forschber. Plön* **2**, 66–9.

—— (1893*v*). Sanguijuelas de la peninsula iberica. *Ann. Soc. Hist. nat. Espan.* **22**, 243–58.

—— (1893*w*). *Supplément à la notice sur les titres et travaux scientifiques.* Paris.

—— (1894*a*). Viaggio di Leonardo Fea in Birmania e regioni vicine. 57. Hirudinées. *Ann. Mus. Stor. nat. Genova* (2) **14**, 113–18.

—— (1894*b*). Révision des Hirudinées du Musée de Dresde. *Abh. Ber. Zool. Mus. Dresden* 1892–93, No. 4.

—— (1894*c*). Hirudinées de l'Italie continentale et insulaire. *Boll. Mus. Zool. Anat. Torino* **9**, No. 192.

—— (1894*d*). Courtes notices sur les Hirudinées. XIX. Sur les *Branchellion* des mers d'Europe. *Bull. Soc. zool. France* **19**, 85–8.

—— (1896*a*). Courtes notices sur les Hirudinées. XX. Hirudinées de la Prusse orientale. *Bull. Soc. zool. France* **21**, 118–20.

—— (1896*b–e*). Courtes notices sur les Hirudinées. (*b*) XXI. Sur la (*Glossiphonia?*) *scutifera* Young, 1894. *Bull. Soc. zool. France* **21**, 137; (*c*) XXII. Hirudinées de l'île Borkum. *Bull. Soc. zool. France* **21**, 137–8; (*d*) XXIII. Hirudinées de Terre-Neuve et des îles adjacentes. *Bull. Soc. zool. France* **21**, 138–40; (**e**) XXIC. Présence de la *Glossiphonia complanata* en Amérique. *Bull. Soc. zool. France* **21**, 140.

—— (1896*f*). Description de quelques Hirudinées asiatiques. *Mém. Soc. zool. France* **9**, 316–30.

—— (1896*g*). Hirudineen aus dem Togoland. *Arch. Naturg* **62** I, 49–53.

—— (1896*h*). Viaggio del Dott. A. Borelli nella Republica Argentina e nel Paraguay. 21. Hirudinées. *Boll. Mus. Zool. Anat. Torino* **11**, No. 263.

—— (1896*i*). Campagnes de l'Hirondelle et de la Princesse Alice. Hirudinées. *Bull. Soc. zool. France* **21**, 196–8. (Reprinted posthumously in 1936, *Res. camp. Sci. Monaco* **96**, 76–8.)

—— (1897*a*). Hirudineen Ost-Afrikas. Die thierwelt Ost-Africas 4, Lfg. 2, Berlin (Dietr. Reimer).

—— (1897*b*). Hirudinées des Indes néerlandaises. *Weber's Zool. Ergebn. in Niederl. Ost-Indian*, **4**, 332–56.

—— (1897*c*). Hirudinées du Musée de Leyde. *Not. Zool. Mus. Leyden* **19**, 73–113.

—— (1898). Nouveau type d'Hirudinée (*Torix mirus*). Bull. Sci. France Belg. **28**, 339–44.

—— (1899*a*–*d*). Courtes notices sur les Hirudinées. (*a*) XXV. Sur la *Clepsine maculosa* Rathke, 1862; (*b*) XXVI. Sur la *Clepsine polonica* Lindenfeld et Pietruszynski, 1890; (*c*) XXVII. Sur les genres *Liostoma* Wagler et *Haementeria* de Filippi; (*d*) XXVIII. Sur le *Liostomum Ghilianii* (F. de Filippi). *Bull. Soc. zool. France* **24**, 181–9.

—— (1899*e*). Apropos de Sangsues fixées dans le pharynx. *Arch. Parasis., Paris* **2**, 142–4.

—— (1900). Hirudineen. *Erg. Hamb. Magalhaens. Sammelreise* 1892/93, Hamburg, 3.

—— (1905). Hirudineen aus Montenegro. *SB. Böhm. Ges. Wiss. Prag*, 1905, No. 12, 1–3.

—— (1908). Hirudinées. In H Gadeau de Kerville. *Voyage Zoologique en Khroumirie (Tunisie)*, V–VI 1906, pp.307–10. Paris.

—— (for biography see: 1908. *Chanteclair* No. 11, 3 pp.).

—— (1917). Monographie des Hémadipsines (Sangsues terrestres). *Bull. Soc. Path. exot.* **10**, 640–75.

Blest, A.D. *et al.* (1983). The cytoskeleton of microvilli of leech photoreceptors. A stable bundle of actin microfilaments. *Cell Tissue Res.* **234**(1), 9–16.

Blinn, D.W. and Dehdashti B. (1984). The nocturnal feeding behaviour of *Erpobdella punctata* (Hirudinoidea) in a near thermally constant environment. 47th Annual Meeting of the American Society of Limnology and Oceanography, Vancouver, B.C. Abst. p. 9. (= *E.montezuma*).

Blyth, R.I.K. (1950). War on leeches. *Br. med. J.* **ii**, 1058.

Boardman, W. (1933). Leeches. *Aust. Mus. Mag.* **v**(2), 64–9.

Bobin, G. (1949). Remarques cytologiques sur les cellules graisseuses de *Glossossiphonia complanata* L. *Bull. Soc. zool. France* **74**(4)(5), 300–7.

—— (1950). Sur les cellules à sphérules colorées et leur parenté avec les cellules adipeuses chez *Glossossiphonia complanata* L. *Arch. Zool. exp. gene.* **87**, 69–94.

Boehm, G. (1947). Uber eine rotfluoreszierende, als Porphyrin anzusehende Substanz in den Augen von *Hirudo medicinalis*. *Experientia, Basel* **3**(6), 241.

Bohl, M. (1973). Prophylaxe und Therapie von Fischkrankheiten. *Münchener Beit. Abwasser. Fisch. Flussbiol.* **21**, 52–65. [Control of leech infestation.]

Bojanowski, C. (1862). Beobachtungen über die Blutkrystalle. *Z. wiss. Zool.* **12**, 312–32. [Plate XXX, Fig. 1: crystals from gut of *Hirudo*.]

Bolsius, H. (1889). Recherches sur la structure des organes segmentaires des Hirudinées. *Cellule* **5**, 367–436.

—— (1894). See Bolsius, H. (1894*c*) in Autrum (1939*c*).

—— (1897). Unpaired gland of *Haementeria*. *J.R. microsc. Soc.* 1897(2), 125–6.

—— (1899). See Bolsius, H. (1899*b*) in Autrum (1939*c*).

—— (1900). See Bolsius, H. (1900*b*) in Autrum (1939*c*).

Bora, S.C., and Bhatia, M.L. (1958). Nephridial system of the land leech *Haemadipsa zeylanica agilis* (Moore). *Proc. Indian Sci. Congr.* **45**, 343–4.

Borgstrom, R., and Halvorsen, O. (1972). [New records of fish leeches from Norway.] *Fauna, Oslo* **25**(1),31–4. [English summary.]

Boroffka, I. (1968). Osmo- und Volumenregulation bei *Hirudo medicinalis*. *Z. vergl. Physiol.* **57**, 348–75. [See also Zerbst-Boroffka, I.]

—— and Hamp, R. (1969). Topographie des Kreislaufsystems und Zirulation bei *Hirudo medicinalis* (Annelida, Hirudinea). *Z. Morph. Tiere* **64**(1), 59–76.

—— Altner, H., and Haupt, J. (1970). Funktion und Ultrastruktur des Nephridiums von *Hirudo medicinalis:* I. Ort und Mechanismus der Primärharnbildung. *Z. Vergl. Physiol.* **66**(4), 421–38.

Borovitzkaia, M. (1949). [On parasitic leeches of the family Ichthiobdellidae occurring in the pallial cavity of Cephalopod Mollusca.] *C. r. Acad. Sci. URSS,* new series **68**(2), 425–7. [*Crangonobdella achmerovi = Platybdella fabricii*.]

Botallo, L. (1577). De curiatione per sanguinis missionem, de incidendae venae, cutis scarificandae et hirudinum applicandarum modo. Lyons. [In W.I.M.H. Lib.]

Bottenberg, H. (1935). Die Blutegelhandlung ein vielseitiges Verfahren des biologischen Medizin. 128 pp. Hipprokrates, Stuttgart. [*Surg. Gen. Cat.* 4th Ser. vol. 2, 1937.]

—— (1943). Neue Gesichtspunkte fur die Blutegelbehandlung. *Munch. Med. Wschr.* **90**, 128.

—— (1947). Nuevos puntos de vista en el tratamiento por sanguijuelas. *Actual. med. Granada* **33**, 100.

—— (1948). *Die Blutegelbehandlung.* Hippokrates-Verlag, Stuttgart.

Botzler, R.G., Wetzler, T.F., and Cowan, A.B. (1973). *Listeria* in aquatic animals. *J. wild. Dis.* **9**(2), 163–70.

Bourcart, N. (date undetermined). Technique d'isolement d'une souche neuve d'*Helobdella algira* à partir d'exemplairus parasites par des trypanosomes d'amphibiens. *Ann. Parasitol. hum. comp.* **30**, 504–5.

Bourne, A.G. (1884). Contributions to the anatomy of the Hirudinea. *Q. Jl microsc. Sci.* **24**, 419–506.

—— (1888). See Autrum (1939*c*).

Bouvet, J. (1968*a*). Notes sur les Hirudinées des Alpes françaises *Trocheta bykowskii* Gedroyc, 1913 (Sous-ordre des Pharyngobdellae, famille des Erpobdellidae). *Trav. Lab. Hydrobiol. Piscic. Un. Grenoble* **57–8** (1956–66), 105–9.

—— (1968*b*). Notes sur les Hirudinées des Alpes françaises. 2. *Dina lineata* O.F. Müller 1774 (Sous-ordre des Pharyngobdellae, famille des Erpobdellidae.) *Trav. Lab. Hyrobiol. Piscic. Un. Grenoble* **59–60**, 203–6.

—— (1977). Notes sur les Hirudinées des Alpes francaises. III. *Erpobdella octoculata* L. (Sous-ordre des Pharyngobdellae, famille de Erpobdellidae). *Trav. Lab. Hydrobiol. Piscic. Un. Grenoble* **66–8**, 89–94.

Bower, S.M., and Woo, P.T.K. (1977*a*). Morphology and host specificity of *Cryptobia catostomi* n. sp. (Protozoa: Kinetoplastida) from white sucker (*Catostomus commersoni*) in southern Ontario. *Can. J. Zool.* **55**, 1082–92.

—— (1977*b*). Division and morphogenesis of *Cryptobia catostomi* (Protozoa: Kinetoplastida) in the blood of white sucker (*Catostomus commersoni*). *Can. J. Zool.* **55**, 1093–9.

Bowling, D., Nichols, J., and Parnas, I. (1978). Destruction of a single cell in the central nervous system of leech as a means of analysing its connexions and functional role. *J. Physiol.* **282**, 169–80. [1979BA:28650.]

Boynton, W.H. (1913). Duration of the infectiveness of virulent Rinderpest blood in the water leech, *Hirudo boyntoni* Wharton. [= *Hirudinaria manillensis.*] *Philipp. J. Sci.* **8B**, 509–21. [see also: 1928, **36B**, 1–35.]

—— (1928). See above and Autrum (1939*c*).

Braconnier-Faymendy, M. (1933*b*). [Studies on *Hirudo* excretion.] Bordeaux. [Possibly a thesis; summary of data and this incomplete reference found in Heidermanns, 1937.] [see also: *C. r. hebd. Séanc. Soc. Biol., Paris* **114**, 705–6.]

Bradbury, S. (1955). A cytological study of the metabolism of iron in the leech, *Glossiphonia complanata. Q. Jl. microsc. Sci.* **96**(2), 169–72.

—— (1956). A histochemical study of the adipose cell of the leech, *Glossiphonia complanata. Q. Jl microsc. Sci.* **97**(4), 499–517.

—— (1957*a*). A histochemical study of the connective tissue-fibres in the leech, *Glossiphonia complanata. Q. Jl microsc. Sci.* **98**(3), 29–45.

—— (1957*b*). A histochemical study of the pigment cells of the leech, *Glossiphonia complanata. Q. Jl microsc. Sci.* **98**(3), 301–14.

—— (1958). A cytological and histological study of the connective tissue fibres of the leech (*H. medicinalis*). *Q. Jl microsc. Sci.* **99**(2), 131–42.

—— (1959). The botryoidal and vaso-fibrous tissue of the leech, *Hirudo medicinalis. Q. Jl microsc. Sci.* **100**(4), 483–98.

—— and Meek, G.A. (1958*a*). The fine structure of the adipose cell of the

leech *Glossiphonia complanata. J. biophys. biochem. Cytol.* **4**(5), 603–7.
—— —— (1958*b*). A study of fibrogenesis in the leech, *Hirudo medicinalis. Q. Jl microsc. Sci.* **99**(2), 143–8.

Bragina, E.V. (1970). [On the parasite fauna of fish in water reservoirs of the Kustanai region (USSR).] In *Rybn. resursy vodoemov Kazakhstana i ikh ispol'z*, Alma-Ata, USSR. 'Nauka', No. 6, pp. 290–7. [In Russian; from *Referativnyi Zhurnal, Biologiya* 8 K8 (1970).] [Two leeches: Kazakhistan.]

Brake, Z. (1958). *Hirudo medicinalis* spermatogeneze, *Lat. P.S.R. Zinat. Akad. Vestis* **3**, 99–106.

Brandersky, G. (1957). Die neuzeitliche Behandlung des akuten Myokardinfarktes. *Wien. Med. Wschr.* 820.

Brandes, G. (1900*a*). See Autrum (1939*c*).

—— (1900*b*). Die Begattung von *Clepsine tessulata. Z. Naturwiss.* **73**, 126–8.

—— (1901). Die Begattung der Hirudineen. *Abl. naturf. Ces. Halle* **22**, 373–82.

—— Brandt, B.B. (1936). Parasites of certain North Carolina Salientia. *Ecol. Monogr.* **6**, 491–532. [*Macrobdella ditetra.*]

Brandt, J.F. (1833). See Autrum (1939*c*).

—— (1836). Remarques sur les nerfs stomatogastriques ou intestinaux, dans les animaux invertébrés. *Ann. sci. nat. Zool.* Series 1, Vol. 5. [Translation of German memoirs inserted in Vol. III of the *Mémoires de l'Académie Impériale des Sciences de Saint-Pétersbourg.*]

Branning, T. (1982). Learning from a giant leech. *Nat. Wildl.* **20**(3), 35–7.

Branson, B.A. (1963). Notes on distribution and leech feeding habits in Oklahoma. *Nautilus* **76**(4), 148–9.

—— and Amos, B.G. (1961). The leech *Placobdella pediculata* Hemingway parasitizing *Aplodinotus grunniens* in Oklahoma. *Southwest Nat.* **6**, 53.

Braude, G.L. (1948). [Observation on insemination in *Hirudo medicinalis.*] *Dokl. Akad. Nauk. SSSR Moscow* **63**(3), 325–8. [In Russian.]

—— (1950). O topographicheskoi anatomii polovogo apparata meditsinskoi piyavki. *Dokl. Akad. Nauks SSR Moscow* **70**(2), 307–10. [In Russian.]

—— (1951*a*). [Investigations on the development of teeth in *Hirudo medicinalis* and *Haemopis sanguisuga.*] *Dokl. Akad. Nauk SSSR Moscow* **79**(6), 1041–4. [In Russian.]

—— (1951*b*). [On the structure of the dental apparatus of *Haemopis sanguisuga.*] *Dokl. Akad. Nauk. SSSR Moscow* **86**, 869–72. [In Russian.]

Brezeanu, G., and Marinescu-Popescu, V. (1964). [Hydrobiologisches Studium des Cerna-Beckens.] *Hidrobiologia, Bucuresti* **5**, 65–94. [In Rumanian; German summary.]

Brien, P. (1947). *Guide des Travaux pratiques de zoologie.* Masson et

Desoer, Paris. [Contains a guide to the dissection of *Hirudo.*]

Brightwell, L.R. (1940). The lore of the leech. *Aquarist, Lond.* **10**, 223–4.

Brinck, P. and Wingstrand, K.G. (1951). The mountain fauna of the Virihaure area in Swedish Lapland. *Lunds Univ. Arssk. N.F.* afd. 2. Bd. 46, **2**, 1–173. [*Glossiphonia concolor.*]

Brinkhurst, R.O. (1982). Evolution in the Annelida. *Can. J. Zool.* **60**, 1043–59.

—— and Jamieson, B.G.M. (1971) *Aquatic oligochaeta of the world.* pp. 860. Oliver and Boyd, Edinburgh.

Brinkmann, A. Jr (1947). Two new antarctic leeches. *Nature, Lond.* **160**, 756.

——(1948). Some new and remarkable leeches from the antarctic seas. *Norske Vid. Akad. Oslo* 3–17. [*Scient. Res. Norw. Antarct. Exped.* 1927–28, Oslo, No. 29, 1–17.]

Bristol, C.L. (1898). The metamerism of *Nephelis.* A contribution to the morphology of the nervous system with a description of *Nephelis lateralis. J. Morph.* **15**, 17–72.

Brockbank, W. (1954). The ancient arts of cupping and leeching — bibliography. In *Ancient therapeutic arts,* pp. 100–2. Heinemann, London.

Brockleman, W.Y. (1969). An analysis of density effects and predation in *Bufo americanus* tadpoles. *Ecology* **50**(4), 632–44. [See also: Natural regulation of density in tadpoles of *Bufo americanus.* Ph.D. thesis, University of Michigan (1968).]

Brodfuehrer, P.D., and Friesen, W.O. (1984). A sensory system initiating swimming activity in the medicinal leech. *J. exp. Biol.* **108**, 341–55.

—— —— (submitted) Identification of neurones in the leech through local ionic manipulations. *J. exp. Biol.* **113**, 455–60.

Bronn, H.G. (1936–39). Hirudinea. *Klassen und Ordnungen des Tierreichs.* Akademische Verlagsgesellschaft, Leipzig. [See Autrum (1936, 1939*a*–*c*); Herter (1936*b*, 1937); Schleip (1939).]

Brook, D.R., and Welch, N.J. (1977). *Marvinmeyeria lucida* (Moore, 1954) (Annelida: Hirudinea) a commensal of *Helisoma trivolvis* (say) (Mollusca: Gastropoda) in Nebraska. *Trans. nebr. Acad. Sci.* **4**, 21–2. [Identification doubtful.] [1978BA:20124.]

Brooks, V.K. (1882). *Handbook of invertebrate zoology.*

Brotskaya, V.A., Zhdanova, N.N., and Semyonova, N.L. (1963). Bottom fauna of the Velikaya Salma and the adjoining regions of Kandalaksha Bay of the White Sea. *Trudy belom. Biol. Sta. M.G.V.* **2**, 159–82. [In Russian; English summary.] [*Mysidobdella borealis* from Kandalaksha Bay, White Sea.]

Brown, J.E., Baugh, R.F., and Hougie, C. (1980). Inhibition of the intrinsic generation of activated factor-X by heparin and hirudin. *Thromb. Res.* **17**, 267–72. [1980BA:58992.]

Brown, R.B. (1968). A fall and winter population of the macro-invertebrate fauna of Lincoln Beach, Utah Lake, with notes on invertebrates in fish stomachs. Unpublished Master's thesis, Department of Zoology, Brigham Young University, Provo, Utah.

Bruces, J.R., Colman, J.S., and Jones, N.S. (1963). *Marine fauna of the Isle of Man.* Liverpool University Press. [Leeches, p.109.]

Bruck, F. (1943). [Leeches as substitutes for venesection.] *Wien med. Wchnschr.* **93**, 422. [In German.]

Bruenko, V.P. (1976). [The feeding of pike *Exos lucius* of the Kremenchug Reservoir, Ukrainian SSR, USSR, during spawning.] *Gidrobiol. Zh.* **12**(1), 121–5.

Brues, C.T. (1939). Studies on the fauna of some thermal springs in the Dutch East Indies. *Proc. Am. Acad. Sci., Boston* **73**(4), 71–95.

Brumpt, É. (1899). De l'accouplement chez les Hirudinées. *Bull. Soc. zool. France* **24**, 221–38. [See also: Galliard (1952); Hackett (1952); Lavier (1951).]

—— (1900*a*). Reproduction des Hirudinées. Formation du cocon chez *Piscicola* et *Herpobdella. Bull. Soc. zool. France* **25**, 47–51.

—— (1900*b*). Reproduction des Hirudinées. *Mém. Soc. zool. France* **13**, 286–430.

—— (1900*c*). Reproduction des Hirudinées. Recherches expérimentales sur la fécondation. *Bull. Soc. zool. France* **25**, 90–3.

—— (1900*e*). Monographie de la *Clepsine* (*Glossiphonia complanata* Linné). In *Zoologie déscriptive des invertébrés*, Vol. 2, pp. 1–51. Paris.

—— (1901*b*). Reproduction des Hirudinées. Existence d'un tissu de conduction spécial et d'aires copulatrices chez les Ichthyobdellides. *C. r. Ass. Franc. Sci.* (1900), part 2, 688–710.

—— (1904, 1907, 1914). See Autrum (1939*c*).

—— (1906). Expériences relatives au mode de transmission des Trypanosomes et des Trypanoplasmes par les Hirudinées. *C. r. Soc. Biol. Paris* **61**, 77–9.

—— (1919). Le Professeur R. Blanchard (28 Feb. 1857–7 Feb. 1919). *Presse Med.* 13 Feb. 1919.

—— (1941). V. Observations biologiques diverses concernant *Planorbis* (*Australorbis*) *glabratus* hôte intermediaire de *Schistosoma mansoni. Ann. Parasitol. hum. comp.* **18**, 9–45. [Experimental study of leeches feeding on snails.]

—— (1949). Hirudinea. *Précis de Parasitologie*, 6th edn, pp. 119, 316–27. Masson, Paris.

Bruun, A.F. (1928*a,b*, 1938*a*). See Autrum (1939*c*).

—— (1939*b*). Freshwater *Hirudinea*. In *The zoology of Iceland*, 2, Part 22, 1–4.

Brykov, D. (1856). [Handbook for breeding, keeping and using leeches.] (incomplete reference)

Bubinas, A.D. (1979). [Feeding of noncommercial fish in the reservoir of the Kaunas hydroelectric station, Lithuanian SSR, USSR, in 1972–1975: 1. Three-spined and nine-spined sticklebacks.] *Liet. Tsr. Mokslu Adad. Darb. Ser. C Biol. Mokslai* **3**, 85–94. [In Russian; Lithuanian summary.]

Buchanan, R.E., and Gibbons, N.E. (1974). *Bergey's Manual of determinate bacteriology*, 8th edn. Williams and Wilkins, Baltimore.

Büchli, K. (1924). Bloedzuigers in de neusholte van eenden. *Tijdschr. Diergeneesk.* **51**, 153–5. [*Theromyzon tessulatum* in nose and in brain.]

Buchner, P. (1965). *Endosymbiosis of animals with plant microorganisms*. Interscience, New York. [Leeches: 433–7 (digestive); 609–11 (excretory).] [This is a revised edition of an earlier book in German, *Endosymbiose der Tiere mit Pflanzichen Mikroorganismen*, pp. 371–5. Verlag Birkhäuser, Basel (1953).]

Buck, H. (date undetermined). Zut Verbreitung einiger Gruppen nederer Susswassertiere in Fleissgewassern Nordwurttembergs. *Jh. Ver. vaterl. Naturk. Wurttemb.* **3**, 153–73.

Buddenbrook, W. von (1956). *Vergleichende Physiologie*, Bd. III. Birkausen, Stuttgart.

Budge, J. (1849). See Autrum (1939*c*).

—— (1850). [First observation of haematocrystals in gut of *Hirudo*.] *Kölnische Zeitg.* No. 300.

Budzynski, A.Z., Olexa, S.A., and Sawyer, R.T. (1981). Composition of salivary gland extracts from the leech, *Haementeria ghilianii*. *Proc. Soc. exp. Biol. Med.* **168**, 259–65.

—— Olexa, S.A., Brizuela, B.S., Sawyer, R.T., and Stent, G.S. (1981*a*). Anticoagulant and fibrinolytic properties of salivary proteins from the leech *Haementeria ghilianii*. *Proc. Soc. exp. biol. Med.* **168**, 266–75.

—— —— —— —— (1981*b*). Anticoagulant and fibrinolytic properties of salivary gland proteins from the leech *Haementeria ghilianii*. *Thrombosis and Haemostasis* **46**, 64[Abstr. presented at VIIIth International Congress on Thrombosis and Haemostasis, Toronto, July, 1981.]

Buldovskii, A.T. (1936). O biologii i prom'slovom ispol'zovanii ussuriiskoi (amurskoi) cherepakhi *Amyda maaki* Brandt. *Tr. Dal'nevost. Phil. Akad. Nauk SSSR* **I**, 62–102.

Bullock, T.H., and Horridge, G.A. (1965). *Structure and function in the nervous system of invertebrates*. Freeman, San Francisco.

Bunker, T.D. (1980). The contemporary use of the medicinal leech. *Injury: The British Journal of Accident Surgery.* **12**, 430–2. [Leeches used to treat periorbital haematoma.]

Buren, G.V. (1952). Der Geistsee. *Mitt. naturf. Ges. Bern. N. F.* **9**, 1–51.

Bürger, O. (1891). Beiträge zur Entwicklungsgeschichte der Hirudineen. Zur Embryologie von *Nephilis*. *Zool. Jb. Anat.* **4**, 697–738.

—— (1894). Neu Beiträge zur Entwicklungsgeschichte der Hirudineen. Zur Embryologie von *Hirudo medicinalis* und *Aulastomum gulo*. *Z. wiss. Zool.* **58**, 440–59. [see also *Zool. Zbl.* 1, 661–4 (1894).]

—— (1902). Weitere Beiträge zur Entwicklungsgeschichte der Hirudineen. Zur Embryologie von *Clepsine*. *Z. wiss. Zool.* **72**, 525–44.

Burghardt, G.M. (1968). Comparative prey-attack studies in newborn snakes of the genus *Thamnophis*. *Behaviour* **33** (1/2), 77–114.

Burgova, M.P., Kurepina, V.G., and Kondrashova, M.N. (1973). [Pyridine nucleotides and flavoproteins of giant neurons of the medicinal leech.] *Biol. Nauki* **16**(18), 47–50. [In Russian.] [1974BA:66594.]

—— —— —— (1975). [Fluorescence of flavoproteins and pyridine nucleotides of the leech nerve cell: III. Effect of potassium chloride solutions.] *Vestn. Leningr. Un. Ser. Biol.* **4**, 149–50. [In Russian; English summary.] [1976BA:14592.]

—— —— and Stidenkina, A.S. (1974). [Flavoprotein and pyridine nucleotide fluorescence of the leech nerve cells: II. The relationship between redox-reactions of Retzius cells.] *Vestn. Leningr. Un. Ser. Biol.* **3**, 70–7. [In Russian; English summary.] [1975BA:14432.]

—— —— —— and Kondrashova, M.N. (1977). [The fluorescence of flavoproteins and pyridine nucleotides of the leech nerve cell. IV. The effect of succinate solutions.] *Vestn. Leningr. Un. Ser. Biol.* **2**, 63–70. [In Russian; English summary.] [1978BA:46376.]

—— Stidenkina, A.S., Emel'yanenko, N.S., and Kondrashova, M.N. (1980). The redox reaction of flavine and pyridine nucleotides of nerve cells of the leech, *Hirudo medicinalis*. *J. evol. Biochem. Physiol.* **16**(3), 187–91 [English transl. of *Zh. evol. Biokhim. Fiziol.*] [1980BA:15464.]

Burn, P.R. (1980). The parasites of smooth flounder, *Liopsetta putnami* (Gill), from the Great Bay Estuary, New Hampshire. *J. Parasitol.* **66**(3), 532–1.

Burreson, E.M. (1975). Biological studies on the hemoflagellates of Oregon marine fish and their potential leech vectors. Ph.D. Dissertation. pp. 143. Oregon State University, Corvallis.

—— (1976*a*). *Aestabdella* gen. n. (Hirudinea: Piscicolidae) for *Johanssonia abditovesiculata* Moore 1952 and *Ichthyobdella platycephali* Ingram 1957. *J. Parasitol* **62**(5), 789–92.

—— (1976*b*). *Trachelobdella oregonensis* sp. n. (Hirudinea: Piscicolidae), parasitic on the Cabezon, *Scorpaenichthys marmoratus* (Ayres), in Oregon. *J. Parasitol.* **62**(5), 793–8.

—— (1977*a*). A new marine leech *Austrobdella californica* n. sp. (Hirudinea: Piscicolidae) from southern California flatfishes. *Trans. Am. microsc. Soc.* **96**(2), 263–7.

—— (1977*b*). *Oceanobdella pallida* n. sp. (Hirudinea: Piscicolidae) from the English sole, *Parophrys vetulus*, in Oregon. *Trans. Am. microsc. Soc.* **96**(4), 526–30.

—— (1977*c*). Two new marine leeches (Hirudinea: Piscicolidae) from the west coast of the United States. *Instituto de Biologia. Publicaciones Especiales. Excerta Parasitológica en memoria del doctor Eduardo Caballero y Caballero. 29 April 1977*, pp. 503–12.

—— (1977*d*). Two new species of *Malmiana* (Hirudinea: Piscicolidae) from Oregon coastal waters. *J. Parasitol.* **63**(1), 130–6.

—— (1979). Structure and life cycle of *Trypanoplasma beckeri*, new species (Kinetoplastida), a parasite of the Cabezon, *Scorpaenichthys marmoratus* in Oregon, USA, coastal waters. *J. Protozool.* **26**(3), 343–7. [1980BA:51576.] [*Malmiana diminuta* vector.]

—— (1981*a*). A new deep-sea leech, *Bathybdella sawyeri*, n. gen., n.sp., from Thermal Vent areas on the Galápagos Rift. *Proc. biol. Soc. Wash.* **94**(2), 483–91.

——(1981*b*). Effects of mortality caused by the hemoflagellate *Trypanoplasma bullocki* on the summer flounder populations in the Middle Atlantic Bight. *Int. Counc. exp. Sea* G **61**, 1–6.

—— (1982). The life cycle of *Trypanoplasma bullocki* (Strout) (Zoomastigophorea: Kinetoplastida). *J. Protozool.* **29**(1), 72–7.

—— (1984). A new species of marine leech (Hirudinea: Piscicolidae) from the north-eastern Pacific Ocean, parasitic on the English sole, *Parophrys vetulus* Girard. *Zool. J. Linn. Soc.* **80**, 297–301.

—— and Allen, D.M. (1978). Morphology and biology of *Mysidobdella borealis* (Johansson) comb. n. (Hirudinea; Piscicolidae) from mysids in the western North Atlantic. *J. Parasitol.* **64**(6), 1082–91

—— and Zwerner, D.E. (1982). The role of host biology, vector biology and temperature in the distribution of *Trypanoplasma bullocki* infections in the lower Chesapeake Bay. *J. Parasitol.* **68**(2), 306–13.

Büsing, K.H. (1951). *Pseudomonas hirudinis*, ein bakterieller Darmsymbiont des Blutegels (*Hirudo officinalis*). *Zentbl. Bakterio. Parasitenk.* **157**(7), 478–84.

—— Döll, W., and Freytag, K. (1953). Die Bakterienflora der medizinischen Blutegel. *Arch. Mikrobiol.* **19**(1), 52–86.

Butler, W.J. (1948). Leeches in ducks. *Mont. Livest. Sanit. Board Rep.* 1939–40, 12.

Buu-Hoi, N.P. (1962). Repellent action of different chemical substances against earth leeches in Vietnam. *C. r. Soc. Biol., Paris* **156**(2), 277–9.

Bychowsky, A. (1921). Über die Entwicklung der Nephridien von *Clepsine sexoculata* Bergm. (= *compl.* Sav.). Ein Beitrag zum Nephridialproblem. *Rev. Suisse Zool.* **29**, 41–131.

Bykhovskaya-Pavlovskaya, I.E. (1949). [Parasitic fauna of *Perca fluviatilis*

L. and the influence of some ecological factors on its alteration.] *Bull. Acad. Sci. Moscow* (*Biol.*) 316–39.

—— *et al.* (1964). See Epshtein, V.M. (1964*b*).

Bywater, J.E.C., and Mann, K.H. (1960). Infestation of a monkey with the leech *Dinobdella ferox. Vet. Rec.* **72**, 955

Caballero, Ed. y C. (1930*a*). Contribución al conocimiento de los hirudíneos de México. *Limnobdella mexicana* R. Blanchard. *Anal. Inst. Biol. Un. Méx.* **1**, 247–51. [see also M.F. Eliezer.]

—— (1930*b*). Revisión de los hirudíneos mexicanos. II. *Haementeria officinalis. Anal. Inst. biol. Un. Méx.* **1**, 319–25.

—— (1931*a*). *Glossiphonia socimulcensis* n. sp. *Anal. Inst. Biol. Un. Méx.* **2**, 85–90.

—— (1931*b*). Batrachobdellinae subfam. nov. *Anal. Inst. Biol. Un. Méx.* **2**, 223–9. [= *Limnobdella mexicana.*]

—— (1931*c*). Potamobdellinae subfam. nov. *Anal. Inst. Biol. Un. Méx.* **2**(3), 325.

—— (1932*a*). *Herpobdella ochoterenai* n. sp. *Anal. Inst. Biol. Un. Méx.* **3**, 33–9.

—— (1932*b*). Algunas sanguijuelas de la región de Tenancingo. *Anal. Inst. Biol. Un. Méx.* **3**, 41–2.

—— (1932*c*). *Limnobdella tehuacanea* Jiminez 1865, Caballero 1931. *Anal. Inst. Biol. Un. Méx.* **3**, 43–7.

—— (1933*a*). *Haemopis profundisulcata* n. sp. *Anal. Inst. Biol. Un. Méx.* **4**, 23–6.

—— (1933*b*). Sanguijuelas del estado de Guanajuato (*Hirudinea*). *Anal. Inst. Biol. Un. Méx.* **4**, 179–85.

—— (1934*a*). Hirudineos de México. Tesis.

—— (1934*b*). *Limnobdella cajali* n. sp. (Hirudinea). *Anal. Inst. Biol. Un. Méx.* **5**, 237–41.

—— (1935). Hirudineos de Mexico XI. *Glossosiphonia* Castle 1900. *Anal. Inst. Biol. Un. Méx.* **6**, 49–51.

—— (1937). Hirudíneos del valle del Mezquital HGO XII. *Anal. Inst. Biol. Un. Méx.* **8**, 181–8.

—— (1940*a*). Nuevos genero y especie de Hirudineo perteneciente a la subfamilia *Haemadipsinae*. XV. *Anal. Inst. Biol. Un. Méx.* **11**(2), 573–83.

—— (1940*b*). Nota sobre la presencia de la *Placobdella rugosa* (Hirudinea: Glossiphoniidae) en las aquas de lago Xochimilco. *Anal. Inst. Biol. Un. Méx.* **11**(1–2), 255–60.

—— (1940*c*). Sanguijuelos del lago de Patzcuaro y descripción de una nueva especie, *Illinobdella patzcuarensis*. XIV. *Anal. Inst. Biol. Un. Méx.* **11**(2), 449–64.

—— (1941). Hirudineos de México. XVI. Nuevos huespedes y localidades para algunas sanguijuelas ya conocidas y descripción de una nueva especie. *Anal. Inst. Biol. Un. Méx.* **12**(2), 747–57.

—— (1952). Sanguijuelas de México. XVIII. Presencia de *Macrobdella decora* (Say, 1824) Verrill, 1872, en el norte del pais, y una nueva desinencia para los ordenes de Hirudinea. *Anal. Inst. Biol. Un. Méx.* **23**(1–2), 203–9.

—— (1955). Hirudineos de México XIX. Presencia de *Pontobdella macrothela* Schmarla, 1861, en aguas marinas del Golfo de México. *Anal. Esc. Nac. Cienc. Biol. Méx.* **18**(3–4), 153–8.

—— (1956). Hirudíneos de México. XX. Taxa y nomenclature de la clase Hirudinea hasta generos. *Anal. Inst. Biol. Un. Méx.* **27**(1), 279–302. [English summary.]

—— (1958). Hirudineos de México. XXI. Descripción de una nueva especie de sanguijuela procedente de las selas Estado de Chiapas. *Anal. Inst. Biol. Un. Méx.* **28**(1–2), 241–5.

—— (1959). Hirudineos de México. XXII. Taxa y nomenclature de la Clase Hirudinea asta generos (Nueva edición). *Anal. Inst. Biol. Un. Méx.* **30**(1–2), 227–42. [English summary.]

—— (1969). John Percy Moore, 1869–1965. *Anal. Inst. Biol. Un. Nat. Auton. Méx.* **40**, Ser. Zool. (2), 319–20.

—— (1970) 1974. Cambio de Nomenclatura. *Anal. Inst. Biol. Un. Nat. Auton, Méx.* **41**, Ser. Cienc. del Mar y Limnol. (1), 155–6.

Cadenat, J. (1953). La collection de parasites de poissons de mer ouest africains de la Station de Biologie Marine de Gorec. *Bull. Inst. franc. Afr. Noire* **16**a(1), 300–1.

Cadwallader. P.L. (1978). First record of *Richardsonianus australis* (Bosisto, 1959) (Hirudinea: Richardsonianidae) taking a blood meal from a fish. *R. Soc. Victoria Proc,* **90**(1/2), 283–6. [On 3 of 130 galaxiid *Galaxias olidus.*] [1979BA:36499.]

Cain, A.J. (1947*a*). Demonstration of lipine in the Golgi apparatus in gut cells of *Glossiphonia*. *Q. Jl microsc. Sci.* **88**(2), 151–7.

—— (1947*b*). The use of nile blue in the examination of lipoids. *Q. Jl microsc. Sci.* **88**, 383–92. [*Glossiphonia complanata.*]

—— (1947*c*). An examination of Baker's acid haematein test for phospholipines. *Q. Jl microsc. Sci.* **88**, 467–78. [*Glossiphonia complanata.*]

Caine, R.L. (1955). A morphologic and taxonomic study of the flagellated enteric protozoa of leeches. Ph.D. thesis, University of Iowa. [University Microfilms, Ann Arbor, Mich.] [*Diss. Abstr. Int.* **15**(8), 1465.]

Cajal, S.R. (1904). Variaciones morfológicas de reticulo nervioso de invertebrados y vertebrados sometidos a la acción de condiciones naturales (nota preventiva). *Trabajos Lab. Invest. biol.* **3**, 287–97.

—— (1908). L'Hypothèse de la continuité d'Apathy résponse aux objections de cet auteur contre la doctrine neuronale. *Travaux Lab. Rech. biol. Un. Madrid* **6**, 21–89.

—— (1911). *Histologie du système nerveux de l'homme et des vertébrés*, Vol. 1. A. Maloine, Paris.

Calabrese, R.L. (1977*a*). The neural control of alternate heartbeat coordination states in the leech, *Hirudo medicinalis*. *J. comp. Physiol.* **A122**(1), 111–43. [1978BA:21702.]

—— (1977*b*). Regeneration of an intersegmental interneuron in the leech. *Neurosci. Abstr.* **3**, 102. (Abstr. 296.)

—— (1979*a*). Neural generation of the peristaltic and nonperistaltic heartbeat coordination modes of the leech, *Hirudo medicinalis*. *Am. Zool.* **19**, 87–102. [1980BA:56834.]

—— (1979*b*). The roles of endogenous membrane properties and synaptic interaction in generating the heartbeat rhythm of the leech, *Hirudo medicinalis*. *J. exp. Biol.* **82**, 163–76.

—— (1980). Control of impulse-initiation sites in a leech interneuron. *J. Neurophysiol.* **44**(5), 878–96.

—— and Peterson, E. (1983). Neural control of heartbeat in the leech, *Hirudo medicinalis*. In: *Neural origin of rhythmic movements*. (ed. A. Roberts and B. Roberts). *Soc. Experimental Biol.*, *Symp.* **37**.

Caldwell, M.J. (1966). The effect of hirudin, a potent anti-thrombin, on the thromboplastin generation test. (Method abstract.) *Proc. Fedn Am. Socs exp. Biol.* **25**(2 Pt. 1), 255.

Calow, P., and Riley, H. (1980). Leeches: an adaptational approach to their ecology and physiology. *J. Biol. Educ.* **14**, 279–89.

—— (1982). Observations on reproductive effort in British erpobdellid and glossiphoniid leeches with different life cycles. *J. Anim. Ecol.* **51**, 697–712.

Calvert, A.C., and Calvert, P.P. (1917). *A year of Costa Rican natural history*. Macmillan, New York.

Camatini, M., and Franchi, E. (1974). Organization of glycerinated muscle fibers and isolated primary microfilaments of oligochaete annelids (*Lumbricus terrestris*) and Hirudinea (*Glossiphonia complanata*). *Boll. Zool.* **41**(4), 465–6. [1977BA:30978.]

—— Ceresa Castellani, L., Franchi, E., Lanzavecchia, G., and Paoletti, L. (1976). Thick filaments and paramyosin of annelid muscles. *J. ultrastruct. Res.* **55**(3), 433–47. [*Glossiphonia complanata.*] [1978BA:55192.]

Cameron, A. (1950). Haematemesis from leeches. *Br. med. J.* **ii**, 679–80.

Cammelli, E., DeBellis, A.M., and Nistri, A. (1974). Distribution of acetycholine and acetycholinesterase activity in the nervous tissue of the frog and of the leech. *J. Physiol.* **242**, 88P–90P.

Campbell, R.A. (1973). Studies on the biology of the life cycle of *Cotylurus flabelliformis* (Trematoda: Strigeidae). *Trans. Am. microsc. Soc.* **92**(4), 629–40. [*Helobdella* can serve as second intermediate host.]

Canning, E.U., Cox, F.E.G., Croll, N.A., and Lyons, K.M. (1973). The natural history of Slapton Ley Nature Preserve: VI. Studies on the parasites. *Field Stud.* **3**(5), 681–718. [1975BA:44925.]

Canova, F. (1940). Grave forms of dyspnea and hemoptysis caused by leeches. *Riv. Med. trop. Med. Indigena* **4**, 148–50.

Capelli, G.M. (1980). Seasonal variation in the food habits of the crayfish *Orconectes propinquus* in the Trout Lake, Vilas County, Wisconsin, USA (Decapoda: Astacidea: Cambaridae). *Crustaceana* (Leiden) **38**(1), 82–6. [Eating leeches.]

Carayon-Gentil, A., and Gautrelet, J. (1938). See Autrum (1939c).

Carbonetto, S., and Muller, K.J. (1977). A regenerating neurone in the leech can form an electrical synapse on its severed axon segment. *Nature, Lond.* **267**, 450–2. [1977BA:1979.]

Cargo, D.G. (1960). Predation of eggs of the spotted salamander, *Ambystoma maculatum*, by the leech *Macrobdella decora*. *Chesapeake Sci.* **1**(3), 119–20.

Carlson, C.A. (1968). Summer bottom fauna of the Mississippi River, above Dam 19, Keokuk, Iowa. *Ecology* **49**(1), 162–9.

Carr, J.F., and Hiltunen, J.K. (1965). Changes in the bottom fauna of western Lake Erie from 1930 to 1961. *Limnol. Oceanogr.* **10**, 551–569.

Carus, J.V. (1885). *Prodromus faunae mediterraneae* I. Stuttgart. [*Trachelobdella lubrica.*]

Caspers, N. (1975). [Caloric values of the most abundant invertebrates of two woodland-brooklets of the 'Naturpark Kottenforst-Ville'.] *Arch. Hydrobiol.* **75**(4), 484–9. [In German; English summary.] [*Glossiphonia complanata; Erpobdella octoculata.*]

Castle, W.E. (1900a). Some North American fresh-water *Rhynchobdellidae*, and their parasites. *Bull. Mus. comp. Zool. Harvard* **36**, 17–64.

—— (1900b). Metamerism of the leech. *Science, NY* **11**, 175.

—— (1900c). The metamerism of the *Hirudinea*. *Proc. Am. Acad. Sci., Boston* **35**, 285–303.

Castro, O.F. (1970a). Aspectos da biologia de *Haementeria gracilis* (Weyenbergh, 1883) (Hirudinea–Glossiphoniidae). In *Resumos 22 Reuniao Anual*, p. 299. Salvador, Bahia, Brasil. (Abstr.)

—— (1970b). Consideracoes sobre o desenvolvimento de *Helobdella anoculis* Weber, 1915 (Hirudinea, Glossiphoniidae). *Reuniad Anual da Sbpc*. 22. Salvador. (Resumos.) Sao Paulo, Sociedade Brasileira para o progresso da Ciencia, p. 300. (Resumo.) (Abstr.)

—— (1970c). Sistemática, ecologia e desenvolvimento de hirudíneos da regiâo norte-ocidental do Estado de Sao Paulo. Sao Paulo, USP. Tese (doutoramento). 190 pp., Faculdade de Filosofia, Ciências e Letras de Sao José do Rio Preto.

—— (1971). Sôbre Hirudíneos no Brasil. *Rev. Ciências. Fac. Ciên. Letr. Votuporanga.* **1**, 1–69. [In Portuguese; English summary.]

Catling, P.M., and Freedman, B. (1980). Food and feeding of sympatric snakes at Amherstburg, Ontario, Canada. *Can. Field-Nat.* **94**(1), 28–33. [1980BA:22391.] [Gartern snake *Thamnophis butleri*, feeds on leeches.]

Catterina, G. (1897). See Autrum (1939c).

Causey, D. (1953). Marine leeches. *Educational Focus* **25**(2), 19–23.

Cekanovskaya, O.V. (1962). (The Aquatic Oligochaeta of the U.S.S.R.) *Opred. Faune. SSSR* **78**, 1ff. [*Agriodrilus vermivorous:* 353, Figs 220–1.] [In Russian.]

Champeau, A. (1966). Contributions à l'étude écologique de la faune des eaux temporisées de la Haute Carmargue. *Archo. oceanogr. Limnol.* **14**, 309–57. [*Hirudo medicinalis* and *Helobdella stagnalis* from temporary pond, la Haute Camargue.]

Chandra, M., and Mukharjee, R.N. (1973). Record of *Paraclepsis praedatrix* Harding, 1924 and *Glossiphonia weberi* Blanchard, 1897 (Annelida: Hirudinea), from a new host, *Rana limnocharis* Weigman. *Current Sci.* **42**(14), 512–13.

—— and Saha, S.S. (1966). Record of *Paraclepsis praedatrix* Harding 1924 (Annelida: Hirudinea) from a new host, *Natrix piscator* (Schneider), the checkered keelback (Reptilia: Serpentes). *J. Bombay nat. Hist. Soc.* **63**(2), 448–9.

Chang, H.C., and Gaddum, J.M. (1933). Choline esters in tissue extracts. *J. Physiol.* **79**, 255–85. [Leeches: 263–7.]

Chang Yin-Pi (1963). [New records of leeches from southern Tibet.] *Acta zool. sinica* **15**(3), 489–90. [In Chinese; English translation.]

Chapeville, F. (1952). [Anticoagulatory mechanism of leech extract.] Thèse de Docteur Vétérinaire, Paris.

Chapman, G. (1958). The hydrostatic skeleton in the invertebrates. *Biol. Rev.* **33**, 338–71.

Chapron, C., and Relexans, J-C. (1971). Connexions intercellulaires et évolution nucléaire au cours de la préméiose ovocytaire. Etude ultrastructurale chez le Lombricien *Eisenia foetida*. *C.r. Acad. Sci. Paris*, **272**, 3307–10.

Charbonnel-Berault, A. (1962). [On hirudin.] Thèse spécialité, Paris.

Chatin, J. (1880). See Autrum (1939c).

Chaudhuri, A.K., and Khalsa, H.G. (1968). Changes in bleeding and coagulation time in human subjects due to the biting by land leech *Haemadipsa sylvestris* Blanchard. *Indian J. Physiol. all. Sci.* **22**(1/2), 18–23.

Chaudhuri, S.K.B. (1971). Common leech as a cause of haemoptysis and epistaxis. *J. Indian med. Ass.* **56**(8), 234.

Chelladurai, J.E. (1934). On a new Indian leech, *Hemiclepsis viridis*, sp. nov. *Rec. Ind. Mus.* **36**, 345–52.

Chemberlin, D.R. (1968). Ecological studies of the leech *Haemopis grandis*. M.S. thesis, Mankato State College, Mankato, Minn.

Chen Ching Mei (1959). [Study of leeches about a major inner province Szechwan fresh water.] [*Zool. Mag.*] **12**, 562–5. [In Chinese.]

—— (1960). [List of the names of the Chinese leeches.] [*Zool. Mag.*] **4**(1), 41–3. [In Chinese.]

Chen, Yi. (1959).]Chinese animal map — segmented animals.] Scientific Press, Peking. [In Chinese.]

—— (1962). An account of the leeches found in Nanking and vicinity with description of a new species, *Placobdella sinensis. Acta zool. sinica* **14**(4), 515–24.

Cherkesova, D.V., Baskova, I.P., Mosolov, V.V., Aldanova, N.A., and Potapenko, N.A. (1978). ['Pseudohirudin' in the body of the medicinal leech.] *Dokl. Acad. Nauk SSSR.* **241**(3), 720–2. [In Russian.]

Chernin, E., Michelson, E.H., and Augustine, D.L. (1956). Studies on the biological control of schistosome-bearing snails. II. The control of *Australorbis glabratus* populations by the leech, *Helobdella fusca*, under laboratory conditions. *Am. J. trop. Med. Hyg.* **5**, 308–14.

—— and Perlstein, J.M. (1971). Protection of snails against miracidia of *Schistosoma mansoni* by various aquatic invertebrates. *J. Parasitol.* **57**(2), 217–19.

Cherry, R.H. and Ager, A.L. (1982). Parasites of American alligators (*Alligator mississippiensis*) in South Florida. *J. Parasitol.* **68**, 509–10.

Chin Ta-hsiung (1941). Notes on leech infestation in man. *Chinese med. J.* **60**, 241–3.

—— (1949). Further notes on leech infestation in man. *J. Parasitol.* **35**(2), 215.

Chouteau, F. (1968). Influence de la pollution industrielle et domestique sur les populations animales de la riviere Isère au cours de sa traversée de la région grenobloise. *Trav. Lab. Hydrobiol. Piscic. Un. Grenoble* **59–60**, 39–63.

Christiansen, M. (1939). *Protoclepsis tesselata* (O.F. Müller), der Entenegel, als Ursache von Krankheit, u.a. Konjunktivitis, bei Gänsen und Enten. *Z. Infektionskrankh. Parasit. Krankh. Hyg. Haustier, Berlin* **55**(1), 75–89. [*Theromyzon tessulatum* on eyes; Europe.]

Chutter, F.M. (1970). Hydrobiological studies in the catchment of Vaal Dam, South Africa, Part 1. River zonation and the benthic fauna. *Int. Rev. ges. Hydrobiol.* **55**, 445–94.

Ciaccio, G. (1957). Essai de conservation d'une souche Cobaye de type O de virus de la fièvre aphteuse sur Sangsue (*H. medicinalis*), hôte non habituel de ce virus. *C.r. Soc. Biol. Paris* **151**, 261–3.

Clark, A.W. (1963). Fine structure of two invertebrate photoreceptor cells. *J. cell Biol.* **19**, 14A.

—— (1965*a*). Microtubules in some unicellular glands of two leeches. *Z. Zellforsch. mikrosk. Anat.* **68**, 568–88.

—— (1965*b*). An electronmicroscopic study of four invertebrate photoreceptors (*Helobdella stagnalis, Placobdella rugosa, Viviparens maleatus, Gyraulus circumstriatus*). *Diss. Abstr.* **25**(12 Pt. 1), 6875. (Abstr. only.)

——— (1967). The fine structure of the eye of the leech, *Helobdella stagnalis. J. cell Sci.* **2**(3), 342–8.

Clark, R.B, and Cowey, J.B. (1958). Factors controlling the change of shape of certain nemertean and turbellarian worms. *J. exp. Biol.* **35**, 731–48.

Claude, A. (1937). Spreading properties of leech extracts and the formation of lymph. *J. exp. Med.* **66**, 353–66.

——— (1940). 'Spreading' properties and mucolytic activity of leech extracts. *Proc. Soc. exp. Biol. Med.* **43**, 684–9.

Claus, C., Grobben, K., and Kühn, A. (1932). [*Ozobranchus quatrefagesi* in mouths of crocodiles, on f.w. turtles, and on Pelicans, Africa.] Lehrbuch der Zoologie 10 Aufl. Berlin, Vienna. [See *Traité de Zoologie*, p. 579. Fig. 418.]

Cleland, J.B. and Johnston, J. (1910). The haematozoa of Australian batrachians, No. 1, Sydney. *Jl R. Soc. NSW* **44**, 252–60.

Clemens, W.A., Rawson, D.S., and McHugh, J.L. (1939). A biological survey of Okanagan Lake, British Columbia. *Bull. fish. Res. B. Can.* **61**, 1–70. [Hirudinea: 60, 68.]

Clifford, H.F. (1969). Limnological features of a northern brown-water stream, with special reference to the life histories of the aquatic insects. *Am. Midl. Nat.* **82**(2), 578–7.

Cliffton, E.E. (1961). Fibrinolytic agents. *Med. Clins N. Am.* **45**, 917–33.

Cline, H.T. (1983). ^{3}H-GABA uptake selectively labels identifiable neurons in the leech central nervous system. *J. comp. Neurol.* **215**(3), 351–8.

Cobbold. P.H. (1971). Uptake of sodium and potassium by the leech, *Hirudo medicinalis. J. Physiol.* **219**, 10–12.

Coggeshall, R.E. (1966). The ganglion–connective junction in the central nervous system of the leech, *Hirudo medicinalis. J. Morph.* **119**(4), 417–24.

——— (1972*a*). 5-Hydroxytryptamine in nerve terminals of the leech. *Anat. Rec.* **172**, 293.

——— (1972*b*). Neuromuscular junctions in the leech. *Proc. Soc. Neurosci.* **2**, ·139.

——— (1972*c*). Autoradiographic and chemical localization of 5-hydroxytryptamine in identified neurons. *Anat. Rec.* **172**(3), 489–98.

——— (1974). Gap junctions between identified glial cells in the leech. *J. Neurobiol.* **5**(5), 463–7. [1975BA:49523.]

——— Dewhurst, S.A., Weinreich, D., and McCaman, R.E. (1972). Aromatic acid decarboxylase and choline acetylase activities in a single identified 5-HT containing cell of the leech. *J. Neurobiol.* **3**, 259–65.

—— and Fawcett, D.W. (1964). The fine structure of the central nervous system of the leech, *Hirudo medicinalis. J. Neurophysiol.* **27**, 229–89.

—— and Yaksta-Sauerland, B.A. (1974). The localization of 5-hydroxytryptamine in chromaffin cells of the leech body wall. *J. comp. Neurol.* **156**(4), 459–70. [1975BA:2510.]

Cohen, M.N., and Sawyer, R.T. (1972). Notes on a marine oligochaete (Annelida: Oligochaeta) from Ft. Johnson, South Carolina. *Bull. S. Carolina Acad. Sci.* **34**, 103.

Coker, R.E., Shira, A.F., Clark, H.W., and Howard, A.D. (1921). Natural history and propagation of freshwater mussels. *Bull. US Bur. Fish.* **37**, 77–181. [Separately issued as *Bur. Fish. Doc.* 893.] [Leeches in mantle cavity.]

Colbo, M.H. (1965). Taxonomy and ecology of the helminths of the American coot in Alberta. M.Sc. thesis, University of Alberta, Edmonton, Alberta. [Ecological relationships of leeches and coots.]

Coleman, C.L. (1975). Studies on the Retzius cell and neuronal 5-HT in the leech *Hirudo medicinalis.* Ph.D. thesis, St. Andrews University, UK.

Coles, J.W. (1974). Survey of Bookham Common: thirty-second year. Vermes: Hirudinea: the medicinal leech. *Lond. Nat.* No. 53, 74–5.

Coles, M.D. (date undetermined). Leeches. *Salm. Trout Mag.* **169**, 172–5.

Collin, A. (1892). See Autrum (1939*c*).

Collins, H.L. (1981). Bait leech: its nature and nurture. *Superior Advisory Notes* No. 12. pp. 4. (Univ. of Minnesota Sea Grant Extension Program, Duluth.)

—— and Holmstrand, L.L. (1984*a*). Indicators of secual maturity in the leech *Nephelopsis obscura* (Annelida: Hirudinea). *Am. Midl. Nat.* **112**(1), 91–4.

—— —— (1984*b*) Early life history and growth of *Nephelopsis obscura,* Verrill, 1872. (Pharyngobdellida: Erpobdellidae). *J. Freshwater Ecology* **2**(6): 549–54.

—— —— and Jesswein, W. (1983). Bait leech, *Nephelopsis obscura,* culture and economic feasibility. Research Report No. 9. 20 pp. Minnesota Sea Grant Program.

Conet, M.A. (1931). *Nosema herpobdellae,* Microsporidie nouvelle parasite des Hirudinées. *Ann. Soc. Sci. Bruxelles* **51**, 170–1.

Conway-Morris, S. (1979). Middle Cambrian polychaetes from the Burgess Shale of British Columbia. *Phil. Trans. R. Soc.* **385B**, 227–74.

—— Pickerill, R.K., and Harland, T.L. (1982). A possible annelid from the Trenton Limestone (Ordovician) of Quebec, with a review of fossil oligochaetes and other annulate worms. *Can. J. Earth Sci.* **19**(11), 2150–7.

Corbett, P. (1967). Leeches. *Nat. Sci. Schools* **5**, 40–7.

Cordero Del, C.M., and Rojo Vazques, F.A. (1979). [Veterinary parasitology in old works of veterinary and related sciences: 2. Brother Miguel Agustin, 17th–18th century.] *Rev. Iber. Parasitol.* **39**(1–4), 209–222. [In Spanish; English summary.]

Cordero, E.H. (1933). Notes sur les Hirudinées. II. *Piscicola platense* n. sp., d'un poisson sud-Americain *Hoplias malabaricus* (Bloch). *Ann. Parasitol. hum. comp.* **11**, 450–62.

—— (1935). Descripción de dos nuevos Hirudíneos del Plata, de los generos *Piscicola y Placobdella. Bol. Un. Nac. La Plata* **18**(6), 13–22.

—— (1936). See Autrum (1939*c*).

—— (1937*a*). Los Hirudíneos del nordeste del Brasil, I. *Ann. Acad. Brasil. Sci.* **9**, 1–26.

—— (1937*b*). Hirudíneos neotropicales y subantárticos nuevos, criticos o ya conocidos del Museo Argentine de Ciencias naturales. *An. Mus. Argentino Ci. Nat., Buenos Aires* **39**, 1–78.

—— (1941). Revisión de las especies de Hirudíneos de la Repúblic Argentina, descriptos por Weyenbergh en 1879 y 1883. *Boll. Acad. Nac. Cienc. Córdoba* **35**(2–3), 182–214.

—— (1946*a*). Notes sur les Hirudinées. III. *Helobdella anoculis* Weber= *Helobdella michaelseni* R. Bl. *Comun. Zool. Mus. Hist. nat. Montevideo* **2**(26), 1–4.

—— (1946*b*). Notes sur les Hirudinées. IV. *Anoculobdella trituberculata* Weber= *Helobdella triserialis* (Em. Bl.). *Comun. Zool. Mus. Hist. nat. Montevideo* **2**(30), 1–9.

Corkum, K.C., and Beckerdite, F.W. (1975). Observations on the life history of *Alloglossidium macrobdellensis* (Trematoda: Macro-deroididae) from *Macrobdella ditetra* (Hirudinea: Hirudinidae). *Am. Midl. Nat.* **93**(2), 484–91.

Cornec, J.P. (1967). Etude des modifications histologiques observées à la suite d'amputations ches *Helobdella stagnalis* (Hirudinée Rhynchobdelle). *Bull. Soc. Zool. Fr.* **92**(4), 779–85.

—— (1971). Sur le comportement des tissus après amputation ches l'Hirudinée Pharyngobdellae *Erpobdella octoculata. ann. Embryol. Morphol.* **4**(3), 269–79.

—— (1978). Evolution histologique et histochimique des téguments maternels dans la zone d'attache des pontes et des jeunes en cours de croissance chez l'Hirudinée glossiphoniide *Helobdella stagnalis* L. *Bull. Soc. Zool. Fr.* **103**(2), 113–24. [In French; English summary.] [1979BA:15813.]

—— (1979). Table de développement d'*Erpobdella octoculata* (Annélide Hirudinée) et description de cas de duplications. *Bull. Soc. zool. Fr.* **104**(2), 137–46. [In French; English summary.] [1980BA:23785.]

—— (1980). Régulation et régénération après amputation de la région postérieur de jeunes Hirudinées de l'espèce *Erpobdella octoculata. Arch. zool. exp. Gen.* **121**(3), 173–81.

—— (1981). Comportement des tissus après amputation de la région antérieure chez les jeunes hirudinées de l'espèce *Erpobdella octoculata.* Cas de régénération antérieure. *Bull. Soc. zool. Fr.* **106**(2), 101–10.

—— and Coulomb-Gay, R. (1975). Détection par l'histochimie des activités phosphatasiques acides et alcalines normales et de leur variations après amputation chez l'Oligochète *Eiseniella tetraedra* (Sav.) et les Hirudinées *Erpobdella octoculata* (L.) et *Helobdella stagnalis* (L.). *C. r. Séanc. Soc. Biol. Paris* **169**(1) 86–93. [In French; English summary.] [1976BA:25924.]

—— Solbani, L., and Turki, M. (1970). Recherches sur la dynamique tissulaire de l'epiderme des annelides. *Ann. Fac. Sci. Marseille* **43B**, 73–80.

Corradetti, R., Moroni, F., and Pepeu, G. (1980). Pharmacological effects of benzodiazepines in the leech, *Hirudo medicinalis,* benzodiazepine and GABA receptors and GABA level. *Pharmacol. Res. Commun.* **12**(6), 581–6.

Corte, L.D. see Della Corte, L.

Costa, O.G. (1856). Paleontologia del regno di Napoli. *Atti Acc. Pontan* **7**, 1–378. [*Hirudella:* 354–5.]

Cott, H.B., (1961). Scientific results of an inquiry into the ecology and economic status of the Nile crocodile (*Crocodylus niloticus*) in Uganda and northern Rhodesia. *Trans. zool. Soc., Lond.* **29**, 211–337.

Couchman, J. and Edwards. J. (in preparation). Isolation and characterization of leech hyaluronidase.

Couteaux, R. (1957). Neurofilaments et neurofibrilles dans les fibres nerveuses de la sangsue. In *Proc. Int. Conf. electron Microscopy,* Stockholm, September, 1956 (ed. F.S. Sjöstrand and Y. Rhodin) pp. 188–90. Almqvist and Wiksells, Uppsala.

—— Carasso, N., and Favard, P. (1975). High and low voltage electron microscopical observations of the neurofibrillar bundles of the leech after silver impregnation. *J. Microsc. Biol. Cell.* **24**(2/3), 283–94. [French summary.] [1977BA:15813.]

Cover, E.C., and Harrel, R.C. (1978). Sequences of colonization, diversity, biomass, and productivity of macroinvertebrates on artificial substrates in a freshwater canal. *Hydrobiologia* **59**(1), 81–95.

Cowie, D.M. (1911). On the hirudin and hirudin immunity. *J. med. Res., St. Louis* **24**, 497.

Cowles, R.P. (date undetermined). A biological survey of the offshore waters of Chesapeake Bay. J.P. Moore reported a leech.

Cozette, P. (1906). Les parasites des poissons. *C. r. Soc. Sav. Paris* 138.

Craig and Faust, E.C. (1964). See Faust, E.C. and Russell, P.F. (1964).

Crellin, J.C. (1969). Leech, honey and tamarind jars. In *Medical ceramics in the Wellcome Institute,* Vol. 1, pp. 127–33. Wellcome Institute Publications, London.

Cresp, J. and Cornec, J.P. (1972). Variations comparées des activités phophatasiques après des amputations pratiquées chez le serpulide

Hydroides norvegica (Gunn) et l'hirudinée pharyngobdelle *Erpobdella octoculata* (Linn.). *Ann. Embryol. Morph.* **5**(4), 297–303. [In French; English summary.]

Crewe, W. and Cowper, S.G. (1973). A leech parasitic on *Bulinus* (Demonstration). *Trans. R. Soc. trop. Med. Hyg.* **67**(1), 65. ['*Helobdella*' feeding in laboratory on *Bulinus truncatus* from an endemic area of schistosomiasis in Egypt.]

Crisp, D.T, and Gledhill, T.A. (1970). A quantitative description of the recovery of the bottom fauna in a muddy reach of a mill stream in southern England after draining and dredging. *Arch. Hydrobiol.* **67**, 502–41.

Cristea, V. (1969). [Leech fauna of Lake Capuso.] *Lucrarile St. Institutulvi Pedagogic din Galati* **3**, 149–58. [In Romanian.]

—— (1970). [Some aspects of the reproductive biology and development of *Erpobdella testacea* (Hirudinea, Pharyngobdellae).] *Stud. Cercet. Biol. Ser. Zool.* **22**(5), 447–53. [In Romanian; English summary.]

—— (1975). [*Dina stschegolewi* (Lukin et Epstein, 1960) (Hirudinea — Erpobdellidae) new for the Romanian fauna.] *Stud. Cercet. Biol. Anim.* **27**(2), 85–8. [In Romanian; English summary.] [1976BA:61081.]

—— and Manoleli, D. (1977). Conspectus des sangsues (Hirudinea) de Roumanie, avec une clé de détermination. *Trav. Mus. Hist. nat. Gr. Antipa* **18**, 23–56. [In French; with English, Romanian, and Russian summaries.]

Croftin, H.D. (1971). A quantitative approach to parasitism. *Parasitology* **62**(2), 179–94.

Cross, W.H. (1976). A study of predation rates of leeches on tubificid worms under laboratory conditions. *Ohio J. Sci.* **76**(4), 164–6. [*Helobdella stagnalis; Erpobdella punctata.*] [1977BA:25606.]

Crothers, J.H. (1966). *Dale Fort marine fauna*, 2nd edn, Suppl. to Vol. 2 of Field Studies. Field Studies Council. [Hirudinea: 44.]

Crowe, M.G.F. (1962). Leeches for leeches (and subsequent correspondence). *Guy's Hosp. Gaz.* **76**, 314–19; 343–4; 366–7.

Crumb, S.E. (1977). Macrobenthos of the tidal Delaware River between Trenton and Burlington, New Jersey. *Chesapeake Sci.* **18**(3), 253–65.

Cuénot, L. (1891). Études sur le sang et les glandes lymphatiques dans la série animale. II. *Arch. Zool. exp. gén.* (2) **9**, 365–475.

——(1931). See Autrum (1939*c*).

Curry, M.G. (1975*a*). Notes on some Louisiana freshwater leeches (Hirudinea), two new to the state. *Am. Midl. Nat.* **93**(2), 509–10.

—— (1975*b*). A new leech (Hirudinea: Glossiphoniidae) for Louisiana with ecological notes. *Ass. Southeast. biol. Bull.* **22**(2), 49. (Abstr.) [*Placobdella montifera.*]

—— (1976). Three leeches (Hirudinea) new to Arkansas with ecological and distribution notes. *Wasmann J. Biol.* **34**(1), 5–8.

—— (1977). Ecology of the leech *Placobdella montifera* (Hirudinea: Glossiphoniidae). Part 1. *Ass. Southeast. biol. Bull.* **24**(2), 45. [1977BA:97234.]

—— (1978). Delaware leeches (Annelida: Hirudinea: Glossiphoniidae): new state records and new molluscan host record for *Placobdella montifera* Moore. *Wasmann J. Bio.* **35**(1), 65–7.

—— (1979). New fresh water unionid clam hosts for three glossiphoniid leeches. *Wasmann J. Biol.* **37**(1–2), 89–92. [1980BA:46401.]

—— and Kennedy, W.G. Jr. (1975). Louisiana turtle leeches with new host record. *Proc. La. Acad. Sci.* **38**, 124.

—— and Leemann, J.E. (1978). Salinity ranges of freshwater glossiphoniid leeches (Annelida: Hirudinea: Glossiphoniidae) in Louisiana. *Ass. Southeast. biol. Bull.* **25**(2), 34. [1978BA:18844.]

—— and Vidrine, M.F. (1976). New fresh-water mussel host records for the leech *Placobdella montifera* with distributional notes. *Nautilus* **90**(4), 141–4.

—— —— (1977). New fresh-water clam host records for the leeches *Placobdella montifera* Moore and *Helobdella stagnalis* L. *Proc. La. Acad. Sci.* **40**, 43–6.

Czechowicz, K. (1961). Studies on the distribution of nucleic acid in the neurosecretory cells of the central nervous system in some leeches. *Zool. Pol.* **2**(2), 91–9.

—— (1963). Neurosecretion in leeches and the annual cycle of its changes. *Zool. Pol.* **13**(3/4), 163–84.

—— (1968). Studies on the participation of the neurosecretory system in the leech in the regulation of water balance. *Zool. Pol.* **18**, 81–95.

—— and Fabianczyk, U. (1979). [Effect of commissurotomy by *Haemopis sanguisuga* on secretory activity of nerve cells.] *Pr. Nauk. Uniw. Slask. Katowicach* **278**, 9–16. [In Polish; English and Russian summaries.] [1980BA:44251.]

—— and Zarzycki, J. (1970). [Studies on the ultrastructure of the nervous system in leeches.] *Folia Morph., Warsaw* **29**, 53–9.

—— Hajnisz, E., Skowronek, M., and Wolna, M. (1980). [Effect of some antibiotics on the neurosecretory system in the leech *Haemopis sanguisuga*. 1. Morphological investigations with the light microscope.] *Pr. Nauk Uniw. Slask Katowicach.* **375**, 173–84.

Czeczuga, B. (1977). Astaxanthin, a dominating carotenoid in four species of leeches. *Bull. Acad. Pol. Sci. Ser. Sci. Biol.* **25**(2), 85–8. [Russian summary.] [1977BA:21002.]

Daborn, G.R. (1976). Colonization of isolated aquatic habitats. *Can. Field-Nat.* **90**, 56–7. [*Marvinmeyeria lucida* on birds.]

Dadswell, M.J. (1979). Biology and population characteristics of the shortnose sturgeon, *Acipenser brevirostrum* Lesueur 1818 (Osteichthys: Acipenseridae), in the Saint John River estuary, New Brunswick, Canada. *Can. J. Zool.* **57**, 2186–210.

Dahl, E., Falck, B., Von Mecklenburg, C., and Myhrberg, H. (1963). Adrenergic sensory neurons in invertebrates. *Gen. Comp. Endocr.* **3**, 693. (Abstr.).

Dahm, A.G. (1962). Distribution and biological patterns of *Acanthobdella peledina* Grube from Sweden (Hirudinea, Acanthobdellae). *Acta Un. Lunds Arsskrift, N. F. Avd. 2* **58**(10), 1–35.

—— (1963). Ivösjöns lägre djurvärld och livsmiljöer. *Skanes Natur, Skanes Naturskyddsförenings Arsskrift.* [Some leeches in south Sweden.]

—— (1964). The leech (Hirudinea) fauna in lotic habitats of the River Mörrams An in Southern Sweden. *Acta Un. Lunds Arsskrift. N. F. Avd. 2* **59**(10), 1–32.

Dall, P.C. (1979*a*). Ecology and production of the leeches *Erpobdella octoculata* L. and *Erpobdella testacea* Sav. in Lake Esrom, Denmark. *Arch. Hydrobiol.*, Suppl. **57**(2), 188–220. [German summary.] [1980BA:72279.]

—— (1979*b*). A sampling technique for littoral stone dwelling organisms. *Oikos* **33**, 106–12.

—— (1979*c*). Distributional relationship and migration of leeches (Hirudinea) in the exposed littoral of Lake Esrom, Denmark. *Oikos* **33**(1), 113–20. [Russian summary.][1980BA:15228.]

—— (1982). Diversity in reproduction and general morphology between two *Glossiphonia* species (Hirudinea) in Lake Esrom, Denmark. *Zoologica Scripta* **11**(2), 127–33.

—— (1983). The natural feeding and resource partitioning of *Erpobdella octoculata* L. and *Erpobdella testacea* Sav. in Lake Esrom, Denmark. *Int. Revue Ges. Hydrobiol.* **68**, 473–500.

—— Heegaard, H. and Fullerton, A.F. (1984). Life history strategies and production of *Tinodes waeneri* (L) (Trichoptera) in Lake Esrom, Denmark. *Hydrobiologia* **112**, 93–104.

Damas, D. (1962). Histologie, cytologie, histochimie du tube digestif de *Glossiphonia complanata* L. Doctorat de 3ème Cycle. *Arch. Zool. exp. Gén.* **101**(2), 41–72.

—— (1964*a*). Succinode hydrogénases ovariennes de *Glossiphonia complanata* L. (Hirudinée, Rhynchobdelle), en rapport avec la phase d'accroissement des oeufs. *C. r. hebd. Séanc. Acad. Sci. Paris* **258**(12), 1895–6.

—— (1964*b*). Structure et rôle du rachis ovarien chez *Glossiphonia complanata* L. (Hirudinée, Rhynchobdelle). *Bull. Soc. zool. Fr.* **89**(2–3), 147–55.

—— (1964*c*). Elaboration de l'acrosome au cours de la spermiogenèse de *Glossiphonia complanata* L. (Hirudinée, Rhynchobdelle). *Bull. Soc. zool. Fr.* **89**(5–6), 669–73.

—— (1965*a*). Présence d'un mucopolysaccharide neutre dans l'acrosome du spermatozoide de *Glossiphonia complanata* L. (Hirudinée, Rhynchobdelle). *Ann. Histochim.* **10**, 11–15.

—— (1965*b*). Mode de nutrition des cellules mâles et femelles de *Glossiphonia complanata* (Hirudinée, Rhynchobdelle) durant la spermatogenèse et l'ovogenèse. Démonstration du 19 mai 1965 à la Société Zoologique de France. [**40**, 337–8.]

—— (1965*c*). Répartition des orthophosphatases alcalines chez une Hirudinée *Glossiphonia complanata* L. *Bull. Soc. zool. Fr.* **90**, 385–91.

—— (1966*a*). Anatomie et histologie des canaux éjaculateurs de *Glossiphonia complanata* L. (Hirudinée, Rhynchobdelle). *Arch. Zool. exp. Gén.* **107**(2), 325–35.

—— (1966*b*). Phénomène neurosecrétoire en rapport avec la reproduction de *Glossiphonia complanata* L. (Hirudinée, Rhynchobdelle). *Bull. Soc. zool. Fr.* **91**, 613–21.

—— (1968*a*). Histochimie des canaux éjaculateurs de *Glossiphonia complanata* (Hirudinée, Rhynchobdelle.) *Ann. Histochim.* **13**, 111–22.

—— (1968*b*). Origine et structure du spermatophore de *Glossiphonia complanata* L. (Hirudinée, Rhynchobdelle). *Arch. Zool. exp. Gén.* **109**(1), 79–86. [English summary.]

—— (1968*c*). Les cellules germinales mâles de *Glossiphonia complanata* L. (Hirudinée, Rhynchobdelle). Origine, évolution et structure. *Bull. Soc. zool. Fr.* **93**(3), 375–85. [English summary.]

—— (1969*a*). Données histochimiques sur la cuticule de *Glossiphonia complanata* L. (Hirudinée, Rhynchobdelle). *Arch. Zool. exp. Gén.* **110**(3), 417–33. [English summary.]

—— (1969*b*). Appareil génital et fécondation par voie hypodermique chez *Glossiphonia complanata* L. (Hirudinée, Rhynchobdelle). Étude histophysiologique. Thèse Fac. Sciences Université Paris. N° CNRS:AO 3587.

—— (1972). Durcissement de la cuticule des mâchoires de *Hirudo medicinalis* (Annélide, Hirudinée) aboutissant aux structures dentaires: étude histochimique et ultrastructurale. *Arch. Zool. exp. Gén.* **113**(3), 401–21.

—— (1973*a*). Etude ultrastructurale des organes de Bayer (complexe épithélio-musculaire) chez l'Hirudinée *Glossiphonia complanata* L. *C. r. hebd. Séanc. Acad. Sci. Paris* **276**, 2545–8.

—— (1973*b*). Ultrastructure de l'épithélium tégumentaire de *Glossiphonia complanata* L. (Hirudinée); cellules épithéliales et organes de Bayer. *Z. Zellforsch.* **134**, 355–65.

—— (1974*a*). Etude histologique et histochimique des glandes salivaires de la sangsue médicinale, *Hirudo medicinalis* (Hirudinée, Gnathobdelle). *Arch. Zool. exp. Gén.* **115**(2), 279–92. [English summary.]

—— (1974*b*). Ultrastructure du spermatozoïde de *Glossiphonia complanata*, L. (Hirudinée, Rhynchobdelle). *C. r. hebd. Séanc. Acad. Sci. Paris* **279**, 1353–6.

—— (1977). Anatomie microscopique et évolution de l'appareil génital femelle de *Glossiphonia complanata* (L.) (Hirudinée, Rhynchobdelle), au cours du cycle annuel. Etude histologique et ultrastructurale. *Arch. Zool. exp. Gén.* **118**(1), 29–42. [English summary.] [1977BA:62939.]

Damerau, G. (1960). [An almost forgotten note of interest on leech therapy.] *Z. Arztl. Fortbild.* **54**, 719–21. [In German.]

Danapala, S.B., and Fernando, C.H. (1958). A record of *Dinobdella ferox* (Blanchard), a leech found in the nasal cavities of a buffalo, with notes on leeches infesting domestic animals in Ceylon. *Ceylon vet. J.* 1–5.

Dang, D.M., and Curtin, C.B. (1978). Initial study of the distribution of leeches in Nebraska. *Trans. Nebr. Acad. Sci.* **5**, 9–14. [1979BA:61111.] [Also, same title *Proc. Nebr. Acad. Sci. Affil. Soc.* **87**, 12 (1977).]

Daniels, B.A. (1973). Observations on the anatomy and biology of the blue crab leech. *Bull. South Carolina Acad. Sci.* **35**, 109.

—— (1975a). Life history of a glossiphoniid fish leech. *Can. Soc. Zool.* Annual Meeting, 8–11 June 1975. Un. of Guelph. (Abstr. p. 42.)

—— (1975b). On the biology of *Actinobdella inequiannulata* (Hirudinea, Glossiphoniidae) parasitic on *Catostomus* spp. in Algonquin Park. Master's thesis. University of Toronto.

—— and Freeman, R.S. (1976). A review of the genus *Actinobdella* Moore, 1901 (Annelida, Hirudinea). *Can. J. Zool.* **54**(12), 2112–17. [1977BA:51838.]

—— and Sawyer, R.T. (1973). Host-parasite relationship of the fish leech *Illinobdella moorei* Meyer and the white catfish *Ictalurus catus* (Linnaeus). *Bull. Ass. Southeast. Biol.* **20**(2), 48.

—— —— (1975). The biology of the leech *Myzobdella lugubris* infesting blue crabs and catfish. *Biol. Bull.* **148**, 193–8.

Danilova, R.D. (1964. [Removal of a leech from the larynx.] *Vestn. Otorinolar.* **26**, 94–5. [In Russian.]

Daniyarov, M.R. (1975). [Parasite fauna of fishes from the spring 'Chilu-Chor Chashma' (Tadzhik SSR) with constant and high water temperature.] *Parazitologiya* **9**(4), 312–14. [In Russian; English summary.] [Water 18–20°C year round; *Piscicola geometra*.] [1976BA:39166.]

Dann. O., and Sucker, H. (1967). Phosphotransacetylase und biogene Hemmstoffe der Acetat-aktivierung bzw. der Acetylphosphatase im Blutegel (*Hirudo medicinalis*). *Justus Lievigs. Ann. Chem.* **705**, 238–53.

—— et al. (1968). [Contraction of leech back muscle with quaternary and tertiary drug bases. II. Acetylcholine.] *Pharmazie* **23**, 69–75. [In German.]

D'Arcy W. Thompson. (1961). *On growth and form*. Abridged edition 346 pp. Cambridge University Press, London.

Darinskaya, V.S., and Darinskii, Yu. A. (1973). [Changes in neuron volume during depolarization and hyperpolarization.] *Arkh. Anat. Gistol. Embriol.* **65**(9), 40–7. [In Russian; English summary.] [1974BA:67982.]

Darinskii, Yu. A. (1973). [Effect of the microelectrode on neuron recording electrophysiological parameters.] *Fiziol. Zh. SSSR IM.I.M. Sechenova* **59**(4), 551–7. [In Russian; English summary.]

—— and Darinskaya, V.S. (1973). [Neuron response to injury and its activation.] *Tsitologiya* **15**(7), 939–43. [In Russian; English summary.] [1974BA:43461.]

Dartevelle, E. (1949). Les invertebrés des environs de Léopoldville. Les Vers. *Zoo. Léo., Léopoldville* **2**, 15–23. [Hirudinea: 17–18.]

Das, S.M., Malhotra, Y.R., and Duda, P.L. (1964). The palearctic elements in the fauna of Kashmir region. *Kashmir Sci.* **1**, 100–11.

Dastugue, G. (1952). [Sensitizing or antagonistic action of some synthetic antihistamine on the denervated leech preparation, comparison with atropine.] *C. r. hebd. Séanc. Soc. Biol. Paris* **146**, 647. [In French.]

Dastych, H. (1967). *Haementeria costata* Fr. Müller) w jeziorze Bytynskim. *Not. Przyr., Poznan* **1**, 14–15.

Dathe, H. (1934, 1935). See Autrum (1939c).

David, O.F. (1970) (1969)). Spontaneous electrical activity in the longitudinal muscle of the leech *Hirudo medicinalis. J. evol. Biochem. Physiol.* **5**(6), 432–7. [English transl. of *Zh. Evol. Biokhim. Fiziol.* **5**, 540–6 (1969).]

—— (1978). [Neural control of somatic muscle function in the earthworm, *Allobophora longa,* and in the leech, *Hirudo medicinalis.*] *Zh. Evol. Biokhim. Fiziol.* **14**(1), 34–42. [In Russian; English abstr.] [English transl., *J. evol. Biochem. Physiol.* **14**(1), 25–31.] [1980BA:4562.]

Davies, R.W. (1971). A key to the freshwater Hirudinoidea of Canada. *J. Fish. Res. Bd Can.* **28**(4), 543–52.

—— (1972). Annotated bibliography to the freshwater leeches (Hirudinoidea) of Canada. *Fish. Res. Bd. Can.,* Tech. Rep. No. 306, pp. 1–15.

—— (1973). The geographic distribution of freshwater Hirudinoidea in Canada. *Can. J. Zool.* **51**(5), 531–45.

—— (1976). Ecology of *Nephelopsis obscura* (Hirudinoidea) (Abstr.) Abstracts of Symposia and Contributed Papers. Annual Meeting of the Canadian Society of Zoologists, Un. of Regina, Saskatchewan, 8–11 June 1976, p. 51.

—— (1978a). The morphology of *Ozobranchus margoi* (Apathy) (Hirudinoidea), a parasite of marine turtles. *J. Parasitol.* **64**(6), 1092–6.

—— (1978b). Reproductive strategies shown by freshwater Hirudinoidea. *Verh. Int. Verein. Limnol.* **20**, 2378–81.

—— (1979). Dispersion of freshwater leeches (Hirudinoidea) to Anticosti Island, Quebec. *Can. Field-Nat.* **93**, 310–13.

—— (1984). Sanguivory in leeches and its effects on growth, survivorship and reproduction of *Theromyzon rude. Can. J. Zool.* **62**(4), 589–593.

—— and Chapman, C.G. (1974). First record from North America of the piscicolid leech, *Ozobranchus margoi*, a parasite of marine turtles. *J. Fish. Res. Bd Canada* **31**(1), 104–6.

—— and Everett, R.P. (1975). The feeding of four species of freshwater Hirudinoidea in southern Alberta. *Verh. Int. Verein. Limnol.* **19**, 2816–27.

—— —— (1977). The life-history, growth and age structure of *Nephelopsis obscura* Verrill, 1872 (Hirudinoidea) in Alberta. *Can. J. Zool.* **55**, 620–7.

—— and Reynoldson, T.B. (1975). Life history of *Helobdella stagnalis* (L.) in Alberta. *Verh. Int. Verein. Limnol.* **19**, 2828–39.

—— —— (1976). A comparison of the life-cycle of *Helobdella stagnalis* (Linn. 1758) (Hirudinoidea) in two different geographical areas in Canada. *J. anim. Ecol.* **45**(2), 457–70. [1976BA:61085.]

—— and Wilkialis, J. (1980). The population ecology of the leech (Hirudinoidea: Glossiphoniidae) *Theromyzon rude. Can. J. Zool.* **58**(5), 913–16.

—— —— (1981). A preliminary investigation on the effects of parasitism of domestic ducklings by *Thermoyzon rude* (Hirudinoidea: Glossiphoniidae). *Can. J. Zool.* **59**(6), 1196–9.

—— —— (1982). Observations on the ecology and morphology of *Placobdella papillifera* (Verrill) (Hirudinoidea: Glossiphoniidae) in Alberta, Canada. *Am. Midl. Nat.* **107**(2), 316–24.

—— Linton, L.R. and Wrona, F.J. (1982). Passive dispersal of 4 species of freshwater leeches (Hirudinoidea) by ducks. *Freshwater Invertebr. Biol.* **1**(4), 40–4. [1983BA:24853.]

—— Reynoldson, T.B., and Everett, R.P. (1977). Reproductive strategies of *Erpobdella punctata* (Hirudinoidea) in two temporary ponds. *Oikos* **29**, 313–19. [Russian summary.]

—— Singhal, R.N. and Blinn, D. W. (1985). *Erobdella montezuma* (Hirudinoidea Erpobdellidae), a new species of freshwater leech from North America (Arizona, U.S.A.). *Can. J. Zool.* **63**(4), 965–9.

—— Wrona, F.J., and Everett, R.P. (1978). A serological study of prey selection by *Nephelopsis obscura* Verrill (Hirudinoidea). *Can. J. Zool.* **56**(4), 587–91. [French summary.]

—— —— and Linton, L. (1979). A serological study of prey selection by *Helobdella stagnalis* (Hirudinoidea). *J. anim. Ecol.* **48**(1), 181–94. [1979BA:52747.]

—— —— —— (1982). Changes in numerical dominance and its effects on prey utilization and inter-specific competition between *Erpobdella punctata* and *Nephelopsis obscura* (Erpobdellidae: Hirudinoidea) — an assessment. *Oikos* **39**, 92–9. [Russian summary.]

—— Linton, L.R., Parsons, W., and Edgington, E.S. (1982). Chemosensory detection of prey by *Nephelopsis obscura* (Hirudinoidea: Erpobdellidae). *Hydrobiologia* **97**, 157–61.

—— Wrona, F.J., Linton, L., and Wilkialis, J. (1981). Inter- and intraspecific analyses of the food niches of two sympatric species of Erpobdellidae (Hirudinoidea) in Alberta, Canada. *Oikos* **37**, 105–11. [Russian summary.]

Davila, H.V., Salzberg, B.M., and Cohen, L.B. (1973). Optical recording of neuronal impulses in the leech central nervous system. *Biol. Bull.* **145**, 431–2. (See also 1973. *Nature New Biol.* **241**, 159.)

Davis, C.C. (1968). Mechanisms of hatching in aquatic invertebrate eggs. *Oceanogr. Mar. Biol. Ann. Rev.* **6**, 325–76.

Dawydoff, C. (1959). Développement des Hirudinées. *Grassé's Traité de Zoologie*, Vol. 5, pp. 665–86.

de Beauchamp, P. (1904, 1905). See Beauchamp, P. de (1904, 1905a,b) in Autrum (1939c).

de Beer, G.R. (1958). *Embryos and ancestors*, 197 pp. Oxford University Press.

Debout, G., and Provost, M. (1981). Le Marais de la Sangsuriére. *Courrier Nat.* **74**, 10–18.

Dechtiar, A.O. (1972a). New parasite records for Lake Erie fish. *Great Lakes Fish Comm. Tech. Rep.* **17**, 1–20.

—— (1972b). Parasites of fish from Lake of the Woods, Ontaria. *J. Fish. Res. Bd. Can.* **29**(3), 275–83.

Degner, E. (1921). See Autrum (1939c).

Deitmer, J.W., and Schlue, W.R. (1981a). Measurements of the intra-cellular potassium activity of Retzius cells in the leech central nervous system. *J. exp. Biol.* **91**, 87–101.

—— —— (1981b). Distribution of intra- and extracellular K^+ in the leech central nervous system studied using double-barrelled ion-sensitive microelectrodes. In: *Progress in enzyme and ion-selective electrodes.* (eds. D.W. Lübbers, H. Acker, R.P. Buck, G. Eisenman, M. Kessler and W. Simon.) pp. 93–9. Springer Verlag, Heidelberg.

—— —— (1981c). Active regulation of intracellular potassium in sensory neurones of the leech central nervous system. *Naturwissenschaften* **68**, 622–3.

—— —— (1983). Intracellular Na^+ and Ca^+ in leech Retzius neurones during inhibition of the $Na^+ - K^+$ pump. *Pflüger's Arch. Eur. J. Physiol.* **397**, 195–201. (See also J. Physiol. **257**, 53P. 1984).

—— —— (1984). Na-dependent changes of intracellular calcium in leech sensory neurones. *Proc. Physiol. Soc.* July 1984. 60 pp.

Dejoux, C., and Saint-Jean, L. (1972). Etude des communités d'invertébrés d'herbiers du lac Tschad: recherches préliminaires. *Cah. ORSTOM (Hydrobiol.)* **6**(1), 67–83.

Dekleva, N.D. (1971). Eksperimentalno izucavanje uloge katecholamina u mehanizmu traumatskog edema mozga sa posebnim osvrtom na ulogu barometarskog pritiska. Ph.D. thesis. [*Inst. za pato-fiziol. Med. Fak. Beograd.* 1–173. [Leech neurophysiology.]

—— Beleslin, B.B., and Mihailović, Lj. T. (1972). Leech ganglion cells:

effect of adrenaline on spontaneous spike activity in hyposmotic medium. *Proc. XXth Inter. Congr. Aviation Space Medicine*, Nice, p. 50.

—— —— (1974). The influence of adrenaline on electrical activity of leech ganglion cells in hyposmotic medium. *Comp. gen. Pharmacol.* **5**(1), 61–5.

Del Campillo,M.C. (1974). Parasitic zoonoses in Spain. *Int. J. Zoonoses* **1**(2), 43–57. [1975BA:51933.] [Human infestation.]

Della Corte, L., and Nistri, A. (1974). Amine oxidase activity in tissues of the leech (*Hirudo medicinalis.*) *Br. J. Pharmacol.* **52**(1), 129P.

Demirhan, A. (1979). [The use of the leech in folkloric medicine and its evaluation.] *Istanbul. Univ. Tip. Fak. Mecm.* **42**(3), 523–9. [1980BA:82041.]

Demshin, N.I. (1968). [Sexually mature *Hirudineatrema oschnarini* n. g, n. sp. (Hirudineatrematinae n. subf.) parasitic in leeches.] *Soobshch dal'nevost. Fil. V. L. Komarova sib. Otdel. Akad. Nauk SSSR* **26**, 41–6. [In Russian.]

—— (1971). [The role of ecology and phylogenetic links in determining the life cycles of helminths developing with the assistance of oligochaetes and leeches.] In *Parazity Vodnykh bexpozvorochnykh zhivotnykh. I Vsesoyuznyi Symposium*, pp. 23–33, Nauka, Vladivostok. [In Russian.]

—— (1975). [*Oligochaeta and Hirudinea as intermediate hosts of helminths.*] Nauka, Novosbirsk, Siberia. [In Russian.] [Aug. 1976 *Helminthological Abstracts:* 3854.]

Dence, W.A. (1943). A leech feeding on *Ligula. J. Parasitol.* **29**, 299–300.

Dendy, A., and Olliver, M.F. (1901). On a New Zealand fresh-water leech (*Glossiphonia* [*Clepsine*] *novaezealandiae*, n. sp.). *Trans. Proc. NZ Inst.* **33**, 99–103.

Denis, M.A. (1933). Les sangsues en médecine. Thése. Lille. no. 34. pp. 72. Société d'Edition du Nord.

Denzer, U. (1935). (=U. Malbrandt) Helligkeits- und Farbensinn bei deutschen Süsswasseregeln. *Zool. Jahrb. Physiol.* **55**, 525–62.

De Priester, W.C. (1924). Paarden en Bloedzuigers. *De Levende Natuur* **28**, 285.

Dequal, L. (1912). Contributo alla conoscenza degli Irudinei italiani. *Arch. zool. Ital.* **5**(1), 1–14.

—— (1916). Irudinei. In Viaggio de Dott. E. Festa nel Darien, nell Ecuador e regioni vicine. *Boll. Mus. Zool. Anat. Torino* **31**(717), 1–20.

—— (1917). Nuovi Irudinei esotici del Museo zoologico di Torino. *Boll. Mus. Zool. Anat. Torino* **32**(724), 1–20.

Derganc, M., and Zdravic, F. (1960). Venous congestion of flaps treated by application of leeches. *Br. J. plast. Surg.* **13**, 187–92.

Deriemer, S.A., and Macagno, E.R. (1980). Positional correlation of synaptic boutons in pairs of mechanosensory cells in the leech. *Biol. Bull.* **159**(2), 483.

—— —— (1981). Light microscopic analysis of contacts between pairs of identified leech neurons with combined use of horseradish peroxidase and Lucifer Yellow. *J. Neurosci.* **1**(6), 650–7.

—— —— and Muller, K.J. (1981). Overlap of interganglionic dendritic fields of mechanosensory cells in leech segmental ganglia. *Biol. Bull.* **161**(2), 341–2.

Derosa, Y.S., and Friesen, W.O. (1981). Morphology of leech sensilla: observations with the scanning electron microscope. *Biol. Bull.* **160**(3), 383–93.

De Roth, G.C. (1953). Some parasites from Maine freshwater fishes. *Trans. Am. microsc. Soc.* **72**, 49–50.

De Silva, P.H.D.H. See Silva, P.H.D.H. De.

Despotov, B. (1964). [Collected works. I.P. Pavlov de Plovdiv.] Bulgaria.

—— (1965*a*). Procédés anticoagulants et fibrinolytiques d'alimentation de la sangsue pharmaceutique et leur signification therapeutiques dans les thromboses. *Folia Med., Plovdiv* **7**, 291–5. [In French; Russian summary.]

—— (1965*b*). Sur l'action anticoagulante locale de la sangsue pharmaceutique dans les atherothromboses et la maladie hypertonique. *Folia Med., Plovdiv* **7**, 327–32. [In French; Russian Summary.]

—— (1966). État anticoagulant du sang absorbé du reservoir epigastrique de la sangsue pharmaceutique. *Folia Med., Plovdiv* **8**, 18–22. [In French; Russian summary.]

Desser, S.S. (1973). A description of intraerythrocytic schizonts and gametocytes of a haemogregarine of the snapping turtle *Chelydra serpentina. Can. J. Zool.* **51**(4), 431–33.

—— (1976). The ultrastructure of the epimastigote stages of *Trypanosoma rotatorium* in the leech *Batracobdella picta. Can. J. Zool.* **54**(10), 1712–23. [1977BA:21355.]

—— and Weller, I. (1977). Ultrastructural observations on the body wall of the leech, *Batracobdella picta. Tissue Cell* **9**(1), 35–42. [1977BA:38889.]

Dev, B. (1963). Structure of the nephridia of the Indian cattle leech *Hirudinaria granulosa* (Savigny) with remarks on their nephridial microflora. *Experimentia* **19**, 402–3.

—— (1964*a*). Alkaline phosphatase in the nephridial bladder of the Indian leech *Hirudinaria granulosa* (Savigny). *Curr. Sci.* **33**, 114–15.

—— (1964*b*). Stone formation (lithiasis) in the nephridial bladders of the common Indian cattle leech *Hirudinaria granulosa* (Savigny). *Nature, Lond.* **202**, 208.

—— (1964*c*). Excretion and osmoregulation in the leech, *Hirudinaria granulosa* (Savigny). *Nature, Lond.* **202**, 414.

—— (1964*d*). On the histochemical localization of the alkaline phosphates and its physiological significance in the testicular nephridia of the Indian leech, *Hirudinaria granulosa* (Savigny). *J. Histochem. Cytochem.* **12**, 311–14.

—— (1964*e*). Histochemical localization of adenosinetriphosphatase in the Indian leech *Hirudinaria granulosa* (Savigny). *Z. Zellforsch.* **63**, 913–20.

—— (1964*f*). Histochemical localization of alkaline phosphatase in the pre-testicular nephridia of *Hirudinaria granulosa* (Savigny). *Acta histochem.* **19**, 271–4.

—— (1964*g*). On the functional anatomy of the nephridia of the Indian leech, *Hirudinaria granulosa* (Savigny). *Acta anat.* **59**, 327–32.

—— (1964*h*). On the enzymatic patterns of alkaline phosphatase and adenosinetriphosphatase in the nephridia of the Indian leech: *Hirudinaria granulosa* (Savigny) and their physiological significance. *Ann. Histochim.* **9**, 371–6.

—— (1965*a*). The cytochemical localization of the alkaline phosphatase in the testicular nephridia of the Indian leech, *Hirudinaria granulosa*, with special reference to the inner lobe. *Q. Jl microsc. Sci.* **106**, 31–5.

—— (1965*b*). Studies on excretion in the common Indian leech, *Hirudinaria granulosa* (Savigny). *Acta biol. hung.* **15**, 425–30.

—— (1965*c*). On the atrophy of the nephridia of the Indian leech, *Hirudinaria granulosa* (Savigny) and its bearing on the enzymatic make up of the nephridia. *Tsitologiya Moskva* **7**, 719–22.

—— (1968). Histochemical localization of alkaline phosphatase in the testicular nephridia of the land-leech, *Haemadipsa* sp. *Acta anat.* **69**, 287–91.

—— (1969*a*). Pattern of urinary out-flow, quantitative relationship between ammonia and urea, excreted by leeches, with comments on their nephridial microflora. *Symp. Recent Trends in Animal Physiol., Gorakhpur*, 12–13 Feb. 1969, pp. 12–13.

—— (1969*b*). Further studies on lithiasis (stone formation) in the nephridia bladders of the Indian cattle-leech, *Hirudinaria granulosa* (Savigny). *38th Annual Session of the Nat. Acad. of Sciences, India, Ranchi*, 17–19 March 1969, p. 66.

—— and Mishra, G.C. (1971). Histoenzymology of the salivary complex, and mechanism of biting of the medicinal leech, *Poecilobdella granulosa* (Savigny, 1822). I. Adenosine triphosphatase. *Acta morph. neerl. scand.* **9**, 117–24.

—— —— (1972*a*). On the mode of blood-sucking by the Indian medicinal leech, *Poecilobdella granulosa* (Savigny, 1822) and its functional correlation with histoenzymatic and colorimetric studies of biting complex and coordinators. *Acta histochem.* **42**, 15–28.

—— —— (1972*b*). Histoenzymology of the salivary complex, of the Indian cattle leech, *Poecilobdella granulosa* (Savigny, 1822). I. Acetyl cholinesterase, *41st Ann. Sess. Nat. Acad. Sci. India, Varanasi*, Feb. 24–26, p. 92.

—— —— (1972*c*). Distribution of alkaline phosphatase in different rectal zones of the common Indian leech, *Poecilobdella granulosa* (Savigny, 1822). *41st Ann. Sess. Nat. Acad. Sci. India, Varanasi*, Feb. 24–26, p. 93.

—— —— (1972*d*). On the histoenzymatic localization of acid phosphatase in the cerebral and sub-pharyngeal ganglia of the common Indian leech, *Poecilobdella granulosa* (Savigny, 1822). *41st Ann. Sess. Nat. Acad. Sci. India, Varanasi*, Feb. 24–26, p. 94.

—— —— (1976). A simple new technique to demonstrate the presence of giant nerve-cells (Retzius' cells) in the subpharyngeal and segmental ganglia of the common Indian cattle leech, *Poecilobdella granulosa. Curr. Sci.* **45**(15), 556–7.

—— —— (1977*a*). Alkaline phosphatase and its implications in the nervous system of Indian cattle leech, *Poecilobdella granulosa. Cell. molec. Biol.* **22**(2), 191–6. [French summary.] [1979BA:41816.]

—— —— (1977*b*). On the distribution of 5'-nucleotidase and its functional significance in the nervous system of Indian cattle leech, *Poecilobdella granulosa. Cell. molec. Biol.* **22**(2), 197–202. [French summary.] [1979BA:41815.]

—— —— (1977*c*). Effect of decerebration on the enzymatic system of stomach and intestine of Indian cattle leech, *Poecilobdella granulosa. Z. Parasitenkd* **52**(1), 97–102. [1977BA:62940.]

Diamond, L.S. (1958). A study of the morphology, biology and taxonomy of **trypanosomes of Anura. Ph.D. thesis. University of Minnesota.**

Dick, R.I. (1959). Preliminary notes on the relationships existing between the leech, *Marsupiobdella africana*, the river crab, *Potamon perlatus*, and the platanna, *Xenopus laevis*, as observed in specimens taken from the Kromboom River in the Cape Province of the Union of South Africa. *J. scient. Soc. Cape Town.* **2**, 47–9.

Dickinson, M.H., and Lent, C.M. (1984). Feeding behaviour of the medicinal leech, *Hirudo medicinalis* L. *J. Comp. Physopol.* **154**, 449–55.

Dickinson, W.L. (1890). See Autrum (1939*c*).

Didenko, A.V. (1977*a*). [The effect of synaptic activation on RNA synthesis in Retzius cells of the medicinal leech.] *Fiziol. Zh. SSSR IM. I.M. Sechenova* **63**(10), 1482–6. [In Russian.] [1978BA:71042.]

—— (1977*b*). [Effect of temperature on bioelectric activity of Retzius neurons in the medicinal leech.] *Fiziol. Zh. SSSR IM.I. M. Sechenova* **63**(11), 1611–15. [In Russian.] [1978BA:21703.]

—— (1979). Influence of inhibitors and stimulators of protein metabolism on resting potential of Retzius neurons of medicinal leech. *Dokl. biol. Sci.* **246**, 960–2. [English translation of *Dokl. Akad. Nauk SSSR, Ser. Biol.*] [1980BA:36083.]

—— Bazanova, I.S., Evdokimov, S.A., and Merkulova, O.S. (1972). [Correlation between the bioelectric unit activity of the leech Retzius cells and the fluorescence of the RNA acridine derivatives of the same neurons.] *Fiziol. Zh. SSSR IM. I. M. Sechenova* **58**(10), 1569–78. [In Russian; English summary.] [1974BA:61150.]

—— —— and Kazanskii, V.V. (1973). [Correlation between RNA content and membrane potential of Retzius neurons of the leech.] *Fiziol. ZH. IM. I. M. Sechenova* **59**(6), 955–6. [In Russian.] [1974BA:2699.]

—— and Kazanskii, V.V. (1979). [Analysis of the connection between neuron size and bio-electric activity.] *Fiziol. Zh. SSSR IM. I. M. Sechenova* **65**(8), 1231–4. [In Russian.]

—— and Sergeev, N.A. (1974). [Some characteristics of the bioelectric activity of Retzius neurons of the medicinal leech.] *Fiziol. Zh. SSSR. IM. I. M. Sechenova* **60**(12), 1888–90. [In Russian.] [1975BA:43710.]

Dienske, H. (1968). A survey of the metasoan parasites of the rabbit fish, *Chimaetra monstrosa* L. (Holocephali). *Neth. J. sea Res.* **4**, 32–58. [Page 50, *Calliobdella nodulifera* on the head of fish; see Olsson (1896).]

Diesing, C.M. (1850). Systema Helmintha, Vol. 1, Sect. 1: Vindobonae Mollia. [In Latin.] Reprinted by Hafner, New York (1960).

Dilbs and Anderson (1964). [Leeches in humans in Ceylon.] [See Keegan (1968), p. 7.]

Dimitru, V., and Somnes, G.O. (1931). Action thérapeutique de l'hirudine dans les phlébites, la septicémie, et dans quelques affections de nature microbienne. *Presse méd.* **39**, 1359.

Dinland, F., and Bottenberg, H. (1935). See Dinland and Bottenberg (1935*b*) in Autrum (1939*c*).

Dioni, W. (1967). Investigación preliminar de la estructura basica de las asociaciones de la micro y meso fauna de las raices de las plantas flotantes. *Acta zool. Lilloana* **23**, 111–37. [English summary.]

Diwany, E.H. (1925). Recherches expérimentales, sur l'histophysiologie comparée de l'appareil digestif des invertebrés hematophages. 1. Les Hirudinées. *Arch. Anat., Strasbourg* **40**, 229–58.

Dobrowolski, K.A. (1958). [Parasites of leeches of Druzno Lake (Parasitofauna of the biocoenosis of Druzno Lake — Part V).] *Acta parasitol. pol.* **6**(5), 179–94. [In Polish; English summary.]

Dogel, V.A. (1945*a*). Vliyanie rasprostraneniya chozyaina na ego parazitophauna. Sravnenie parazitophaun'Kazakhstanskogo i Dal'nevostostochnogo Sazanov. *Izv. Akad. Nauk. Kazakhsk. SSR Ser. Zool.* **4**, 5–8. [In Russian.]

—— (1945*b*). Analiz parazitophaun' osetrov'ch i otsenka ee patogennogo znacheniya. *Izv. Akad. Nauk Kazakhsk. SSR Ser. Biol.* **4**, 9–19. [In Russian.]

—— and Akhmerov, A. Kh. (1946). Parazitophauna p'b Amura i ee zoogeographicheskoe zhachenie. *Tr. Ubil. Nauchn. Sessii Leningr. Gos. Un. Sektsiya Biol. Nauk.* 171–8. [In Russian.]

—— and Bauer, O.N. (1955). Bor'ba s parazitaph'mi zabolevaniyami p'b v prudov'ch. *Khozyaistvakh, Moscow-Leningrad* 1–86. [In Russian.]

—— and B'Khovskii, G.E. (1934). Phauna parazitov p'b Aral'skogo Morya. *Parazitol. Sb. Zool. Inst. Akad. Nauk SSSR* **4**, 241–346. [In Russian.]

—— —— (1939). Fish parasites of the Caspian Sea.] *Tr. Po Kompleksnomu Izuch. Kasp. Morya Akad. Nauk SSSR* **7**, 1–152. [In Russian.] [*Piscicola caspica.*]

—— and Bogolepova, I.I. (1957). [Parasites of Baikal Fishes.] *Tr. Baikal. Limnolog. Sta.* **15**, 446–50. [In Russian.]

—— and Smirnova, K.V. (1949). Parazitophauna p'b ozera Baikal i ee zoogeographicheskoe znachenie. *Vestn. Leningr. Un.* **7**, 12–34. [In Russian.]

—— and Petruchevskii, G.K. (1939). Parazitophauna p'b Nevskoi rub'. *Tr. Petergophsk. Biolog. Inst.* **10**, 366–434. [In Russia.]

—— —— and Polanski, Y.I. (eds.) (1961). *Parasitology of fishes.* Oliver and Boyd, London. [*Piscicola geometra*, a serious pest to hatchery fish.] [See: Bauer (1961).]

Smirnova, K.V., and Roznachenko, L.K. (1945). Parazit' prom'slov'ch p'b ozera Zaisan. *Izv. Akad. Nauk. Kazakhsk. SSR Ser. Biol.* **4**, 31–7. [In Russian.]

Dolley, J.S. (1933). *Am. Midl. Nat.* **14**, 202.(incomplete reference)

Dollfus, R.P. (1953). Parasites animaux de la morue atlanto-arctique *Gadus callarias* L. *Encycl. Biol.* **43**, 1–428.

—— (1964). Sangsue tentaculifère de la peau d'un téléostéen du genre *Chaenichthys* J. Richardson 1844. *Bull. Mus. nat. Hist.* **36**(6), 831–48.

—— (1971). Hirudinea. In *Marion and Prince Edward Islands* (ed. E. van Zinderen Bakker Sr *et al.*). Balkena, Cape Town.

Dombrowski, H. (1953). Die Nahrungsmenge des Fischegels *Piscicola geometra* L. (Zugleich ein Beitrage zur Physiologie des Blutes des Karpfens *Cyprinus carpio* L.) *Biol. Zbl., Leipzig* **72**(5–6), 311–14.

Domermuth, R.B., and Reed, R.J. (1980). Food of juvenile American shad, *Alosa sapidissima*, juvenile blueback herring, *Alosa aestivalis*, and pumpkinseed, *Lepomis gibbosus*, in the Connecticut River below Holyoke Dam, Massachusetts. *Estuaries* **3**(1), 65–8.

Dordillon, I.R. (1904). *Grammaire et Dictionnaire de la Langue des Iles Marquises.* [Gives a native name *toke omo toto* for 'sangsue'.]

Dorier, A. (1951). Présence de *Glossiphonia heteroclita* sub. sp. *hyalina* dans la cavité palléale de *Limnaea stagnalis. Trav. Lab. Hydrobiol., Grenoble* 46–7.

Dregol'skaya, I.N. (1963). [The possibility of using the heat resistance of cells as species criteria in coelenterates and worms.] *SB Rabot Inst. Tsitol. Akad. Nauk SSSR* **6**, 134–44. [In Russian; English summary.]

Dresscher, Th. G.N., and Engel, H. (1946). De Medicinale Bloedzuiger in Nederland. *Natuurhist. Maanblad.* **35**(7–8), 47–9. [See also Engel.]

———— ———— (1947a). De medicinale bloedzuiger in Nederland. *Pharmaceutische Weekblad, Amsterdam* **82**(39/40), 555–64.

———— ———— (1947b). Een nieuwe diersoort voor het Nederlands gebied. *Levende Nat., Amsterdam* **50**(8/9), 105–7.

———— ———— (1948). Hirudinea of the genus *Helobdella* from Curacao and Venezuela. *Stud. Fauna Curacao, Aruba, Bonaire Venezuelan Islands* **15**, 87–8.

———— ———— (1949). De medicinale bloedzuiger. *Natuurhist. Maandbl.* **38**, 20.

———— ———— (1953). Clef des Hirudinées de Hollande. *De Zoetwater Bloedzuigers. Hybi-Commissie der N. J. N.*

———— ———— (1955). *Trocheta bykowskii* Gedroyć, 1912, in the Netherlands. *Beaufortia* **5**(43), 11–13.

———— and Higler, L.W.G. (1982). De Nederlandse Bloedzuigers (Hirudinea). *Wet. Meded. K. Ned. Natuurh. Veren.* No. 154, 64 pp.

———— ———— and Middelhoek, A. (1960). De Nederlandse bloedzuigers (Hirudinea). *Wet. Meded. K. Ned. Natuurh. Veren.* **39**, 1–60.

———— ———— and Van Der Spoel, S. (1966). Neue Fundorte und Variabilitat der *Xerobdella anulata* Autrum, 1958 (Hirudinea, Haemadipsidae). *Beaufortia* **13**(162), 213–19. (Series of Miscellaneous Publications). Zoological Museum, Amsterdam.

Drewes, C.D., Landa, K.B., and McFall, J.L. (1978). Giant nerve fibre activity in intact, freely moving earthworms. *J. exp. Biol.* **72**, 217–27.

———— and Pax, R.A. (1974). Neuromuscular physiology of the longitudinal muscle of the earthworm, *Lumbricus terrestris*. III. Mapping of motor fields. *J. exp. Biol.* **60**, 469–75.

Druga, A., Lantos, T., Varga, G., and Doman, V. (1966). The action of heparin, Li heparin and hirudin on reversion motion and vacuole formation of *Paramecium-multimicronucleatum* arising upon potassium-ion effect. *Acta biol. acad. sci. hung.* **17**(4), 408. (Abstr.)

Drummond, D.C. (1952). The medicinal leech (*Hirudo medicinalis*) in Yorkshire — another record. *Naturalist*, Hull. 1952. p. 158.

Dubinina, M.N. (1945). [Influence of hibernation on the parasites of fishes in the hibernating grounds of Volga delta.] *Mag. Parasitol., Moscow* 104–25. [In Russian.]

Du Buisson, M., and Van Den Berghe, L. (1927). See Autrum (1939c).

Duchesne, P.Y. (1969). Phénomènes neurosécrétoires au niveau du ganglion caudal de la sangsue *Hirudo medicinalis*. *C. r. Séanc. Soc. Biol.* **163**(4), 998–1000.

DuFour, P. (1950). 'Got a black eye? You can still get leeches.' *Times-Picayune* (La.) 20 July 1950.

Dukina, V.V., Tvorovskii, V.S., and Epstein, V.M. (1974). [Finding of the fish leech *Caspiobdella fadejewi* on shrimps (mysids) in the Pechenezhsk Water Reservoir.] *Zool. Zh.* **53**(9), 1413–14. [In Russian; English summary.][English transl.: **53**, 1431.][1975BA:61456.][On 10 per cent of all *Paramysis lacustris.*]

Durchon, M. (1967). *L'Endocrinologie des Vers et des Mollusques.* Masson, Paris. [Hirudinea.] [See also: *Ann. Sci. Nat. Biol. Anim.* [II] **13**, 427–51.]

Dussumier (incomplete reference) (1827). [Letter recording collecting rhynchobdellid leeches on the Malabar Coast, India.] *Mém. Mus. d'Hist. Nat.* **15**, 379.

Dyk, V. (1941). Patalogichỳ vỳznam rybi pijavky. *Zvěrolokařsk. Obz.* **24**, 1–4.

—— (1954). [Little known parasites on the fishes of South Moravia.] *Cas. Mar. Mus.* **123**, 39–45. [English summary.]

—— (1963). Der gemeine Fishegel (*Piscicola geometra*). *Angewandte Parasitol.*, IV, 2. *Merkbl. angew. Parasitenk.* Schädlinsbek, Merkbl. 7, Jena.

Dykova, I., and Lom, J. (1978). Histopathological changes in *Trypanoplasma borelli* infection in goldfish. *J. Protozool.* **25**, 36A.

Eaglestone, A.A., and Lockett, T.A. (1964). Rockingham leech jar. In *The Rockingham Pottery, Rotherham*, p. 127.

Earp, B.J., and Schwab, R.L. (1954). An infestation of salmon fry and eggs. *Prog. Fish-Cult.* **16**, 122–4.

East, C.F.T., and Bain, C.W.C. (1931). *Recent advances in cardiology*, 2nd edn, pp. v–x, 1–353. Churchill, London. [Relief of venous engorgement of liver by application of six leeches along right costal margin.]

Ebelova, L.M. (1968). [Amitotic division of nerve cells of the CNS of invertebrates as a reaction to mechanical trauma.] *Tsitol. Genet.* **2**(1), 54–62. [In Russian.]

Eberhardt, L.L., Meeks, R.L., and Peterle, T.J. (1971). Food chain models for DDT kinetics in a freshwater marsh. *Nature, Lond.* **230**, 60–2.

Ebrard, E. (1857). *Nouvelle Monographie des Sangsues Médicinales.* Paris.

Eckert, R. (1963). Electrical interaction of paired ganglion cells in the leech. *J. gen. Physiol.* **46**, 573–87.

Eddy, S., and Hodson, W.C. (1950). Leeches. Class Hirudinea. In: *Taxonomic keys to the common animals of the North Central States.* pp. 32–6. Burgess Publishing Company. Minneapolis, Minnesota.

Egorova, T.A., Stidenkina, A.S., and Burgova, M.P. (1979). [Extracellular effect of hydrogen ions on the redox of pyridine nucleotides and flavoproteins of intact nerve cells.] *Vestn. Leningr. Un. Biol.* **3**, 64–9. [In Russian; English summary.] [1980BA:58074.]

Ehinger, B., Falck, B., and Myhrberg, H.E. (1968). Biogenic amines in *Hirudo medicinalis. Histochemie* **15**, 140–9.

Ehlers, E. (1869). Uber fossile Wurmer aus dem lithographischen Schiefer in Bavern. *Palaeontographica* **17**, 145–75.

Ehrenberg, F. (1950). [Modern therapy with leeches.] *Berl. med. Ztschr.* **1**, 183. [In German.]

Eibl-Eibesfeldt, I. (1955). Über die Abwehr von Pferdeegeln (*Haemopis sanguisuga* L.) durch Frösche und Molche. 3. *Tierpsychologie* **12**, 175.

Eichler, D.U., and Eichler, W. (1941). Das Vorkommen von Egeln bei Vögeln. *Mitt. Vereins. Sächs. Orn.* **6**, 115–60.

Einstein, E.B. (1978). 'The leech gatherers.' Aquatint by George Walker, 1814. Yale Medical Library, New Haven. Clements C. Fry Collection. *J. Hist. Med.* **33**, 214, illus.

Eisenstadt, T.B., see Aisenstadt, T.B.

Elder, D. (1979). An epigenetic code. *Differentiation* **14**(3), 119–22. [1980BA:29910.]

Elder, H.Y., and Owen, G. (1967). Occurrence of 'elastic' fibres in the invertebrates. *J. Zool., Lond.* **152**, 1–8.

Elder, J.F. (1974). A study of the metabolism of the carnivorous leech *Mooreobdella microstoma* (Family Erpobdellidae). M.Sc. thesis, University of Tulsa.

—— and Rogers, S.H. (1977). Carbohydrate metabolism in the leech *Mooreobdella microstoma* (Erpobdellidae). *Proc. Okla. Acad. Sci.* **57**, 32–8.

—— and Sawyer, R.T. (1975). Notes on the ecology and occurrence of the leech *Mooreobdella microstoma* (Erpobdellidae) in Tulsa County, Oklahoma. *Proc. Okla. Acad. Sci.* **55**, 48.

Eliezer, M.F. (1966). Helmintos de la colección Eduardo Caballero en el Museo de Historia Natural de la Ciudad de Mexico. *Acta zool. méx.* **8**(6), 1–2.

Elkan, E., and Reichenbach-Klinke, H.H. (1974). *Color atlas of the diseases of fishes, amphibians and reptiles.* H. Publications. [1975BA:11622.]

Elliot, E.J. (in press). Chemosensory stimuli in leech feeding behaviour: an important role for NaCl and arginine. *J. Comp. Physiol.*

Elliott, E.J., and Muller, K.J. (1980). Long term survival of anucleate glial segments during nerve regeneration. *Neurosci. Abstr.* **6**, 679.

—— —— (1981). Long-term survival of glial segments during nerve regeneration in the leech. *Brain. Res.* **218**, 99–113.

—— —— (1982). Synapses between neurons regenerate accurately after destruction of ensheathing glial cells in the leech (*Hirudo medicinalis*). *Science* **215**(4537), 1260–2. [see also: *Neruosci. Abstr.* **7**, 260.] [1982BA:17814.]

—— *et al.* (1983). Sprouting and regeneration of sensory axons after destruction of ensheathing glial cells in the leech nervous system. *J. Neurosci.* **3**(10), 1994–2006.

Elliott, G.F.S., Laurie, M., and Murdoch, J.B. (1901). *Fauna, flora and geology of the Clyde Area, Glasgow.* [Three marine leech species noted.]

Elliott, J.M. (1971*a*). The distances travelled by drifting invertebrates in a Lake District stream. *Oecologia* **6**, 350–79.

—— (1971*b*). Upstream movements of benthic organisms in a Lake District stream. *J. Anim. Ecol.* **40**, 235–52.

—— (1973*a*). The life cycle and production of the leech *Erpobdella octoculata* (L.) (Hirudinea: Erpobdellidae) in a Lake District stream. *J. anim. Ecol.* **42**, 435–48.

—— (1973*b*). The diel activity pattern, drifting and food of the leech *Erpobdella octoculata* (L). (Hirudinea: Erpobdellidae) in a Lake District stream. *J. anim. Ecol.* **42**, 449–59.

—— and Mann, K.H. (1979). *A key to the British freshwater leeches with notes on their life cycle and ecology*. Freshwater Biological Association. Scientific Publication No. 40, pp. 1–72.

—— Mugridge, R.E.R. and Stallybrass, H.G. (1979) *Haementeria costata* (Hirudinea: Glossiphoniidae), a leech new to Britain. *Freshwater Biol.* **9**, 461–5.

—— and Tullet, P.A. (1982*a*). *Provisional atlas of the freshwater leeches of the British Isles*. Freshwater Biological Association. Occasional Publication No. 14. 32 pp.

—— —— (1982*b*). Leech parasitism of waterfowl in the British Isles. *Wildfowl* **33**, 164–70.

—— —— (1984). The status of the medicinal leech *Hirudo medicinalis* in Europe and especially in the British Isles. *Biological Conservation* **29**, 15–26.

Elliott, W.T. (1917). *Glossiphonia* (leech) destroying *Limnaea peregra*. *Proc. Malac. Soc., Lond.* **12**, 307.

Ellis, A.E. (1977). Perils of the deep. *Conchologist Newsl.* **44**, 310. [Attacks by '*Hirudo*' on divers in ponds.]

Emden, M. Van (1929). Bau und Funktion des Botryoidgewebes von *Herpobdella atomaria* Carena. *Z. wiss. Zool.* **134**, 1–83.

Engel, H., and Dresscher, Th. G.N. (1944*a*). Wat weten wij van de bloedzuigers (Hirudinea) welke in Nedernland voorkomen? *Lev. Natuur* **49**(1), 8–10. [See also Dresscher.]

—— —— (1944*b*). Bloedzuigers. *Lev. Natuur* **49**(5), 59.

Enno van Gelder, J.G. (1928). Hat zwemmen van bloedzuigers en alen. *Lev. Natuur* **32**, 107–10.

Epps, G. (1982). The brain biologist and the mud leech. *Science 82* **3**(1), 34–41. [see B. Zipser.]

Epshtein, V.M. (1954). [Certain particular freshwater leeches.] *Zool. Zhurn.* **33**(3), 549–54. [See also Lukin; Parukhin.]

—— (1957*a*). [A new species of leech from the Amur Basin.] *Zool. Zhur.* **36**(9), 1414–17. [In Russian; English summary.] (Transl.)

—— (1957*b*) K phaune p'b'ikh piyavok presn'kh vod SSSR. *Tez. Dokl. Nauchn. Konpher. Vsesouzn. Obsch. Gel'mintolog., Posvyaschennoi 40 Godovschine Velikoi Oktyabr'skoi Sotsialisticheskoi Revolutsii, 11– 15 Dekabrya*, Moscow, Vol. 2, pp. 150–1. [In Russian.]

—— (1959). The systematic position, life history and origin of the endemic leech *Trachelobdella torquata* (Grube). *Dokl. Akad. SSSR Biol. Sci.* **125**(4), 935–7. [English transl. **125**, 393–5.]

—— (1960). Geographicheskoe rasprostranenie presnovodn'kh r'b'ikh piyavok na territorii SSSR. *sb. 'Prodlem' Parazitologii', Tr. 3-i Nauchn. Konpherenstii Parazitol. UK SSR, Kiev* 413–14. [In Russian.]

—— (1961*a*). [On the external morphology, life history and taxonomic position of the endemic Baikal leech, *Codonobdella truneata* Grube.] *Dokl. Akad. Nauk. SSSR. Biol. Sci.* **139**(4), 1108–11. [In Russian; English Transl. **139**, 734–6.]

—— (1961*b*). [A new species of fish leech, *Piscicola fadejewi* n. sp., and some suppositions as to the origin of this species.] *Dopovidi Akad. Nauk. Ukrain. RSR* **12**, 1644–8. [In Russian; English summary.] [Transl. 3711, by RTS, UK).

—— (1961*c*). A review of the fish leeches (Hirudinea, Piscicolidae) from the northern seas of the S.S.S.R. *Dokl. Akad. Nauk. SSSR Biol. Sci.* **141**(6), 1508–11. [Transl. **141**, 1121–4.]

—— (1962*a*). A survey of fish leeches (Piscicolidae) from the Bering and Okhotsk Seas and from the Sea of Japan. *Dokl. Akad. Nauk. SSSR Biol. Sci. Sect.* **144**(5), 1181–4. [Transl. **144**, 648–51.]

—— (1962*b*). Class: Leeches *Opredelitel' Parasitov Presnovodn'kh R'b SSSR Moscow-Leningrad* 617–26. [In Russian.]

—— (1963). [Fish leeches of freshwater and seas of the USSR.] *Avtoreph. Diss.* [Report on a dissertation.] Kharkov 1–16.

—— (1964*a*). Towards the zoogeographical characterization of the fish leeches of the Amur Basin. *Dokl. Akad. Nauk SSSR* **159**(5), 1178–82. [English transl. **159**, 907–9.]

—— (1964*b*). Annelida: Hirudinea. In *Key to parasites of freshwater fish of the USSR* (ed. I.E. Bykhovskaya-Pavlovskaya *et al.*). Office of Technical Services (TT64–11040), US Dept. Commerce, Washington, DC [Hirudinea: pp. 727–39.] [Originally in Russian: *Akad. Nauk SSSR Opred. po Faune SSSR Opred. po Faune SSSR* 1–776 (1962).]

—— (1965). [On the systematic position, distribution and origin of the Caspian endemic leech *Piscicola caspica* Selensky (Hirudinea, Piscicolidae)]. *Zool. Zhurn.* **44**(12), 1858–61. [Technical Translation, National Research Council of Canada, Ottawa.]

—— (1966*a*). *Acanthobdella livanowi* sp. n. — A new species of leech (Archihirudinea) from Kamchatka. *Dokl. Akad. Nauk SSSR* **168**(4), 955–8. [Transl. *Dokl. Biol. Sci.* 450–3.]

—— (1966*b*). [*Caspiobdella tuberculata*, gen. et sp. nov., a new genus and leech species (Hirudinea; Piscicolidae) from the Caspian Sea.] *Helminthologia* **7**, 151–4. [In Russian; English summary.]

—— (1967*a*). [Regularities of the geographical distribution of the marine fish leeches (Hirudinea, Piscicolidae.] *Zool. Zh.* **46**(5), 680–91. [In Russian, English summary.] [Transl. Scitran, Santa Barbara, Ca., for the EPA (1975).]

—— (1967*b*). [On relations and geographical distribution of fish leeches (Hirudinea, Piscicolidae) of the genus *Carcinobdella* Oka, 1910.] *Zool.*

Zh. **46**(11), 1648–54. [In Russian; English summary.] [Transl. UDC 595.143.2, by National Sciences Library, National Research Council of Canada, Ottawa; and Dr M.C. Meyer.]

—— (1968*a*). [Revision of the genera *Oxtonostoma* and *Johanssonia* (Hirudinea: Piscicolidae).] *Zool. Zh.* **47**(7), 1011–21. [In Russian.] [Transl. P.G. Rossbacher, Dept. Modern Languages, Oregon State University.]

—— (1968*b*). [Zoogeographical analysis of fish leeches in the Antarctic and revision of the genus *Trachelobdella* Diesing, 1850.] (Abstr.) [*All Union Conference (5th) on the Diseases and Parasites of Fish and Aquatic Invertebrates*, 29 Nov.–4 Dec. 1968, pp. 137–8, *Leningrad.*] Izdat.

—— (1968*c*). [Leeches.] In *Opredelitel' Phaun' Chernogo i Azovskogo Morei*, pp. 394–405. Kiev. [In Russian.]

—— (1968*d*). [Leeches.] In *Atlas Zhivotn'kh Kaspiiskogo Morya*, pp. 113–17. Moscow. [In Russian.]

—— (1968*e*). Hirudinea. In *Identification key to the fauna of the Black and Azov Seas*, Vol. 1: *Free-living invertebrates* (ed. F. Mordukhai-Boltovskoi.) *Akad. Nauk Ukrain SSR Inst. Biol. Yushn. Morei Kiev* 394–404. [In Russian.]

—— (1969). [Revision of genera *Piscicola* and *Cystobranchus* (Hirudinea: Piscicolidae).] In *Problems in Parasitology, Proceedings of the VIth Scientific Conference of Parasitologists of the UKSSR* (ed. A.P. Markevich) Kiev, Vol. 2, pp. 286–7. (Trans.)

—— (1970*a*). [Unevenness of the evolutionary rate of organ systems and the taxonomic principles of fish leeches (Hirudinea: Piscicolidae).] [*All-Union Conference (1st) on Parasites and Diseases of Marine Animals, Sevastopol*], 1970, Kiev, pp. 140–2. Izdat. (Transl.)

——(1970*b*). [Bipolar distribution of marine fish leeches (Hirudinea: Piscicolidae).] [*All-Union Conference (1st) on Parasites and Diseases of Marine Animals, Sevastopol*, 1970], Kiev, pp. 143–6. Izdat. (Transl.)

—— (1970*c*). [Fish leeches (Hirudinea: Piscicolidae) in the Antarctic seas from the collection of the Zoological Institute of the Academy of Sciences of the USSR.] [*All-Union Conference (1st) on Parasites and Diseases of Marine Animals, Sevastopol*, 1970], Kiev, pp. 146–9. Izdat. (Transl.)

—— (1972). [The taxonomy of the antarctic leech *Pontobdella rugosa* (Piscicolidae.] *Zool. Zh.* **51**(8), 1142–6. [Technical Translation, National Research Council of Canada, Ottawa.]

—— (1973*a*). [Diagnoses of the genera *Calliobdella*, *Trachelobdella*, *Limnotrachelobdella* and *Baicalobdella* (Hirudinea, Piscicolidae) and determination of the taxonomic value of the diagnostic characteristics.] *Zool. Zh.* **52**(3), 332–41. [Transl. P.G. Rossbacher, Dept. Modern Languages, Oregon State University.]

—— (1973*b*) [The taxonomic position, mode of life and geographic distribution of *Levinsenia rectangulata*.] *Parazitologiya* **7**(3), 286–92. [Transl. for NERC-Library, EPA, by Leo Kanner Associates, Redwood City, Ca.]

—— (1973c). [New information on the structure, geographic distribution and hosts of the marine tropical leech *Trachelobdella lubrica* (Piscicolidae).] *Parazitologiya* **7**, 427–35. [Transl. P.G. Rossbacher, Dept. of Modern Languages, Oregon State University.]

—— (1974). [New information on Antarctic fish leeches descending from tropical elements.] [*All-Union Conference (VI) on the Diseases and Parasites of Fish and Aquatic Invertebrates*, 3–5 April, 1974], pp. 301–5. Tezesy Doknadov.

—— (1982). [Numbers of leeches in Northwest Pacific ecosystems.] *Hydrobiol. J.* **18**(3), 93. [In Russian.]

—— and Korotaeva, V.D. (1973). [Generic diagnosis of *Trachelobdella* Diesing, 1850 (Hirudinea, Piscicolidae) and the geographic distribution of the species of this genus.] *Izv. Tikhookean. Nauchno-Issled. Inst. Rybn. Khoz. Okeanogr* **87**, 178–84. [Trans. Translation Bureau (JS), Multilingual Services Division, Fisheries and Marine Service, Arctic Biological Station, Ste. Anne de Bellevue, Quebec, Canada, Translation Series No. 3277.]

Epure, E.X. (1945). *Cystobranchus respirans* (Troschel), un rare Ichthyobdellide trouvé en Roumanie. *Bul. Soc. Stinte Cluj, Timisoara* **9**(4), 557–63.

—— (1947). Die Fauna der Hirudinen im Banat. *Ann. Sci. Univ. Jassy* **30**(2), 29–37. [Also Vol. 9 (1943).]

Ergens, R. (1962). Helmintofauna ryb dvou jihoceskych rybnicních soustav II. Trematoidea, Monogenoidea, Nematoda, Acanthocephala a Hirudinea. *Cs. Parasitol.* **9**, 167–91.

Erickson, J.E. (1976). New host and distribution records for *Piscicolaria reducta* Meyer 1940. *J. Parasitol.* **62**(3), 409.

—— (1978). Parasites of the Banded Darter, *Etheostoma zonale* (Pisces: Percidae). *J. Parasitol.* **64**(5), 899.

Erickson, R.C. (1948). Life history and ecology of the canvasback, *Nyroca valisineria* (Wilson) in southwestern Oregon. Ph.D. thesis, Iowa State College, Ames, Iowa. [*Theromyzon* in nose, buccal cavity and on body.]

Ernst, C.H. (1971). Seasonal incidence of leech infestation on the painted turtle, *Chrysemys picta*. *J. Parasitol.* **57**(1), 32.

—— (1976). Ecology of the spotted turtle, *Clemmys guttata* (Reptilia, Testudines, Testudinidae), in Southeastern Pennsylvania. *J. Herpetol.* **10**(1), 25–33. [*Placobdella parasitica*.][1976BA:18900.]

—— and Ernst, E.M. (1977). Ectoparasites associated with neotropical turtles of the genus *Callopsis* (Testudines, Emydidae, Batagurinae). *Biotropica* **9**(2), 139–42. [1978BA:3635.]

Ertlová, E. (1967). [Occurrence of *Cystobranchus respirans* (Troschel) in the Czechoslovakian area of the Danube River.] *Biológia Bratisl.* **22**(5), 386–8. [In Czechoslovak.]

Erwin, D.N. (1975). Effects of neuroaminidase upon the serotonin receptors of Retzius cells in the leech. Ph.D. dissertation, University of Oklahoma.

—— and Theis, R.E. (1975). Effect of neuraminidase and ethylene-diaminetetra-acetic acid (EDTA) upon the serotonin receptors of Retzius cells in the leech. *Fedn. Proc. Fedn. Am. Socs. exp. Biol.* **34**, 359.

Etziony, M.A. (1962). A note on the ancient and modern diagnosis and treatment of hemoptysis due to leeches. *Bull. Hist. Med.* **36**, 529–31.

Evsin, V.N., and Ivanov, N.O. (1979). The summer feeding of brook trout, *Salmo trutta*, in the Pulonga River (Kola Peninsula). *Vopr. Ikhtiol.* **19**(6), 1098–104. [In Russian.]

Ewers, W.H. (1974). *Trypanosoma aunawa* sp. n. from an insectivorous bat, *Miniopterus tristris*, in New Guinea, which may be transmitted by a leech. *J. Parasitol.* **60**(1), 172–8.

Ewert, J.P., and Traud, R. (1979). Releasing stimuli for antipredator behaviour in the common toad *Bufo bufo* (L.). *Behaviour* **68**(1/2), 170–80. [1979BA:13178.]

Eyres, J.P., and Pugh-Thomas, M. (1978). Heavy metal pollution of the River Irwell (Lancashire, U.K.) demonstrated by analysis of substrate materials and macroinvertebrate tissue. *Environ. Pollut.* **16**(2), 129–36. [*Erpobdella octoculata* tissue analysis.] [1979BA:18954.]

Fahy, E. (1974). [Reference to Irish leeches.] *Ir. nat. J.* **18**, 9–12.

Faivre, E. (1854). Note sur les oeufs parasites de la Sangsue médicinale. *Soc. Biol. Mém., Paris*, **I**, 143–5.

—— (1855). Observations histologiques sur le grand sympathique de la Sangsue médicinale. *Ann. Sci. Nat. Ser. Zool.* **IV**, 249–61.

—— (1856). Études sur l'histologie comparée du système nerveux chez quelques Annélides. *Ann. Sci. Nat. Ser. Zool.* **V**, 337–74; **VI**, 16–82.

Faller, A. (1951). *Fauna Republicii Populare Romane*, Indrumator, part iA, *Protozoare, Viermi, Arthropode*. Bucarest.

—— (1963). Elektronenmikroskopische Untersuchung der glatten Muskelzellen im Neurilemma des Zentralnervensystems von *Hirudo medicinalis* L. In *Comptes rendus de l'Union Libre des Anatomistes des Universités Suisse, Lausanne*, 1963. *Acta Anat.* **55**(4), 399. (Abstr.)

—— (1964). Zur Frage der Struktur der glatten Muskelzellen von *Hirudo medicinalis* L. *Z. Zellforsch. Mikrosk. Anat.* **63**(6), 799–815.

Falls, D.L., and Macagno, E.R. (1981). A quantitative study of the branching pattern of T cell processes in *Hirudo medicinalis*. *Biol. Bull.* **161**(2), 342–3.

Fänge, R. (1979). Protozoan infections (haemogregarines, trypanosomes) of the blood of gadoid fish, *Melanogrammus aeglefinus* (haddock) and *Gaidropsaurus cimbrius* (four-bearded rockling). *Acta zool., Stockholm* **60**(3), 129–37.

Farber, I.C., and Grinvald, A. (1983). Identification of presynaptic neurons by laser photostimulation. *Science.* (Incomplete reference)

Faust, E.C. (1940). Leech infestation. In *Modern medical therapy in general practice*, Vol. 2. Wood, Baltimore.

—— and Russell, P.F. (1964). Leeches. In *Craig and Faust's Clinical parasitology*, 7th ed, pp. 704–12. Lea and Febiger, Philadelphia.

Fawcett, D.W. (1966). On the occurrence of a fibrous lamina on the inner aspect of the nuclear envelope in certain cells of vertebrates. *Am. J. Anat.* **119**(1), 129–46.

—— and Coggeshall, R.E. (date undetermined). The fine structure of the central nervous system of *Hirudo medicinalis. Anat. Rec.* **145**, 362.

Fenton, J.W. (1981). (Hirudin possibly involving multiple binding states). *Ann. N.Y. Acad. Sci.* **267**, 469–95.

Fermond, C. (1984). See Autrum (1939c).

Fernàndez, J.H. (1968). A light and electron microscopical study of the nervous system of the leech and the snail. An analysis of neural and glial relationships, transport pathways and degenerative and regenerative changes that follow lesions. *Diss. Abstr.* **29B**, 12 Pt. 1, 4485.

—— (1975). Structure and arrangement of synapses in the CNS of the leech under normal and experimental conditions. *Neurosci. Abstr.* **I**, 662.

—— (1976a). Organization of fiber pathways and synaptic sites in the CNS of the leech *Macrobdella decora. Anat. Rec.* **184**(3), 401.

—— (1976b). Development of the nervous system in the horse leech. *Neurosci. Abstr.* **II**, 634.

—— (1978). Structure of the leech nerve cord: distribution of neurons and organization of fiber pathways. *J. comp. Neurol.* **180**(1), 165–91. [BA:39968.]

—— (1980). Embryonic development of the glossiphoniid leech *Theromyzon rude:* characterization of developmental stages. *Devel. Biol.* **76**(2), 245–62.

—— (1982). [Organization and development of the nervous system of the leech.] *Arch. Biol. Med. exp.* **15**, 229–43. [In Spanish.]

—— and Fernandez, M.S.G. (1974). Morphological evidence for an experimentally induced synaptic field. *Nature, Lond.* **251**, 428–30.

—— and Olea, N. (1982). Embryonic development of glossiphoniid leeches. In: *Developmental biology of freshwater invertebrates.* (eds. F.W. Harrison and R.R. Cowden) pp. 317–61. Alan R. Liss, Inc., New York.

—— and Stent, G.S. (1980). Embryonic development of the glossiphoniid leech *Theromyzon rude:* structure and development of the germinal bands. *Devl. Biol.* **78**(2), 407–34.

—— and Stent. G.S. (1982). Embryonic development of the hirudinid leech *Hirudo medicinalis:* structure, development and segmentation of the germinal plate. *J. Embryol. exp. Morph.* **72**, 71–96. [1983BA:34220.]

Fernando, C.H. (1960). The Ceylonese fresh water crabs (Potamidae). *Ceylon J. Sci. biol. Sci.* **3**, 191–222.

Ferraguti, M. See also separate Bibliography on Branchiobdellida, p. 801.

—— (1983). Annelida-Clitellata. In: *Reproductive biology of invertebrates.* Vol. II (ed. K.G. Adyiodi and R.G. Adyiodi). pp. 343–76. Wiley, London.

—— (1984). The comparative ultrastructure of sperm flagella central sheath in Clitellata reveals a new autapomorphy of the group. *Zoologica Scripta* **13**.

Ferrari, G., and Rindi, G. (1952). Attività catalasica di alcune specie di irudinei. *Boll. Soc. ital. Biol. sper.* **28**, 1510–12.

Fett, M.J. (1978). Quantitative mapping of cutaneous receptive fields in normal and operated leeches, *Limnobdella. J. exp. Biol.* **76**, 167–79. [1979BA:28658.]

Fialkowski, W. (1979). Ecology of leeches (Hirudinea) in the organically polluted part of Drwinka stream. *Acta hydrobiol.* **21**(4), 475–88.

Filippi, F. de (1849). Sopra un nuovo genere (*Haementeria*) di Annelidi della famiglia delle Sanguisughe. *Mem. Acc. Sci. Torino 2* **10**, 1–14.

Films on the Hirudinea. See: Malecha, J. (1976*b*); Westheide, W. (1978, 1980, 1981*a,b*).

Fioravanti, R., and Fuortes, M.G.F. (1972). Analysis of responses in visual cells of the leech. *J. Physiol.* **227**, 173–94.

Fischer, A., and Weigelt, K.R. (1975). Strukturelle Beziehungen zwischen jungen Oocyten und somatischen Zellen bei den Anneliden *Platynereis* und *Piscicola. Verh. Dt. Zool. Ges.* **67**, 319–23. [In German; English summary.] [1976BA:2800.]

Fischer, E. (1966*a*). Cytochemical investigation on the muscles of *Haemopis sanguisuga* L. *Acta biol. acad. sci. hung.* **17**, 393. (Abstr.)

—— (1966*b*). Lokalisation der Phosphorylase im Pferdeegel *Haemopis sanguisuga* L. *Acta histochem.* **25**(5/8), 371–4.

—— (1967*a*). [Histochemical examination of food storage in the organism of leeches.] *Bull. Teachers' Training Inst., Pécs, Ser. Biol.* **11**, 33–45. [In Hungarian.]

—— (1967*b*). Einige zytochemische Eigenschaften der Muskeln des Pferdeegels (*Haemopis sanguisuga* L.). *Zool. Anz.* **178**, 326–33.

—— (1967*c*). Zytochemische Studien über die Spermien und die Spermiogenese des Pferdeegels *Haemopis sanguisuga* L. *Zool. Anz.* **179**(3/4), 298–302.

—— (1968*a*). Histochemische und heilungsphysiologische Studien an der Korperdecke des Pferdeegels (*Haemopis sanguisuga* L.). (Incomplete reference)

—— (1968*b*). Histochemical examinations of the botryoidal and vasofibrous tissues of horseleech (*Haemopis sanguisuga* L.). *Acta. biol. acad. sci. hung.* **19**, 495–6. (Abstr.)

—— (1968*c*). Experimentally induced changes of the glycogen contents in the muscular and nervous system of the horse leech, *Haemopis sanguisuga* L.: a histochemical study. *Acta biol. acad. sci. hung.* **19**(4), 455–63.

—— (1969). Morphological background of the regulation of nephridial activity in the horse leech (*Haemopis sanguisuga* L.). Study of nephridial innervation by means of esterase reaction. *Acta. biol. acad. sci. hung.* **20**(4), 381–7.

—— (1970*a*). Histochemical examination of the botryoidal tissue and its secretory phase in the horseleech (*Haemopis sanguisuga* L.). *Acta biol. acad. sci. hung.* **21**(3), 281–92.

—— (1970*b*). Lokalisation der Leucyl-Aminopeptidase (LAP) beim Blutegel (*Hirudo medicinalis* L.) und beim Pferdeegel (*Haemopis sanguisuga* L.). *Acta histochem.* **37**(1), 170–5.

—— (1970*c*). [Histological, histochemical and histophysiological study of the botryoidal and vaso-fibrous tissues of *Haemopis sanguisuga* L.] *Bull. Teachers' Training Inst., Pécs, Ser. Biol.* **14**, 53–68. [In Hungarian.]

—— (1972*a*). Oxidation and detoxification of aromatic compounds in the botryoidal tissue of the horseleech, *Haemopis sanguisuga* L. *Acta biol. acad. sci. hung.* **23**(3), 289–97.

—— (1972*b*). [Histological, histochemical and histophysiological examination of the chloragogen and botryoidal tissues of Annelids.] Thesis, Pécs. [In Hungarian.]

—— and Klujber, L. (1970). [Protein und sulfatiertes Mucopolysaccharid in den Botryoidgranula des Pferdeegels (*Haemopis sanguisuga* L.).] *Acta histochem.* **38**(2), 241–9. [In Russian; English and French summaries.]

—— Lovas, M., and Nemeth, P. (1976*a*). Zinkporphyrin-Pigmente im Botryoidgewebe von *Haemopis sanguisuga* L. und ihre Lokalisation mit der Diaminobenzidin-H$_2$O$_2$ Reaktion. *Acta histochem.* **55**(1), 32–41. [English summary.]

—— —— —— (1976*b*). An experimental analysis of the vasofibrous tissue in the leech, *Haemopis sanguisuga* L. *Zool Anz.* **197**(5/6), 289–99. [German summary.] [1977BA:27051.]

Fish, T.D., and Vande Vusse, F.J. (1976). *Hirudicolotrema richardsoni* gen. et sp. n. (Tremadota: Macroderoididae) from Minnesota hirudinid leeches. *J. Parasitol.* **62**(6), 899–900. [1977BA:45399.]

Fitzgerald, S.W.R. (1983). Studies on the defence mechanisms of *Arenicola marina* L. (Polychaeta) and the coelomocytes of other annelids (Polychaeta, Hirudinea). Hirudinea: Vol. 1. Ph.D. thesis. University College of Swansea, U.K.

Fitzpatrick-McElligott, S., and Stent, G.S. (1981). Appearance and localization of acetylcholinesterase in embryos of the leech *Helobdella triserialis. J. Neurosci.* **1**(8), 901–7.

Fjeldsa, J. (1971). [Leeches (Hirudinea) and snails (Gastropoda) from freshwater in Nordland and Troms.] *Fauna* **24**, 41–8. [In Norwegian.]

—— (1972). Records of *Theromyzon maculosum* (Rathke 1862): Hirudinea, in N. Norway. *Norw. J. Zool.* **20**(1), 19–26.

—— and Raddum, G.G. (1973). [Three limnic invertebrate species new to Iceland, found in Myvatn in 1969.] *Náttúrufraedingurinn* **42**(1–2), 103–13. [In Icelandic; English summary.] [*Theromyzon maculosum.*]

Flacke, W., and Yeoh, T.S. (1968a). The action of some cholinergic agonists and anticholinesterase agents on the dorsal muscle of the leech. *Br. J. Pharmacol. Chemother.* **33**, 145–53.

—— —— (1968b). Differentiation of acetycholine and succinylcholine receptors in leech muscle. *Br. J. Pharmacol. Chemother.* **33**, 154–61.

Flerov, B.A. (1977). [Physiological action mechanisms of toxic substances and adaption of aquatic animals to them.] *Gidrobiol. Zh.* **13**(4), 80–6. [In Russian; English summary.] [1978BA:74087.]

Fletcher, P., and Forrester, T. (1975). The effect of curare on the release of acetylcholine from mammalian motor nerve terminals and an estimate of quantum content. *J. Physiol.* **251**(1), 131–44.

Flint, R.W., and Merckel, C.N. (1978). Distribution of benthic macroinvertebrate communities in Lake Erie's eastern basin. *Int. Ver. Theor. Angew. Limnol. Verh.* **20**, 240–51.

Floravanti, R., and Fuortes, M.G.F. (1972). Analysis of responses in visual cells of the leech. *J. Physiol.* **227**(1), 173–94.

Flössner, D. (1976). [Biomasse und Produktion des Macrobenthos der Mitten Saale.] *Limnologica* **10**(1), 123–53. [1976BA:30465.] [*Erpobdella octoculata*, a dominant species in the lower, more polluted sections.]

Fomina, M.S., and Tereshkov, O.D. (1970). [On neuron background activity of the leech ganglion.] *Biol. Nauk.* **77**(5), 27–30. [In Russian.]

—— (1972). Electrical transmission between symmetrical neurons in leech ganglia. *Neurosci. behav. Physiol.* **5**, 91–6. [Transl. *Fiziol. Zh. SSSR Sechenova* **57**, 532–8 (1971); in Russian.]

Font, W.F., and Corkum, K.C. (1977). Distribution and host specificity of *Alloglossidium* in Louisiana. *J. Parasitol.* **63**(5), 937–8.

Fontaine, M.T. (1978). Contribution to the zoogeographical study of the Sambre-et-Meuse district, Belgium, and particular study of a stream. *Ann. Soc. R. Zool. Belg.* **108**(3/4), 159–70. [In French; English summary.] [Biology of *Erpobdella octoculata.*]

Forbes, S.A. (1890a). An American terrestrial leech. *Bull. Ill. Lab.* **3**, 118–22.

—— (1890b). An American terrestrial leech. *Am. Nat.* **24**, 646–9.

Forrester, D.J., and Sawyer, R.T. (1974). *Placobdella multilineata* (Hirudinea) from the American alligator in Florida. *J. Parasitol.* **60**(4), 673.

Forrester, T. (1966). An isometric leech muscle preparation with stable sensitivity for the assay of acetylcholine. *J. Physiol.* **187**, 12P.

Forselius, S. (1952). Blodigeln (*Hirudo medicinalis* L.) i Norden. *Svensk Faun. Revy.* **3**, 67–79.

Forsyth, D.J. (1977). Limnology of Lake Rotokana and its outlet stream. *N. Z. Jl mar. freshwat. Res.* **11**(3), 524–40. [1979BA:14602.] [The leech '*Helobdella*' sp, 1.3/m²; mean pH 2.1.]

Foucher, G., Henderson, H.R., Maneau, M., Merle, M., and Braun, F.M. (1981). Distal digital replantation: one of the best indications for microsurgery. *Internat. J. Microsurgery* **3**, 265 [14 case histories involving leeches to treat venous congestion.]

Fox, F. (1827). [Description of Glass Leech.] *Lancet* **12** (incomplete reference).

Fox, J.G., and Ediger, R.D. (1970). Nasal leech infestation in the rhesus monkey. *Lab. Anim. Care* **20**(6), 1137–8.

França, C. (1908). See França, C. (1908*a,b*) in Autrum (1939*c*).

—— (1915). Le *Trypanosoma inopinatum*. *Arch. Protisenkd.* **36**, 1–12.

Françoise, Ph. (1886). Contribution a l'étude du système nerveux central des Hirudinées. *Tabl. Zool.* **1**, 121–227.

Frank, E., Jansen, J.K.S., and Rinvik, E. (1975). A multisomatic axon in the central nervous system of the leech. *J. comp. Neurol.* **159**(1), 1–14.

Frankenberg, G. von (1947). Beobachtungen über der Fressakt des Vielfrass-Egels (*Haemopis sanguisuga*). *Natur Volk* **77**(1–3), 16–19.

Franz, F. (1903). See Autrum (1939*c*).

Franz, H. (1954). Hirudinea. In *Die Nordost-alpen im Spiegel ihrer Landtierwelt. Umfassend: Fauna, Faunengeschichet, Lebensgemein-schaften und Beeinflussung der Tierwelt durch den Menschen.* **I**, pp. 208–10. Innsbruck. [*Xerobdella*.]

Franzen, A. (1956). On spermiogenesis, morphology of spermatozoon and biology of fertilization among invertebrates. *Zool. Bidrag Fran Uppsala* **31**, 355–482. [*Erpobdella octoculata*.]

—— (1962). See separate bibliography on Branchiobdellida.

Frazer, B.M., and Lent, C.M. (1977*a*). Leydig cells: interneurons in leech CNS. *Am. Zool.* **17**, 875 (Abstr.)

—— —— (1977*b*). Cellular properties of the monoamine neurons in the leech CNS. *Neurosci. Soc.* **3**, 253 (Abstr.)

—— —— (1977*c*). Connectivity of the monoamine neurons in leech CNS. *Am. Zool.* **17**, 875.

Freeman, W.H., and Bracegirdle, B. (1971). *Atlas of invertebrate structure.* Dover, New York.

Freitag, R., Fung, P., Mothersill, J.S., and Prouty, G.K. (1976).

Distribution of benthic macroinvertebrates in Canadian waters of Northern Lake Superior. *J. Great Lakes Res.* **2**(1), 177–82.

French, A.S., and DiCaprio, R.A. (1975). The dynamic electrical behaviour of the electrotonic junction between Retzius cells in the leech. *Biol. Cybernet.* **17**(3), 129–35. [1975BA:8777.]

Freytag, K. (1953) Untersuchungen über den Aufbau der Cuticula von *Hirudo medicinalis* L. *Z. Zellforsch.* **39**, 85–93.

Friant, S.L., Patrick, R., and Lyons, L.A. (1980). Effects of nonsettleable biosolids on stream organisms. *J. WPCF.* **52**(2), 351–63.

Frič, A., and Vávra, V. (1903). Seznam nižší zvireny v uvodi Labe u Podebrad pozorovane s vyobrazenimi. Pijavky (Hirudinei). Vyzkum zvireny ve vodach ceskych. *Archiv. Pro. Prirodoved. Vyzkum Cech.* **11**, 93–4.

Fridman, G.M. (1948). [Benthos of Lake Sevan, Armenia.] *Tr. Sevansk. Gidrobiol. St.* **10**, 7–39. [In Russian.]

—— (1950). [Fauna of Lake Sevan, Armenia.] *Tr. Sevansk. Gidrobiol. St.* **11**, 7–92. [In Russian.] [*Erpobdella octoculata, Helobdella stagnalis Glossiphonia complanata.*]

Friede, R. (1955). Der Kohlenhydratgehalt der Glia von *Hirudo* bei verschiedenen Funktionszustanden. *Z. Zellforsch. micros. Anat.* **41**, 509–20.

Friedrich, G., and Thome, K.N. (1976). Limnologic geologic excursion in the territory of the lower Erft River, West Germany. *Decheniana* **129**, 268–72.

Friend, H. (1922). The annelids of Iceland and the Faroes, *Nature, Lond.* **110**, 342.

Friesen, W.O. (1980). Models, modeling and the leech swimming rhythm. *Behavioural and Brain Sciences* **3**, 546.

—— (1981). Physiology of water motion detection in the medicinal leech. *J. exp. Biol.* **92**, 255–75.

—— (1985). Neuronal control of leech swimming movements: interactions between cell 60 and previously described oscillator neurons. *J. comp. Physiol.* **156**, 231–42.

—— and Dedwylder, R.D. (1978). Detection of low-amplitude water movements: a new sensory modality in the medicinal leech. *Neurosci. Abstr.* **4**, 380.

—— Poon, M., and Stent, G.S. (1976). An oscillatory neural network generating a locomotory rhythm. *Proc. natn. Acad. Sci. USA* **73**(10), 3734–8.

—— —— —— (1978). Neuronal control of swimming in the medicinal leech. IV. Identification of a network of oscillatory interneurons. *J. exp. Biol.* **75**, 25–43.

—— and Stent, G.S. (1977). Generation of a locomotory rhythm by a neural network with recurrent cyclic inhibition. *Biol. Cybernet.* **28**(1), 27–40. [1978BA:71041.]

—————— (1978). Neural circuits for generating rhythmic movements. *A. Rev. Biophys. Bioeng.* **7**, 37–61.

Frieswijk, J.J. (1957). A leech-avoidance reaction of *Physa fontinalis* (L.) and *Physa acuta* Drap. *Basteria* **21**(3), 38–45.

Fritz, H. (1980). [Use of eglins in medicine.] In: *Protein degradation in health and disease. Ciba Foundation Symposium* **75**, 351–79.

—— and Krejci, K. (1976). Trypsin-plasmin inhibitors (bdellins) from leeches. *Meth. Enzymol.* **45**, 797–806. [1977BA:7576.]

—— Förg-Brey, B., Schiessler, H., Arnhold, M., and Fink, E. (1972). Characterization of a trypsin-like proteinase (acrosin) from boar spermatozoa by inhibition with different protein proteinase inhibitors. Part 2. Inhibitors from leeches, soybeans, peanuts, bovine colostrum and sea anemones. *Z. Physiol. Chem.* **353**(6), 1010–12.

—— Gebhardt, M., Meister, R., and Fink, E. (1971). Trypsin-plasmin inhibitors from leeches. Isolation, amino acid composition, inhibitory characteristics. In *Proceedings of the International Research Conference on Proteinase Inhibitors, Munich, 1970* (ed. H. Fritz and H. Tschesche pp. 271–80.) Walter de Gruyter, Berlin.

—— Oppitz, K.H., Gerbhardt, M., Oppitz, I., and Werle, E. (1969). Über das Vorkommen eines Trypsin-Plasmin-Inhibitor in Hirudin. *Z. Physiol. Chem.* **350**, 91–2.

Fuchs, P.A., Henderson, L.P., and Nicholls, J.G., (1982). Chemical transmission between individual Retzius and sensory neurons of the leech in culture. *J. Physiol.* **323**, 195–210. [1982BA:17817.]

—— Nichols, J.G., and Ready, D.F. (1918). Membrane properties and selective connexions of identified leech neurons in culture. *J. Physiol.* **316**, 203–23.

Fühner, H. (1981*a,b*). See Autrum (1939*c*).

Fukuda, T.R. (1937*a–d*). [Ca^{++} ions and skin resting potentials.] *J. Fac. Sci., Tokyo Imp. Un.* **4**(Part 3), 333, 348, 359, and 369.

Fukui, T. (1926). See Autrum (1939*c*).

Fukumoto, T. (1939). [Leech muscle in assav for acetylcholine.] *J. Chiba med. Soc.* **7**, 127.

Fukuoka, G. (1945). Biological and physiological studies of mountain leech from Kasugayama Nara Prefecture. *Physiol. ecol. contr. Hydrol. exp. Sta. Kyoto. Un.* **27**, 1.

Fuld, E., and Spiro, K. (1904). Der Einfluss einiger gerinnungshemmender Agentien auf das Vogelplasma. *Beitr. Chem. Physiol. Path.* **5**, 171–90.

Furukawa, Y., Ikeda, M., and Ohkubo, T. (1963). [Leech muscle in assay for acetylcholine.] *Japan J. vet. Sci.* **25**, 33.

Gadler, R.M., and Olson, Jr., A.C. (1984). Eye anomalies in the leech *Dina anoculata* (Erpobdellidae) *Proc. helminth. Soc. Wash.* **51**(2), 363–4.

Gale, W.F. (1973). Predation and parasitism as factors affecting *Sphaerium transversum* (Say) populations in Pool 19, Mississippi River. *Res. Pop. Ecol.* **14**(2), 169–87. [Japanese summary.] [*Glossiphonia complanata.*]

—— (1975). Bottom fauna of a segment of pool 19, Mississippi River, near Fort Madison, Iowa, 1967–1968. *Iowa St. J. Res.* **49**(4 Part 1), 357–72. [*Glossiphonia complanata, Helobdella stagnalis,* and *Erpobdella punctata* were abundant with densities up to 68 000/m² in areas with shelter.] [1975BA:42441.]

Galliard, H. (1952). Le Professeur Emile Brumpt (1877–1951). *Ann. Parasitol.* **27**, 1–46.

—— (1968). Leeches as possible vectors of mammalian trypanosomes. (Letter.) *Trans. R. Soc. trop. Med. Hyg.* **62**(2), 298.

Gallico, E. (1934). Contributo alla conoscenza della fauna del lago di Mantova. *Boll. Zool.*, Anno V, N.5.

Gallimore, J.R. (1964). Taxonomy and ecology of helminths of grebes in Alberta. M.Sc. thesis, University of Alberta, Edmonton, Alberta. [Ecological relationship of leeches and grebes.]

Galun, R. (1975). The role of host blood in the feeding behavior of ectoparasites. In: *Dynamic aspects of host–parasite relationships.* (ed. A Zuckerman) Vol. 2, pp. 132–62.

—— and Kindler, S.H. (1966). Chemical specificity of the feeding response in *Hirudo medicinalis* (L.). *Comp. Biochem. Physiol.* **17**(1), 69–73.

—— —— (1968). Regulation of feeding in leeches. *Experientia* **24**, 1140.

—— Kosower, E.M., and Kosower, N.S. (1969). Effect of methyl phenyldiazenecarboxylate (azoester) on the feeding behaviour of blood sucking invertebrates. *Nature, Lond.* **224**, 181–2.

Garavaglia, C., Lamia Donin, C.L., and Lanzavecchia, G. (1974). Ultrastructural morphology of spermatozoa of Hirudinea. *J. submicrosc. Cytol.* **6**(2), 229–44. [Italian summary.] [*Hirudo medicinalis, Erpobdella octoculata.*] [1975BA:25538.]

Garcia-Más, I. (1979*a*). Ultraestructura de la cuticula bucofaringae de *Dina lineata* (O.F. Müller, 1774) (Hirudinea, Erpobdellidae). *Bol. R. Soc. Española Hist. Nat.* (*Biol.*) **77**, 361–71. [In Spanish; English summary.]

—— (1979*b*). Anatomia, histologia y ultraestructura del tubo digestivo de *Dina lineata* (O.F. Müller, 1774) (Hirudinea, Erpobdellidae). *II.* Estomago. *Morfologia normal y Patalógica Sec. A.* **3**, 509–25. [In Spanish; English Summary.]

—— (1980*a*). Anatomia, histologia y ultraestructura del tubo digestivo de *Dina lineata* (O.F. Müller, 1774) (Hirudinea, Erpobdellidae). 1. Región bucal y faringe. *Bol. R. Soc. Espanola Hist. Nat.* (*Biol.*) **78**, 59–76. [In Spanish; English summary.]

—— (1980*b*). Anatomia, histologia y ultraestructura del ṭubo digestivo de *Dina lineata* (O. F. Müller, 1774) (Hirudinea, Erpobdellidae). III. Intestino. *Morfologia normal y patológica. Sec. A.* **4**, 485–500. [In Spanish; English summary.]

—— (1981). Estudio histologico y ultraestructural de los musculos intrinsecos del tracto digestivo de *Dina lineata* (O.F. Müller, 1774) (Hirudinea, Erpobdellidae). *Morfologia normal y patológica Sec. A.* **5**, 87–96. [In Spanish; English summary]

Gardner, C.R. (1976). The neuronal control of locomotion in the earthworm. *Biol. Rev.* **51**, 25–52.

—— and Walker, R.J. (1982). The roles of putative neurotransmitters and neuromodulators in annelids and related invertebrates. *Prog. Neurobiol.* **18**, 81–120.

Gardner-Medwin, A.R., Jansen, J.K.S., and Taxt, T. (1973). The 'giant' axon of the leech. *Acta physiol. scand.* **87**, 30A–1A.

Gardeniers, J.J.P. (1966). De invloed van waternerontreiniging op de fauna van de beken van het stroomgebied van de Dommel. Doctoraal Verslag, Kath. Univ. Nijmegen. [Leeches in Holland.]

Gash, R., and Stephen, L.G. (1972). Helminthes fauna of Centrarchidae from two stripmine lakes and a stream in Crawford Country, Kansas, *Trans. Kansas Acad. Sci.* **75**(3), 236–44.

Gasic, G.J., Viner, E.D., Budzynski, A.Z., and Gasic, G.P. (1983). Inhibition of lung tumor colonization by leech salivary gland extracts from *Haementeria ghilianii*. *Cancer Research* **43**, 1633–36.

—— —— and Gasic T.B. (1983). Inhibition of experimental metastases in mice by leech salivary gland extracts and their mechanisms of action. Special Symposium on Cancer Invasion and Metastasis. Houston, Texas. 28 February – 3 March, 1983.

—— *et al.* (1984). In: *Treatment of metastasis: problems and prospects* (ed. K. Hellmann, G. Nicolson, and L. Milas). Taylor and Francis, London.

Gaskell, J.F. (1914). The chromaffine system of annelids and the relation of this system to the contractile vascular system in the leech, *Hirudo medicinalis. Phil. Trans. R. Soc. Lond. B* **205**, 153–211.

—— (1919). Adrenalin in annelids. *J. gen. Physiol.* **2**, 73–85.

Gates, G.E., and Moore, J.E. (1970). The freshwater and terrestrial Annelida. Appendix II. In *Fauna of Sable Island and its zoogeographic affinities.* pp. 1–45. National Museum of Natural Sciences, Ottawa. Publications in Zoology, No. 4.

Geddes, D.C. (1963). *Trocheta subviridis* — a further Yorkshire record. *Naturalist, Hull* **887**, 120.

Gedroyć, M. (1915). Pijawki (Hirudinea) Polski. Studyum monograficzne. *Pozpr. Wiad. Muz. Dzied., Lwów* **1**(1–2), 176–90.

Gee, N.G., and Wu, C.F. (1926). A list of some Chinese leeches. *China J. Sci.* **4**, 43–5.

Gee, W. (1913). The behavior of leeches with especial reference to its modifiability. *Un. Calif. Publ. Zool.* **11**, 197–305.

Geiler, H. (1973). *Haemopis sanguisuga* L. Vielfrass oder Pferdeegel. *Grosses zoologisches Praktikum*, 11b. Fischer, Stuttgart.

Geissbuhler, J. (1938). Beitrage zur Kenntnis der Uferbiozonosen des Bodensees. *Mitt. thurg. naturf. Ges. Frauenfeld.* **31**, 3–74.

Gentry, G. (1938). [Leech design in dresses.] *Progr. Med.*, Suppl. illustré, Apr. 2, 65, p. 30.

Gerasimov, V.D. (1967). Electrical properties and connections of CNS giant nerve cells of *Hirudo medicinalis.* In *Symposium on neurobiology of invertebrates* (ed. J. Salanki) pp. 285–92. Plenum Press, New York. Hungarian Academy of Sciences, Budapest.

—— and Akoev, G.N. (1967*a*). [Features of the giant neuron electrical activity of the leech, *Hirudo medicinalis,* in various saline solutions.] *J. evol. biochem. Physiol.* **3**, 234–40. [In Russian.]

—— —— (1967*b*). [Electrical reactions of different neurons of the leech in a calcium-free solution.] *Dokl. Akad. Nauk SSSR* **172**(2), 494–7. [In Russian.]

—— —— (1967*c*). Effect of various ions on the resting and action potentials of the giant nerve cells of the leech, *Hirudo medicinalis. Nature, Lond.* **214**, 1351–3.

Gerd, S.V. (1946). Obzor gidrobiologicheskikh issledovanii ozer Karelii. *Tr. Karelo-Phinskogo Otd. V.N.I.O.R.Kh.* **2**, 29–132. [In Russian.]

—— and Sokolova, V.A. (1965). Piyavki ozer Karelii. 5b. *Phauna Ozer Karelii. Bespozvonochn'e Moscow-Leningrad* 82–4. [In Russian.]

Gerebtzoff, M.A. (1969*a*). Recherches histoenzymologiques sur les cellules gliales géante de la chaîne nerveuse ventrale de la sangsue *Hirudo medicinalis. C. r. Séanc. Soc. Biol.* **162**(11), 2029–31.

—— (1969*b*). [Comparative histochemical study of metabolic potentiation of neurons and glia in ventral nerve chain of the leech *Hirudo medicinalis.*] *Arch. int. Physiol.* **77**, 532–5.

—— (1970). Recherches histochimiques et histoenzymologiques sur la synergic metabolique entre neurons et névroglia dans la chaîne nerveuse ventrale de la sangsue, *Hirudo medicinalis. Bull. Acad. R. Méd. belg.* **10**, 337–61.

Gerlach, A. *et al.* (1975). [Leeches in the respiratory system.] *Laryngol. Rhinol. Otol.* **54**(2), 123–32. [In German.]

Gerneck, R. (1933). *Entwicklungsphysiologische Untersuchungen an Tubifex- und Clepsine-Embryonen.* Würzburg.

Gersch, M., and Richter, K. (1961). Experimentelle Untersuchungen des physiologischn Farbwechsels von *Piscicola geometra* (Hirudinea). *Zool. Jb. (Allg. Zool.)* **69**(3), 273–84.

Gerschenfeld, H.M. (1973). Chemical transmission in invertebrate central nervous systems and neuromuscular junctions. *Physiol. Rev.* **53**(1), 1–119. [Leeches included.]

Gheorghiu, M.G. (1933). Sur le pigment de *Protoclepsis tesselata* (O.F. Müller). *Bull. Soc. Chim. Biol., Paris* **15**, 552–4.

Ghosh, J.M., Johnson, P., and Nayer, C.K.G. (1963). On the occurrence of the leech *Ozobranchus branchiatus* (Menzies, 1791) in India (Gulf of Kutsch). *J. Bombay nat. hist. Soc.* **60**(2), 469–71.

Gibelli, A. (1963). [Research on the anticoagulant and fibrinolytic action in vivo of hirudin.] *Rivista Pat. Clin.* **18**, 505–19. [In Italian.]

Gibson, R.N., and Tong, L.J. (1969). Observations on the biology of the marine leech *Oceanobdella blennii. J. mar. Biol. Ass. UK* **49**(2), 433–8.

Gilderhus, P.A. (1979). Effects of granular 2,5-dichloro-4-nitrosalicylanilide (Bayer 73) on bethnic macroinvertebrates in a lake environment. *Great Lakes Fish Comm. Tech. Rep.* **34**, 1–5.

Gilkes, M. (1957). Leech bite of the cornea. *Br. J. Ophthal.* **41**, 124.

Gillette, R. (1977). Leech hunt: success in the swamps. *Los Angeles Times* 2–4 October 1977. Vol. XCVI, p. 1ff.

—— (1979). In search of the giant leech. *Adventure Travel* August 1979, 14–22.

Gils, J.B.F. van (1942). De Bloedzuiger als vriend en als vijand. *Ned Tijdschr. Geneesk.* **1**, 22.

Gilyarov. M.S., Lukin, E.I., and Perel', T.S. (1969). [The first terrestrial leech in fauna of the USSR, *Orobdella whitmani* Oka (Hirudinea, Herpobdellidae), a tertiary relict of forests of the southern maritime region.] *Dokl. Acad. Nauk SSSR* **188**(1), 235–7.

Gintsburg, G.I. (1966). [Incorporation of some amino acids into the egg cell during the oogenesis of *Glossiphonia complanata* L.] *Zh. Obshch. Biol.* **27**(5), 575–82. [In Russian; English summary.]

—— (1967*a*). [Autoradiographic study of the nucleic acids biosynthesis during the oogenesis of *Glossiphonia complanata* L.] *Zh. Obshch. Biol.* **28**(3), 322–34. [In Russian; English summary.]

—— (1967*b*). [Peculiarities of the participation of the nucleus of oocytes in the biosynthesis of nucleic acids.] *Struktura in funcii Kletochnogojadra, Iznat nauka* 209–13. [In Russian.]

—— (1971). Incorporation of H^3-thymidine into the cytoplasm of oocytes during various periods of their growth. *Sov. J. devl Biol.* **2**, 63–9. [Transl. from *Ontogenez* **2**(1), 79–87.]

Girard, C. (1850). On a new generic type in the class of worms. *Proc. Am. Ass. Adv. Sci.* **4**, 124–5.

Gislen, T., and Brinck, O. (1950). Subterranean waters on Gotland with special regard to the lummelunda current II. Environmental conditions, plants and animal life, immigration problems. *Acta Un. Lund.* (new series) **46**(6), 1–81.

Giusseppina, F. (1950). Ricerche sulla presenza di roboflavina in alcune specie de 'Irudinei'. *Boll. Soc. ital. Biol. sper.* **26**(2), 144–6.

Glassman, A.B., and Bennett, C.E. (1978). Response of the alligator, *Alligator mississippiensis* to infection and thermal stress. In *US Department of Energy Symposium, Augusta*, 2–4 Nov. 1977. (ed. J.H.

Thorp and J.W. Gibbons) pp. 691–702. Nat. Tech. Information Cen. U.S. Department of Commerce, Springfield, Va.

—— Holbrook, T.W., and Bennett, C.E. (1979). Correlation of leech infestation (*Placobdella multilineata*) and eosinophilia in alligators (*Alligator mississippiensis*). *J. Parasitol.* **65**(2), 323–4. [1980BA:17644.]

Glick, S., and Ritz, N.D. (1957). Hypochronic anemia secondary to leeching. *New Engl. J. Med.* **256**, 409.

Globus, A., Lux, H.D., and Schubert, P. (1973). Transfer of amino acids between neuroglia cells and neurons in the leech ganglion. *Exp. Neurol.* **40**(1), 104–13.

Glover, J.C., and Kramer, A.P. (1982). Serotonin analog selectively ablates identified neurons in the leech embryo. *Science* **216**, 317–19 [see also 1981, *Neurosci. Abstr.* 7, 2.]

Glusa, E., and Markwardt, F. (1980). Adrenaline-induced reactions of human platelets in hirudin plasma. *Haemostasis* **9**(3), 188–92. [1980BA:17416.]

Gobert, J.G., Bertrand, J.Y. Adelet-Suarez, B., and Savel, J. (1977). About some abnormalities of the genital tract in the leech, *Hirudo medicinalis*. *C. r. hebd. Séanc. Soc. Biol. Paris* **171**(3), 534–42. [1978BA:59067.]

Goddard, E.J. (1908). Contribution to our knowledge of Australian Hirudinea Part I. *Proc. Linn. Soc. New S. Wales* **33**, 320–42.

—— (1909*a*). Contribution to our knowledge of Australian Hirudinea. Part II. *Proc. Linn. Soc. New S. Wales* **33**, 854–66.

—— (1909*b*). Contribution to our knowledge of Australian Hirudinea. Part III. *Proc. Linn. Soc. New S. Wales* **34**, 467–86.

—— (1910*a*). Contribution to our knowledge of Australian Hirudinea. Part IV. With a note on a parasitic entoproctus polyzoon. *Proc. Linn. Soc. New S. Wales* **34**, 721–32.

—— (1910*b*). Contribution to a knowledge of Australian Hirudinea. Part V. Leech metamerism. *Proc. Linn. Soc. New S. Wales* **35**, 51–68.

—— (1910*c*). Contribution to a knowledge of Australian Hirudinea. Part VI. The distribution of Hirudinea, with special reference to Australian forms, and remarks on their affinities, together with reflections on zoogeography. *Proc. Linn. Soc. New S. Wales* **35**, 69–76.

—— (1914). On the significance of the somitic constitution, body form and genital apertures in the Hirudinea, in reference to the Arthropoda. *Trans. R. Soc. S. Afr.* **4**, 148–56.

—— and Malan, D.E. (1912). The South African Hirudinea. Part I. *Ann. S. Afr. Mus.* **11**, 307–19.

—— —— (1913). Contribution to a knowledge of South African Hirudinea. On some points in the anatomy of *Marsupiobdella africana*. *Trans. R. Soc. S. Afr.* **3**, 249–54.

Gogava, M.V., and Batsikadze, Ch. M. (1973). [Aspects of the longitudinal muscle action of the leech.] *Soobsch Akad. Nauk Gruz. SSR* **72**(2),

465–8. [In Georgian; Russian and English summaries.] [1974BA:38268.]

Golding, D.W. (1974). A survey of neuroendocrine phenomena in non-arthropod invertebrates. *Biol. Rev.* **49** (no. C.), 161–224. [Hirudinea, 187–8.]

—— and Whittle, A.C. (1978). Neurosecretion and related phenomena in Annelids. *Int. Rev. Cytol.* **50** (Supplement), 189–302.

Goldner, M.M., and Rosen, G.M. (1977). Effects of spin labelled acetylcholine analogs on cholinergic receptors of the leech. *Res. Commun. Chem. Path. Pharmacol.* **16**(3), 393–7. [1977BA:21008.]

Goldsmid, J.M. (1977). Leech infestation of man in Rhodesia. *Trans. R. Soc. trop. Med. Hyg.* **71**(1), 86. [Reports '*Hirudo*' *sjoestedi* and *H. intermedia* from two women in Zimbabwe.]

Golubev, A.I. (1968*a*). [Submicroscopic structure of peripheral nerves of the leech, *Hirudo medicinalis* L.] *Arkh. Anat. Gistol. Embriol.* **54**(2), 38–45. [In Russian.]

—— (1968*b*). [Certain peculiarities of the ultrastructure of the ganglion cells of the medicinal leech.] *Sb. Aspir. Rabot, izd. KGU*, Str. 137–9. [In Russian.]

—— (1970). [The structure and function of the neuroglia in the nervous system of the leech (*Hirudo medicinalis*) (electron microscope data.] *Voprosy evolyutsionnoi morfologgii i biogeografii.* Izdatel'stvo Kazanskogo Universiteta, 23–39. [In Russian.] [The glia of packets, neuropile and connective are ultrastructurally distinguishable.]

Gomez Tersol, A., Talavera, P.A. and Verdue, R.M. (1976). Contribución al conocimiento de los invertebrados de la zona de Revolcadores. In: *Comunicaciones obre el Carst en la provincia du Murcia*, No. 2. Servicio de Investigación y Defensa de la Naturaleza de la Exma Diputación Provincial. [*Hirudo* in Spain]

Gonçalves, M.D.G.R., and Pellegrino, J. (1967). Predatory activity of *Helobdella triserialis* (Blanchard, 1849) upon *Biomphalaria glabrata* under laboratory conditions. *J. Parasitol.* **53**(1), 30.

Gondko, R., Leyko, W., Majdak, M., and Wojtas, F. (1979). Adenine nucleotide content of human blood stored in the alimentary tract of medicinal leech (*H. medicinalis* L.). *Acta Univ. Lodz.* (Ser. II) **23**, 123–7. [Polish and Russian summaries.]

Gondran, M. (1955). Remarques sur les glandes tégumentaires de *Glossossiphonia complanata* L (Hirudinée, Rhynchobdelle). *Arch. Zool. exp. Gén.* **92**(3), 93–115.

Gooding, R.H. (1972). Digestive processes of haematophagous insects. I. A literature review. *Quaest. Entomol.* **8**, 5–60.

Goodrich, E.S. (1899). On the communication between the coelom and the vascular system in the leech, *Hirudo medicinalis*. *Q. Jl microsc. Sci.* **42**, 477–95.

—— (1945). The study of the nephridia and genital ducts since 1895. *Q. Jl microsc. Sci.* **86**, 113–392.

Gopalakrishnan, V. (1968). Diseases and parasites of fishes in warm-water ponds in Asia and the Far East. *FAO Fish. Rep.* (1966, publ. 1968), No. 44(5), pp. 319–43.

Goreglyad, Kh. S. (1955). *Parazit' I Vrediteli P'b*, p. 1–286. Moscow.

Gosset, (1923). Presence de sangsues dans le larynx et le pharynx d'un cheval. *Rec. Med. Vet.* **99**, 480–1.

Gouck, H.K. (1966). Protection from ticks, fleas, chiggers, and leeches. *Arch. Dermatol.* **93**(1), 112–13.

—— Taylor Jr, J.D., and Barnhart, C.S. (1967). Screening of repellents and rearing methods for water leeches. *J. econ. Entomol.* **60**(4), 959–61.

Goulart, A.D. (1963). A hirudofauna do municipio de Porto Alegre (State of Rio Grande do Sul, Brazil). *Iheringia Zool.* **29**, 1–7.

—— (1967). Presenca de *Helobdella obscura* Ringuelet, 1942, e *Helobdella duplicata* var. *tuberculara* Ringuelet, 1958, no Rio Grande do Sul, Brasil. *Iheringia Zool.* **35**, 3–5.

Gould, G.M., and Pyle, W.L. (1898). *Anomalies and curiosities of medicine.* Bell, New York.

Graber, M., and Euzéby, J. (1975). Lutte biologique contre les mollusques vecteurs de trématodoses humaines et animales. Role predateur possible d'*Hirudo medicinalis* Linné (Gnathobdelliformes: Hirudinea) à l'egard de *Biomphalaria glabrata* Say. *Bull. Soc. Sci. Vet. Med. comp. Lyon* **77**(5), 325–8.

Graber, V. (1884). See Autrum (1939c).

Gradwell, N. (1972). Behaviours of the leech, *Placobdella*, and transducer recordings of suctorial pressures. *Can. J. Zool.* **50**(10), 1325–32.

Graf, A., (1893). See Autrum (1939c).

——(1894). See Graf, A. (1894a) in Autrum (1939c).

—— (1899). Hirudineenstudien. *Acta Acad. Leop., Halle* **72**, 215–404.

Graf, L., Patty, A., Barabas, E.B., and Bagdy, D. (1973). On the NH₂-terminal residue of hirudin. *Biochim. biophys. Acta* **310**(2), 416–17. [1974BA:39329.]

Gräfner, G., and Baumann, H. (1974). Blutegelbefall beim Wassergeflügel. *Angew. Parasitol.* **15**(3), 121–4. [*Theromyzon tessulatum.*] [1975BA:51043.]

Graham, L.C. (1966). The ecology of helminths in breeding populations of Lesser Scaup (*Aytha affinis* Eyton) and Ruddy Ducks (*Ozyura jamaicensis* Gmelin). M.Sc. thesis, University of Alberta, Edmonton, Alberta. [Ecology relationships of leeches and ducks.]

Grant, H.C. (1968). Marine and freshwater biology. A note on the freshwater leeches (Hirudinea) of the Lincoln area. *Trans. Lincs. Nat. Un.* **17**, 39.

Grant, J.S. (1958). An experiment with leeches. *Malayan nat. J.* **13**, 31. [Placed buffalo leeches with toads whereupon both toads and leeches died, the toads from loss of blood apparently and the leeches perhaps because they could not prevent coagulation of the blood.]

Granzow, B., Freed, E., and Kristan, W.B. (1980). Long-term behavioral and neuronal changes induced by section of interganglionic connections in leeches. *Neurosci. Abstr.* **10**, 686.

Grassé, P.P. (ed.) (1959). *Traité de zoologie.* Classes des Oligochetes et Hirudinées. 5 (Part I), pp. 224–713. Masson et Cie, Paris. [See Harant and Grassé (1959); Dawydoff (1959).]

Gratia, A. (1920*a,b*, 1921). See Autrum (1939*c*).

Gratiolet, P. (1862). See Autrum (1939*c*).

Grätz, E., and Autrum, H. (1935). Vergleichende Untersuchungen zur Verdauungsphysiologie der Egel II. Die Fermente der Eiweissverdauung bei *Hirudo* und *Haemopis.* *Z. vergl. Physiol.* **22**, 273–83.

Gray, E.G., and Guillery, R.W. (1963*a*). An electron microscopical study of the ventral nerve cord of the leech. *Z. Zellforsch. Mikroscop. Anat.* **60**, 826–49.

—— —— (1963*b*). On the nuclear structure in the ventral nerve cord of the leech *Hirudo medicinalis. Z. Zellforsch.* **59**, 738–45.

Gray, J., Lissmann, H.W., and Pumphrey, R.J. (1938). The mechanism of locomotion in the leech (*Hirudo medicinalis* Ray). *J. exp. Biol.* **15**, 408–30.

Gray, L.J., and Ward, J.V. (1978). Environmental effects of oil shale mining and processing. Part II — the aquatic macro-invertebrates of the Piceance Basin, Colorado, prior to oil shale processing. EPA–600/3–78–097, Environmental Research Laboratory, US EPA, Duluth, Minnesota.

Gray, P. (1952). *Handbook of basic microtechnique.* The Blakiston Company, New York.

Green, C. *et al.* (1983). The medicinal leech. *J.R. Soc. Hlth* **103**(1), 42–4.

Greene, K.L. (1974). Experiments and observations on the feeding behaviour of the freshwater leech *Erpobdella octoculata* (L.) (Hirudinea: Erpobdellidae). *Arch. Hydrobiol.* **74**(1), 87–99. [1975BA:30507.] [No endogenous rhythms.]

Greeze, V.N. (1953). Parazitophauna europeiskikh sigov, akklimatizirovann'kh v Sibiri. *Tr. Vsesouzn. Gidrobiol. Obsch.* **5**, 254–7. [In Russian.]

—— (1957). Osnovn'e chert' gidrobiologii ozera Taim'r. *Tr. Vsesouzn. Gidrobiol. Obsch.* **8**, 183–218. [In Russian.]

Griffiths, R.B. (1974). Parasites and parasitic diseases. In *The husbandry and health of the domestic buffalo* (ed. W.R. Cockrill) pp. 236–75. FAO, Rome. [Brief mention of leeches.]

Grimas, V. (1967). On the fauna in the lower reaches of River Viskan, Southern Sweden. *Rep. Inst. freshwat. Res. Drottingholm, Sweden* **47**, 87–97.

Grinvald, A., Hildesheim, R., Farber, I.C., and Anglister, L. (1982).

Improved fluorescent probes for the measurement of rapid changes in membrane potential. *Biophys. J.* **39**(3), 301–8. [1983BA:15952.]

Groves, E.W. (1975). Survey of Bookham Common. Vermes: Hirudinea: a further note on the medicinal leech. *Lond. Nat.* **54**, 45–6.

Grube, A.E. (1871). Beschreibungen einiger Egelarten. *Arch. Naturg.* **37**, 87–121.

Gruffydd, L.D. (1965). Notes on a population of the leech *Glossiphonia heteroclita*, infesting *Lymnaea pereger*. *Ann. Mag. nat. Hist.* **13**(8), 151–4.

Gruninger, T.L., Murphy, C.E., and Britton, J.C. (1977). Macroparasites of fish from Eagle Mountain Lake, Texas. *Southwest Nat.* **22**(4), 525–35.

Gruzova, M.M. (1975). [The karyosphere in oogenesis.] *Tsitologiya* **17**(3), 219–37. [In Russian; English summary.] [1976BA:41876.]

—— and Zaichikova, Z.P. (1967). [The karyosphere in oogenesis of the leech *Glossiphonia complanata*.] *Tsitologiya* **9**(4), 388–96.

Grynaeus, T. (1962). De hirudine (Magyar). *Commun. Bibl. Hist. med. Hung.* **26**, 129–55.

Grzhimajlo, I. (1859). [Description of 20 years of experiments on breeding leeches.] *Zapiski Komiteta Akklimat. Zhivotnyx* No. 1. [In Russian.] [See Sineva (1949).]

Guerne, J. De (1892). See Autrum (1939*c*).

Gundrize, A.N. (1972). [Parasites of Mongolian grayling.] *Tr. Nauk No-Issled Inst. Biol. Biofiz. Tomsk Un* **2**, 99–101. [Transl. from *REF Zh. Biol.*, No. 12K8.]

Günther, J. (1971*a*). Mikroanatomie des Bauchmarks von *Lumbricus terrestris* L. *Z. Morphol. Tiere* **70**, 141–82.

—— (1971*b*). On the organization of enteroceptive afferents in the body segments of the earthworm. *Verhandl. Dtsch. Gesell. Zool.* **64**, 261–5.

—— (1976). Impulse conduction in the myelinated giant fibers of the earthworm. Structure and function of the dorsal nodes in the median giant fiber. *J. comp. Neurol.* **168**, 505–32.

Gupta, P.K., and Pant, M.C. (1983). Seasonal variation in the energy content of benthic macroinvertebrates of Lake Nainital, Uttar Pradesh, India. *Hydrobiologia* **99**(1), 19–22. [1983BA:47924.]

Gurvich, V.V. (1961). [On the knowledge of microbenthos fauna and benthopelagic plankton in Kabovska water reservoir.] *Z. birn. Prats zool. Mus.* **30**, 29–39. [In Ukrainian; Russian summary.]

Gvozdev, E.V. (1945). Parazitophauna p'b Nagorno-aziatsckoi podoblasti. *Izv. Akad. Nauk Kazakhsk. SSR Ser. Zool.* **4**, 38–44. [In Russian.]

Haage, P. (1975). Quantitative investigations of the Baltic *Fucus* belt macrofauna. 3. Seasonal variation in biomass, reproduction and population dynamics of the dominant taxa. *Contrib. Askö Lab.* No. 10, 1–84. [Seasonal variations of *Piscicola geometra*.]

Hachlov, L. (1910*a*, 1910*b*). See Hachlov, L. (1910*a*) or (1910*b*) in Autrum (1939*c*).

Hackett, L.W. (1952). In memoriam: Emile Brumpt (1877–1951). *J. Parasitol.* **38**, 271–3.

Haderlie, E.C. (1953). Parasites of the freshwater fishes of northern California. *Un. Calif. Publ. Zool.* **57**(5), 303–440. [*Myzobdella lugubris* on *Ictalurus spp.*]

Hadzi, J. (1954). Edina evropska Kopenska pijavka zivi Sloveniji. *Proteus* **16**, 9.

Hagadorn, I.R. (1958). Neurosecretion and the brain of the rhynchobdellid leech, *Theromyzon rude* (Baird, 1869). *J. Morph.* **102**, 55–90.

—— (1962*a*). Neurosecretory phenomena in the leech, *Theromyzon rude*. *Proc. 3rd Int. Conf. Neurosecretion*, Memoir No. 12, pp. 313–21.

—— (1962*b*). Functional correlates of neurosecretion in the rhynchobdellid leech, *Theromyzon rude*. *Gen. comp. Endocr.* **2**, 516–40.

—— (1966*a*) The histochemistry of the neurosecretory system in *Hirudo medicinalis*. *Gen. comp. Endocr.* **6**(2), 288–94.

—— (1966*b*). Neurosecretion in the Hirudinea and its possible role in reproduction. *Am. Zool.* **6**, 251–61.

—— (1967). Hormonal control of spermatogenesis in *Hirudo medicinalis*. *4th Int. Symp. Neurosecretion, Strasbourg*, pp. 219–28. Springer, New York.

—— (1968). The effect of gonadotrophin lack upon the testes of *Hirudo*. *Am. Zool.* **8**, 754. (Abstr.)

—— (1969). Hormonal control of spermatogenesis in *Hirudo medicinalis*. II. Testicular response to brain removal during the phase of testicular maturity. *Gen. comp. Endocr.* **12**, 469–78.

—— (1970). Fine structure spermatogenesis in *Hirudo*. *Am. Zool.* **10**(4), 523. (Abstr.)

—— (1979). Effect of brain removal on the fine structure of *Hirudo* testis. *Am. Zool.* (Abstr.) (Incomplete reference)

[—— (1982). Irvine Ray Hagadorn (Oct. 19, 1932–Aug. 17, 1981). by Lucchesi, J.C. and Gilbert, L.I. *J. Morph.* **172**, 3–4.]

—— and Nishioka, R.S. (1961). Neurosecretion and granules in neurones of the brain of the leech. *Nature, Lond.* **191**, 1013–14.

—— Bern, H.A., and Nishioka, R.S. (1963). The fine structure of the supra-oesophageal ganglion of the rhynchobdellid leech, *Thermoyzon rude*, with special reference to neurosecretion. *Z. Zellforsch. Mikrosk. Anat.* **58**, 714–58.

Hagiwara, S., and Morita, H. (1962*a*). Electrotonic transmission of spikes between two neurons in a leech. *Fedn Proc. Fedn Am. Socs exp. Biol.* **21**, 361.

—— —— (1962*b*). Electrotonic transmission between two nerve cells in leech ganglion. *J. Neurophysiol.* **25**, 721–31.

Hahn, L. (1945). Über die Spattung der Hyaluronsaure durch Mucopolysaccharasen aus Blutegel und *Cl. perfringens. Arkiv. Kem. Min. Geol.* **19a**(33), 1–10.

Hajduk, D. (1979*a*). [New stations of *Hirudo medicinalis* in Poland.] *Przegl. Zoo.* **23**(4), 325–6. [In Polish; English summary.]

——(1979*b*). Prevalance of *Cystobranchus fasciatus* (Kollar) (Hirudinea) in the river of Biebrza. *Waidomości Parazyt.* **25**(3), 349–51. [In Polish; English summary.]

—— and Hajduk, Z. (1979*a*). Hirudinea of the Wolin National Park lakes. *Wiadomości Parazyt.* **25**(6). 669–72. [In Polish; English summary.]

—— —— (1979*b*). Hirudinea of the Lake Jamno. *Wiadomości Parazyt.* **25**(6). 655–68.

Haka, P., Holopainen, I.J., Ikonen, E., Leisma, A., Paasivirta, L., Saaristo, P., Sarvala, J., and Sarvala, M. (1974). Pääjarven pohjaeläimisto. *Luonnon Tuk.* **78**, 157–73.

Hall, F.G. (1922). The vital limits of exsiccation of certain animals. *Biol. Bull.* **42**, 31–51.

Halvorsen, O. (1964). Fiskeiglen, *Piscicola geometra* (L.) fra Ostfold. *Fauna, Oslo* **17**, 75–8. [English summary.]

—— (1966). [*Cystobranchus mammillatus* found in Norway.] *Fauna, Oslo* **19**(1), 38–40. [In Norwegian; English summary.]

—— (1971*a*). Studies of the helminth fauna of Norway XIX. The seasonal cycle and microhabitat preference of the leech, *Cystobranchus mammillatus* (Malm 1863) parasitizing Burbot, *Lota lota* (L.). *Norw. J. Zool.* **19**(2), 177–80.

—— (1971*b*). Studies of the helminth fauna of Norway XVIII: On the composition of the parasite fauna of coarse fish in the River Glomma, south-eastern Norway. *Norw. J. Zool.* **19**(2), 181–92.

—— (1972). Studies of the helminth fauna of Norway XX. Seasonal cycles of fish parasites in the River Glomma. *Norw. J. Zool.* **20**(1), 9–18.

—— and Andersen, K. (1973). Parasitter i ferskvannsmiljoer, biologi og okologi. *Fauna* **26**, 165–89.

Hammersen, F. (1963). Das Muskelgefuge in der Pharynzwand von *Hirudo medicinalis* und *Haemopis sanguisuga. Z. Zellforsch.* **60**, 797–814.

—— and Pokahr. A. (1972*a*). Elektronenmikroskopische Untersuchungen zur Epithelstruktur im Magen-Darmkanal von *Hirudo medicinalis* L. I. Mitteilung: Das Epithel der Divertikel. *Z. Zellforsch. Mikrosk. Anat.* **125**(3), 378–403.

—— —— (1972*b*). Elektronenmikroskopische Untersuchungen zur Epithelstruktur im Magen-Darmkanal von *Hirudo medicinalis* L. II. Metteilung: Das Epithel des Enddarmes. *Z. Zellforsch. Mikrosk. Anat.* **125**(4), 532–52.

—— and Staudte, H.W. (1967). Zum Feinbau des Endothels bei *Hirudo medicinalis* L. *Acta anat., Basel* **68**, 611. (Abstr.).

—— —— (1969). Beiträge zum Feinbau der Blutgefässe von Invertebraten. 1. Die Ultrastruktur des sinus lateralis von *Hirudo medicinalis* L. *Z. Zellforsch. Mikrosk. Anat.* **100**, 215–50. [English summary.]

—— —— and Moehring, E. (1976). Studies on the fine structure of invertebrate blood vessels. II. The valves of the lateral sinus of the leech, *Hirudo medicinalis* L. *Cell Tissue Res.* **172**(3), 405–24. [1977BA:38887.]

Hamond, R. (1972). The non-polychaetous annelids of Norfolk, England, with additional notes on polychaetes. *Cal. Biol. Mar.* **13**, 341–50.

Hanke, R. (1948*a*). The nervous system and the segmentation of the head in the Hirudinea. *Microentomology* **13**, 57–64.

—— (1948*b*). The nervous system and the segmentation of the head in the Hirudinea. Thesis, School of Biological Sciences, Stanford University, California.

Hansen, J. (1982). Leeches, lancets and fleams. Will bloodletting rejoin the medical mainstream? *Science* **81**, 74–7.

Hansen, K. (1962). Elektronenmikroskopische Untersuchung der Hirudineen-Augen. *Zool. Beitr.*, *Neue Folge* **7**(1), 83–128.

Hansen, L.P. (1975). Lakeiglen, *Cystobranchus mammilatus* (Malm, 1863), funet i Deltaet Glomma-Oyeren. *Fauna, Oslo* **28**(2), 97. [English summary.] [On *Lota lota.*]

Hanumante, M.M., and Kulkarni, G.K. (1978). Effect of thermal acclimation on gametogenesis of the Indian freshwater leech, *Poecilobdella viridis* (Blanchard). *Hydrobiologia* **61**(2), 183–6. [1979BA:9553.]

—— —— and Nagabhushanam, R. (1978). Effect of neurohumors on oxygen consumption of the fresh water leech, *Poecilobdella viridis*. *Comp. Physiol. Ecol.* **3**(2), 108–10.

Harant, H. (1929). Essai sur les Hirudinées. *Arch. Soc. Sci. Montpellier* **10**, 1–76. [Also: (1929) Thèse, Montpellier.]

—— and Grassé, P.P. (1959). Classe des Annélides Achètes ou Hirudinées ou Sangsues. In *Traité de Zoologie* (ed. P.P. Grassé) Vol. 5, Part 1, pp. 471–593. Masson, Paris.

—— and Vernières, P. (1935). Une Hirudinées africaine nouvelle *Granelia naivashae* n. gen., n. sp. *Arch. Soc. Sci. Méd. Biol. Année.* **VIe** No. 4, 219–20.

—— —— (1936). Mission scienifique de l'Omo. Hirudinea. III. Fasc. 27 (Zoologie) *Mém. Mus. hist. nat.* (new series) **4**, 219–26.

Hardiman, J (ed.) (1846). *Publ. Archaeolog. Soc.* No. 15. [Transl. of O'Flaherty's (1684) manuscript on West Connaught; refers to *Hirudo medicinalis* in Lake Mask.]

Harding, W.A. (1908). Note on leeches sent by Dr. E.W.G. Masterman from Palestine. *Parasitology* **1**, 282–3.

—— (1910). A revision of the British leeches. *Parasitology* **3** , 130–201.

—— (1911). Note on a new leech (*Placobdella aegyptiaca*) from Egypt. *Ann. Nat. Hist.* **8**(7), 388–9.

—— (1913). On a new land leech from the Seychelles. *Trans. Linn. Soc. Lond.* **16**, 39–43.

—— (1920). Fauna of the Chilka Lake, Hirudinea. *Mem. Ind. Mus.* **5**, Nr. 7, 509–17.

—— (1924). Descriptions of some new leeches from India, Burma and Ceylon. *Ann. Nat. Hist.* **9**(14), 489–99.

—— (1931). See Autrum (1939*c*).

—— (1932). Report on the Hirudinea: Mr. Omer-Cooper's investigation of the Abyssinian fresh waters (Hugh Scott Exped.). *Proc. zool. Soc. Lond.* Pt. I, 81–6.

—— and Moore, J.P. (1927). Hirudinea. In *The fauna of British India, including Ceylon and Burma*, pp. 1–302. London.

Hare, G.M., and Burt, M.D.B. (1975). Abundance and population dynamics of parasites infecting Atlantic salmon (*Salmo salar*) in Brook Trout, New Brunswick, Canada. *J. fish. Res. Bd Can.* **32**(11), 2069–89. [French summary.] [1976BA:44528.] [*Piscicola punctata.*]

Hariri, M. (1966). The biochemical importance of muco-polysaccharides in wound healing and clinical experiences with Hirudinea. *Med. Welt.* **15**, 822–4.

Harmer, S.F., and Shipley, A.E. (1896). *Cambridge natural history.* Vol. 2, pp. 406–8. Macmillan, London. [Records *Limnatis* proving disconcerting to Napoleon's soldiers at the Nile in 1798.]

Harms, C.E. (1959). Checklist of parasites from catfishes of northeastern Kansas. *Trans. Kansas Acad. Sci.* **62**(4), 262.

—— (1960). Some parasites of catfishes from Kansas. *J. Parasitol.* **36**, 695–701.

Harp, G.L., and Harp, P.A. (1980). Aquatic macroinvertebrates of Wapanocca National Wildlife Refuge. *Proc. Arkansas Acad. Sci.* **34**, 115–17.

Harrel, R.C., and Duplechin, J.L. (1976). Stream bottom as indicators of ecological change. 49 pp. Texas Water Resources Institute. Texas A. and M. University.

Harrison, J.L. (1953*a*). Sexual behaviour of land leeches. *J. Bombay nat. hist. Soc.* **51**(4), 959–60.

—— (1953*b*). Leeches. *Med. J. Malaya.* **8**(2), 180–5.

—— (1954). Notes on land leeches. *J. Bombay nat. hist. Soc.* **52**(2/3), 408–72.

—— Audy, J.R., and Traub, R. (1954). Further tests of repellants and poisons against leeches. *Med. J. Malaya* **9**(1), 61–71.

Harry, H.W., and Aldrich, D.V. (1958). The ecology of *Australorbis glabratus* in Puerto Rico. *Bull. Wld Hlth Org.* **18**(3), 819–32. [Found

Helobdella punctatolineata in association with *A. glabratus* under natural conditions.]

Hartley, J.C. (1962). The life history of *Trocheta subviridis* Dutrochet. *J. anim. Ecol.* **31**, 519–24.

Hartmann-Schroder, G. (1971). *Tierwelt Deutschlands*, Teil 58, Annelida, Borstenwermer, Polychaeta. Jena.

Hartnett, J.C. (1972). The care and use of medicinal leeches in 19th century pharmacy and therapeutics. *Pharm. Med.* **14**(4), 127–38. [Fig. 2: leech jar.]

Hartog, C. den, and Van Rossum, E. (1957). De bloedzuiger *Trocheta bykowskii* in de beneden-rivieren. *Levende Nat.* **60**, 228–33.

Hatherill, G.W.B. (1967). *Diestecostoma mexicanum* infestation of dogs. *Vet. Rec.* **81**, 262.

Hatto, J. (1968). Observations on the biology of *Glossiphonia heteroclita* (L.). *Hydrobiologia* **31**(3/4), 363–84. [German summary.]

Hauck, A.K., Fallon, M.J., and Burger, C.V. (1979). New host and geographical records for the leech *Acanthobdella peledina* Grube 1851 (Hirudinea, Acanthobdellidae). *J. Parasitol.* **65**(6), 989.

Haupt, J. (1974). Function and ultrastructure of the nephridium of *Hirudo medicinalis* L. II. Fine structure of the central canal and the urinary bladder. *Cell tissue Res.* **152**(3), 385–401. [German summary.] [1976BA:20034.]

Haupstein, P. (1934). Die Blutegelbehandlung der Thrombophlebitis (Thrombose) nach Operationen und im Wochenbett. *Med. Welt* **8**, 1723.

Haustein, K.O., and Markwardt, F. (1965). Über den Einfluss von Hemmstoffen der Blutgerinnung and Fibrinolyse auf das Sanarelli-Schwartzman-Phänomen. *Thrombos. Diathes. Haemorrh.* **13**, 60–4.

—— —— (1972. Versuche zur pharmakologischen Beeinflussung des lokalisierten Sanarelli-Schwartzman-Phänomen. *Folia Haematol.* **97**, 104–8.

Havet, J. (1900). Structure du système nerveux des annélides *Nephelis, Clepsine, Hirudo, Lumbriculus, Lumbricus* (Méthode de Golgi). *Cellule* **17**, 65–137.

Hawkey, C. (1966). Plasminogen activator in saliva of the vampine bat *Desmodus rotundus*. *Nature* **211**(5047), 434–5.

Hawkins, R.I., and Hellmann, K. (1966). Investigations on a plasminogen activator (Rhokinase) in two blood-suckers, *Rhodnius prolixus* Stal and *Hirudo medicinalis*. *Br. J. Haem.* **12**(1), 86–91.

Haycraft, J.B. (1884*a*). On the action of a secretion obtained from the medicinal leech on the coagulation of the blood. *Proc. R. Soc. Lond.* **36**, 478–87. (28 March) (see also June 7, 1884. *J. Physiol.* **5**, no. III).

—— (1884*b*). Über die Einwirkung eines Secretes des officinellen Blutegels auf die Gerinnbarkeit des Blutes. *Arch. exp. Path.* **18**, 209–17. (24 September)

Hayunga, E.G., and Grey, A.J. (1976). *Cystobranchus meyeri* sp. n. (Hirudinea: Piscicolidae) from *Catostomus commersoni* Lacépède in North America. *J. Parasitol.* **62**(4). 621–7. [1977BA:8815.]

Heath, H. (1925). See Autrum (1939c).

Hecht, G. (1929). See Autrum (1939c).

Hechtel, F.O.P. (1985). Systematics of the jawed leeches of Hong Kong (Hirudinea: Arhynchobdellida). Honours thesis, University College of Swansea, UK.

Heckman, C.W. (1974). The seasonal succession of species in a rice paddy in Vientiane, Laos. *Int. Rev. ges. Hydrobiol. Hydrograph.* **59**(4), 489–507. [*Hirudinaria manillensis.*]

Heckman, R. (1971). Parasites of Cutthroat Trout from Yellowstone Lake, Wyoming. *Prog. Fish-culturist* **33**(2), 103–6.

—— and Farley, D.G. (1973). Ectoparasites of the Western Roach from two foothill streams. *J. wildl. Dis.* **9**(3), 221–4.

Heide, M. (1967). [Leech therapy is still justified.] *Landarzt* **43**, 136–8. [In German.]

Heidermanns, C. (1937). Exkretion und Exkret-Stoffwechsel der Wirbellosen. *Tabulae Biol.* **14**, 209–73. [Summary of literature on excretion of all invertebrates including leeches to date — all secondary.]

Heisler, A. (1937). Aus Forschung und Erfahrung. Landarzt und Naturheilverfahren. Blutegel. *Hippokrates* **50**, 1236.

Heldt, T.J. (1961). Allergy to leeches. *Henry Ford Hosp. med. Bull.* **9**(4), 498–519. [Possible case of local hypersensitivity or infection to a *Placobdella*-like bite.]

Hemingway, E.E. (1912). The leeches of Minnesota. Part II: The anatomy of *Placobdella pediculata. Geol. nat. hist. Surv. Minnesota Zool.* Nr. 5, 29–63.

Henderson, H.P., Matti, B., Laing, A.G., Morelli, S. and Sully, L. (1983). Avulsion of the scalp treated by microvascular repair: the use of leeches for post-operative decongestion. *Br. J. plast. Surg.* **36**(2), 235–9.

Henderson, L.P. (1983). The role of 5-hydroxytryptamine as a transmitter between identified leech neurons in culture. *J. Physiol.* **339**, 309–24. [see also *Neurosci. Abstr.* 1981.]

Henderson, T.B., and Strong Jr, P.N. (1972). Classical conditioning in the leech *Macrobdella ditetra* as a function of CS and UCS intensity. *Conditional Reflex* **7**(4), 210–15.

Hendricks, A.C., Wyatt, J.T., and Henley, D.E. (1971). Infestation of a Texas red-eared turtle by leeches. *Tex. J. Sci.* **22**(2/3), 247.

Hendrickson, J.R. (1958). The Green Sea Turtle, *Chelonia mydas* (Linn.) in Malaya and Sarawak. *Proc. zool. Soc., London* **130**, 524.

Henry, L.M. (1948). The nervous system and segmentation of the head in Annulata. *Microentomology* **13**, 27–48.

Hensley, G.H., and Nahhas, F.M. (1975). Parasites of fishes from the Sacramento-San Joaquin Delta, California. *Calif. Fish Game* **61**(4), 201–8. [*Myzobdella lugubris* on *Ictalurus catus.*] [1976BA:38759.]

Henson, P.P. (1922). Rugby Mollusca. *Rep. Rugby Sch. nat. hist. Soc.* **56**, 23. [Leeches mentioned.]

Hergenrader, G.L., and Lessig, D.C. (1980). Eutrophication of the salt valley reservoirs, 1968–73. III. The macroinvertebrate community: its development, composition and change in response to eutrophication. *Hydrobiologia* **75**(1), 7–25.

Herlant-Meewis, H. (1964). Regeneration in Annelids. *Adv. Morphogenesis* **4**, 155–215.

—— (1977). [Amines and neurosecretory phenomena in Annelids.] *Bull. Cl. Sci. Acad. R. Belg.* 5E. Ser. **63**(3), 240–7. [*Erpobdella octoculata.*] [1978BA:2804.]

Herman, R.L. (1970). Leeches. *Trop. Fish Hobby* **18**, 92–4.

Hermann, E. (1875). See Autrum (1939c).

Herrmann, S.J. (1968a). Systematics, distribution and ecology of Colorado Hirudinea. *Diss. Abst.* **29**(5), 1878–9.

—— (1968b). Zoogeography of Colorado Hirudinea. *J. Colo. -Wyo. Acad. Sci.* **6**, 8.

—— (1970a). New host record for *illinobdella moorei* Meyer, 1940. *Southwest Nat.* **15**(2), 261–8.

—— (1970b). Total residue tolerances of Colorado Hirudinea. *Southwest Nat.* **15**(2), 269–72.

—— (1970c). Systematics, distribution and ecology of Colorado Hirudinea. *Am. Midl. Nat.* **83**(1), 1–37.

Herter, K. (1928b). Reizphysiologische Untersuchungen an dem Egel *Hemiclepsis marginata* O.F. Müll. *Verh. Dt. Zool. Ges.* **32**, 154–60.

—— (1928c). Bewegungsphysiologische Studien an dem Egel *Hemiclepsis marginata* O.F. Müll. mit besonderer Berücksichtigung der Thermokinese. *Z. vergl. Physiol.* **7**, 571–605.

—— (1928d). Reizphysiologie und Wirtsfindung des Fischegels *Hemiclepsis marginata* O.F. Müll. *Z. vergl. Physiol.* **8**, 391–444.

—— (1929a). Über Geotaxis und Phototaxis deutscher Egel. *Verh. Dt. Zool. Ges.* **33**, 72–82.

—— (1929b). Vergleichende bewegungsphysiologische Studien an deutschen Egeln. *Z. vergl. Physiol.* **9**, 145–77.

—— (1929c). Temperaturversuche mit Egeln. *Z. vergl. Physiol.* **10**, 248–71.

—— (1929d). Reizphysiologisches Verhalten und Parasitismus des Entenegels *Protoclepsis tesselata* O.F. Müll. *Z. vergl. Physiol.* **10**, 272–308.

—— (1929*e*). Studien über Reizphysiologie und Parasitismus bei Fisch- und Entenegeln. *SB. Ges. naturf. Fr., Berlin* 142–84.

—— (1930). Fang, Pflege und Zucht der deutschen Süsswasseregel. *Handbuch der biologische Arbeitsmetoden herausgegeben von E. Abderhalden*, Lief. 338, Abt. IX, Teil 2, 2 Hälfte, pp. 1577–1618. Berlin.

—— (1932). Hirudineen. Egel. In *Schulze's Biologie der Tiere Deutschlands*, Lief. 35, Teil 12b, pp. 1–158. Berlin.

—— (1935). Hirudinea. In *Grimpe's Die Tierwelt der Nord- und Ostsee*, 6c₂, pp. 45–106. Leipzig.

—— (1936). Die Physiologie der Hirudineen. In *Bronn's Klassen und Ordnungen des Tierreichs*, 4. Bd., 3. Abt., 4. Buch, Teil 2, 2. Lief, pp. 123–319. Leipzig.

—— (1937). Die Ökologie der Hirudineen. In *Bronn's Klassen und Ordnungen des Tierreichs*, 4. Bd. 3. Abt., 4. Buch., Teil 2, pp. 321–496. Leipzig.

—— (1942). Untersuchungen über den Temperatursinn von Warmbluten Parasiten. *Z. Parasitenk.* **12**, 552–91.

—— (1968). *Die medizinische Blutegel und seine Verwandten*. A. Ziemsen Verlag, Wittenberg Lutherstadt. Die Neue Brehm-Bücherei.

Hertz, L., and Nissen, C. (1976). Differences between leech and mammalian nervous systems in metabolic reactions of potassium ion as an indication of differences in potassium homeostasis mechanisms. *Brain Res.* **110**(1), 182–8. [1977BA:10800.]

Hess, W.N. (1925), Photoreceptors of *Lumbricus terrestris*, with special reference to their distribution. *J. Morph.* **41**(1), 63–93. [Leeches: 85–6.]

Hesse, R. (1896). See Autrum (1939*c*).

—— (1897). Untersuchungen über die Organe der Lichtempfindung bei niederen Tieren. III. Die Sehorgane der Hirudineen. *Z. wiss. Zool.* **62**, 671–707.

—— (1902). Untersuchungen über de Organe der Lichtempfindung bei niederen Tieren. VIII. Weitere Tatsachen. Allgemeines. *Z. wiss. Zool.* **72**, 565–656.

Hessel, R. (1881). Artificial culture of medicinal leeches and of species of *Helix. Bull. US fish Comm.* 264.

—— (1884). Leech culture. *Bull. US fish Comm.* **4**, 175–6.

Hewitt, G.C., and Hine, P.M. (1972). Checklist of parasites of New Zealand fishes and of their hosts. *NZ Jl mar. freshwater Res.* **6**(1/2), 69–114.

Hexaeb, B.M. see Nekhaev, V.M.

Heyligers, P.C. (1965). The soil fauna in the osier beds of the Brabantse Biesbosch. In *Biosoziologie, Bericht über das Internationale Symposium in Stolzenan/Weser 1960 der Internationalen Vereinigung für Vegetationskunde*. Herausgegeben von R. Tüxen, pp. 199–210. Verlag

Dr W. Junk, Den Haag. [*Haemopis sanguisuga*, in soil fauna, Brabantse Biesbosch, Netherlands.]

Hickman, V.V. (1946). *Pontobdella tasmanica* nom. nov. (Hirudinea). *Pap. Proc. R. Soc. Tasmania* 27.

Hicks, F.J., and Threlfall, W. (1973). Metazoan parasites of salmonids and coregonids from coastal Labrador. *J. fish Biol.* **5**, 399–415.

Hiestand, W.A., and Singer, J.I. (1934). Certain factors influencing the respiratory metabolism of the leech (*Hirudo medicinalis*). *Proc. Indiana Acad. Sci.* **43**, 205–10.

Hiester, L. (1759). *A general system of surgery.* (Incomplete reference)

Higashi, A. *et al.* (1968). [Studies on the land leech, *Mesobdella* sp. found in the nostril of a man. 1. Ecological observations.] *J. Otolaryng. Jap.* **71**, 300–4. [In Japanese.]

Higler, L.W.G. (1977). Macrofauna-cenoses on *Stratiotes* plants in Dutch broads. *Verh. Rijksinst. Natuurbeheer* **2**, 1–86. [*Hirudo* in Holland.]

Higuti, K. (1967). Membrane potential and neuromuscular transmission of the obliquely striated muscle of the leech. *Igaku Kenkyu* **37**, 129–47. [*Hirudo nipponia.*]

Hii, J.L.K., Kan, S.K.P., and Yong, K.S.A. (1978). A record of *Limnatis maculosa* (Blanchard) (Hirudinea: Arhynchobdellida) taken from the nasal cavity of man in Sabah, Malaysia. *Med. J. Malaysia* **32**(3), 247–8. [BA:42094.]

Hildebrand, J.G., Barker, D.L., Herbert, E., and Kravitz, E.A. (1971). Screening for neurotransmitters: a rapid neurochemical procedure. *J. Neurobiol.* **23**, 21–46. [Leeches.]

Hillgarth, N., and Kear, J. (1979). Diseases of sea ducks in captivity. *Wildfowl* **30**, 135–41.

Hilsenhoff, W.L. (1963). Predation by the leech, *Helobdella stagnalis*, on *Tendipes plumosus* (Diptera: Tendipedidae) larvae. *Ann. ent. Soc. Am.* **56**, 252.

—— (1964). Predation by the leech, *Helobdella nepheloidea*, on larvae of *Tendipes plumosus* (Diptera: Tendipedidae). *Ann. ent. Soc. Am.* **57**, 139.

—— (1967). Ecology and population dynamics of *Chironomus plumosus* (Diptera: Chironomidae) in Lake Winnebago, Wisconsin. *Ann. ent. Soc. Am.* **60**, 1183–94.

Hiraro, H. (1932). Experimental inquiry into the possibility of transmission of yaws by leeches. *Philippine J. Sci.* **47**, 463–6.

Hirata, K., Ohsako, N., and Mabuchi, K. (1968). Transverse membrane system in the visual cell of the leech, *Hirudo nipponia. J. Electronmicrosc.* **17**, 355. (Abstr.)

Hirschler, J. (1907). See Autrum (1939c).

Hirst, G.K. (1941). Effect of polysaccharide-splitting enzyme on streptococcal infection . *J. exp. Med.* **73**, 493–505. [From leech extract.]

Hoare, C.A. (1932). On protozoal blood parasites collected in Uganda. With an account of the life cycle of the crocodile haemogregarine. *Parasitology* **24**, 210–24.

Hobbs, H.H. Jr, and Figueroa, A.V. (1958). The exoskeleton of a freshwater crab as a microhabitat for several invertebrates. *Virginia J. Sci.* (new series) **9**, 395–6.

Hoepple, R. (1959). *Parasites and parasitic infections in early medicine and science, pp. 173–81 and 300–7.* University of Malaya Press, Singapore. [Leeches in Chinese medicine.]

—— and Tang, C.C. (1941). [Use of leeches.] *Chinese med. J.* **59**, 359–78.

Hoffman, G.L. (1967). Hirudinea. *Parasites of North American freshwater fishes*, pp. 288–98. University of California, Berkeley.

—— (1979). Helminthic parasite. In: *Principal diseases of farm-raised catfish.* (ed. J.A. Blumb) pp. 40–58. Southern Cooperative Series No. 225. Alabama Agricultural. Experimental Station Bulletin (Auburn University).

—— and Meyer, F. (1974). Parasites of freshwater fishes, a review of their control and treatment. *T.F.H.P.* pp. 142–5.

Hoffmann, J. (1955*a*). Faune hirudinéenne du Grand-Duché de Luxembourg. *Arch. Inst. Grand-Ducal Luxemb. Sect. Sci. nat. Phys. Math.* (new series) **22**, 175–211.

—— (1955*b*). Signalement d'une importante station de *Hirudo medicinalis* L. au Grand-Duché de Luxembourg. *Arch. Inst. Grand-Ducal Luxemb. Sect. Sci. nat. Phys. Math.* (new series) **22**, 213–22.

—— (1955*c*). Quelques caractères éthologiques de la Piscicolidae: *Cystobranchus respirans* Troschel. *Arch. Inst. Grand-Ducal Luxemb. Sect. Sci. nat. Phys. Math.* (new series) **22**, 223–5.

—— (1956). Contributions a l'étude des spécifités morphologiques et éthologiques de la Piscicolidée: *Cystobranchus respirans* Troschel 1850. *Arch. Inst. Grand-Ducal Luxemb. Sect. Sci. nat. Phys. Math.* (new series) **23**, 209–40.

—— (1959). A propos d'une 'maladie de captivité' des *Cystobranchus. Arch. Inst. Grand-Ducal Luxemb. Sect. Sci. nat. Phys.* (new series) **26**, 237–43.

—— (1960). Notules hirudinologiques. *Arch. Inst. Grand-Ducal Luxemb. Sect. Sci. nat. Phys. Math.* **27**, 285–91.

—— (1961). Sur la présence de *Trocheta subviridis* Dutrochet, 1817, au Luxembourg. Ajoute à la Faune hirudinéenne du Grand-Duché de Luxembourg. *Arch. Inst. Grand-Ducal Luxemb. Sect. Sci. nat. Phys. Math.* (new series) **28**, 333–7.

Hoffman, R.L. (1964). A new species of *Cystobranchus* from southwestern Virginia (Hirudinea: Piscicolidae). *Am. Midl. Nat.* **72**(2), 390–5.

Höhn-Ochsner, W. (1966). Das Mooreservat Chrutzelried bei Gfenn-Dübendorf ZH:seine Pflanzen und Tierwelt. *Vjschr. Nat. ges. Zürich* **111**, 339–342. [*Erpobdella octoculata* from marshy ground, Zurich.]

Holmes, P.F. (1965). The natural history of Malham Tarn. *Fld. Stud.* **2**(2), 199–223. [*Glossiphonia complanata:* ecological distribution.]

Holmgren, E. (1904). See Autrum (1939c).

Holmquist, C. (1973). Some Arctic limnology and the hibernation of invertebrates and some fishes in sub-zero temperatures. *Arch. Hydrobiol.* **72**(1), 49–70.

—— (1974). A fish leech of the genus *Acanthobdella* found in North America. *Hydrobiologia* **44**(2/3), 241–5.

—— (1975). Lakes of northern Alaska and northwestern Canada and their invertebrate fauna. *Zool. Jb. Syst.* **102**, 334–484. [Hirudinea: 415–21.]

Homblé, A.G. (1975). Bloedzuigers. *Ons Heem, Hooimaand* **29**, 154. [History of leeching.]

Honio, J. (1936). See Honjo, J. (1936) in Autrum (1939c).

Horn, G. (1798). *An entire new treatise on leeches.* London.

Hornbostel, H. (1941). Über die Bakteriologischen Eigenschaften des Darmsymbiontem beim medizinischen Blutegel (*Hirudo officinalis*), nebst Bemerkiungen zur Symbiosefrage. *Zentr. Bateriol.* **148**, 36–47.

Horst, R. (1892). See Autrum (1939c).

Horton, R.G. (1962). Further observations on leeches. *Starfish* **13**, 20–2.

Hotz, H. (1938). *Protoclepsis tesselata* (O.F. Müller). Ein Beitrag zur Kenntnis von Bau und Lebensweise der Hirudineen. *Rev. Suisse Zool.* **45** (Suppl.), 1–380.

House, M.R. (ed.) (1979). *The origin of major invertebrate groups.* pp. 515. Systematics Association Special Volume No. 12. Academic Press, London.

Howard, R.D. (1978). The evolution of mating strategies in bullfrogs, *Rana catesbeiana. Evolution* **32**(4), 850–71. [1979BA:64831.] [*Macrobdella decora* chief predator on eggs.]

Howden, H.F., Martin, J.E.H., Bousfield, E.L., and McAllister, D.E. (1970). Fauna of Sable Island and its zoogeographical affinities: a compendium. *Nat. Mus. nat. Sci. Ottawa Publ. Zool.* **4**, 1–45.

Hraba, M. (1966). [Bleeding from the upper respiratory and digestive tract due to parasites.] *Cesk. Otolaryng.* **15**, 109–11. [In Czechóslovák.]

Hrabe, S. (1936). Pijavka lekarska (Vyskyt na jizni Morave). *Věda Přírodiní* **17**, 124. [*Hirudo medicinalis.*]

—— (1954). Pijavky-Hirudinea. In: "Klíč Zvireny CSR I" (ed. Hrabe *et al.*) pp. 321–3.(Incomplete reference).

Huber, J.C. (1890–91). Die Blutegel im Alterthum: historisch-therapeutische Studien. *Dt. Arch. Klin. Med.* **47**, 522–31.

Hudson, J., and Hudson, M.D. (1841). *A treatise on the medicinal leech, containing remarks on the history, diseases and management of them. Together with the observation of an eminent physician on sanguisuction.* Hull.

Hugghins, E.J. (1972). Parasites of fishes in South Dakota. *S. Dakota exp. Stat. Bull.* **484**, 1–73.

Hulse, A.C. (1976). Carapacial and plastral floral and fauna of the Sonora Mud Turtle, *Kinosternon sonoriense* le Conte (Reptilia, Testudines, Kinosternidae). *J. Herpetol.* **10**(1), 45–8. [1976BA:18922.] [*Placobdella ornata*, in Arizona.]

Hulsebosch, C.E, and Bittner, G.D. (1981). Regeneration of axons and nerve cell bodies in the central nervous system of annelids. *J. comp. Neurol.* **198**(1), 77–88.

Hussein, M.A., and El-Shimy, N. (1982). Description of *Barbronia assiuti* n. sp. (Hirudinea) from Assiut, Egypt. *Hydrobiolgia* **94**, 17–24.

Husson, R. (1963). Considérations sur la reproduction des sangsues du genre *Trocheta*. In *Proc. 16th Int. Cong. Zool.*, Vol. 16, p. 30. (Abstr. only.)

Hutton, R.F., and Sogandares-Bernal, F. (1959). Notes on the distribution of the leech, *Myzobdella lugubris* Leidy, and its association with mortality of the blue crab, *Callinectes sapidus. J. Parasitol.* **45**(4), 384.

Iles, C. (1959). The larval trematodes of certain fresh water molluscs. I. The furcocercariae. *Parasitology* **49**, 478–504.

—— (1960). The larval trematodes of certain fresh water molluscs. II. Experimental studies on the life cycle of two species of furcocercariae. *Parasitology* **50**, 401–17.

Inamura, H. (1979). Studies of feeding mechanism in the blood sucking leech, *Hirudo nipponia*. 2. *Zool. Mag. Tokyo* **88**(4), 646. (Abstr.) [In Japanese.]

Ingram, D.M. (1975). Some Tasmanian Hirudinea. *Pap. Proc. R. Soc. Tasmania* **91**, 191–232.

Invertebrate Red Data Book. (1983). See Wells *et al.* (1983).

Ionescu-Varo, M., and Grigoriu, A. (1965). La neurosécrétion chez *Trocheta bykowskii* Gedroyć (Glossiphoniidae). *Trav. Mus. Hist. nat. 'Grigore Antipa'*, Bucharest **5**, 99–105.

Isakhanyan, G.S. (1982). [Substantiation of some methods of reflex therapy of chronic ischemic heart disease.] *Zh. Eksp. Klin. Med.* **22**(1), 53–6. [1983BA:55021.] [In Russian, with Armenian and English summaries.] ['Normalization of electrophysiological indices and general state of patient observed after treatment with leeches'.]

Ishikawa, A., Hafter, R., Seemüller, U., and Gräff, J.M. (1979). The effect of hirudin on the generalized Schwartzman reaction in rabbits. *Thromb. Haemostasis* **42**(1), 150 [1980BA:25841.]

—— —— —— —— (1980). The effect of hirudin on endotoxin induced disseminated intravascular coagulation (DIC). *Thromb. Res.* **19**(3), 351–8.

Isossimov, V.V. (1926). Zur Anatomie des Nervensystems der Lumbriculiden. *Zool. Jahrb.* (*Anat.*) **48**, 365–404.

—— (1949). [Lumbriculidae of Lake Baikal — A Monographic Treatment.] Doctoral dissertation, 1948. University of Kazan. [In Russian; *Agriodrilus vermivorus:* 183.]

—— (1962). [The Oligochaetes of the family Lumbriculidae of Lake Baikal.] *Trans. Limn. Inst., Acad. Sci. USSR., Siberian Sect.* 1(21)(1), 3. [In Russian.]

Ito, T. (1936a). Zytologische Untersuchungen über die Ganglienzellen des japanischen medizinischen Blutegels, *Hirudo nipponia*, mit besonderer Berücksichtigung auf die 'dunkle Ganglienzelle'. *Okajimas Folia anat. Jap.* **14**, 111–70.

—— (1936b). Zur Zytologie der Gliazellen in der Bauchganglionkette des japanischen medizinischen Blutegels. *Hirudo nipponia. Fol. anat. japon.* **14**, 389–408.

Iuga, G.V. (1931). Les pigments de la *Glossosiphonia paludosa. Arch. Zool. exp. gen.* **71**, 1–97.

Ivanov, A.V., Polyanskii, U.I., and Strelkov, A.A. (1958). *Klass Hirudinea* — piyavki. V Kn. *Bol'shoi Praktikum Po Zoologii Bespozvonochn'ch, I.M.* 502–32. [In Russian.]

Ivanov, G.I. (1968). [Horse leech as a foreign body in otorhinolaryngologic organs.] *Zh. Ushn. Nos. Gorl. Bolez.* **28**, 87–8. [Probably *Limnatis* sp.]

Ivlev, S.V. (1981) [Effect of some transmitters on mechanosensory neurons in the leech, *Hirudo medicinalis.*] *Zh. Evol. Biokhim. Fiziol.* **17**(3), 313–15. [In Russian; English summary.]

—— and Lenkov, D.N. (1982). [The efect of microsuperfusion of glutamic acid, aspartic acid and γ-aminobutyric acid on Retzius cells and AE motoneurons of the leech, *Hirudo medicinalis. Zh. Evol. Biokhim. Fiziol.* **18**(5), 465–70. [In Russian; English summary.][1983BA: 57429.]

Iwata, K.S. (1940a). Spreading of action potential within skin complex of leech. *Jap. J. Zool.* **8**, 433–41.

—— (1940b). Seat of action potential and of resting potential in leech skin. *Jap. J. Zool.* **8**, 443–7.

Izuin, K., and Kawamoto, T. (1960). On cases of parasitism of the nasal cavity of man by leeches in Japan. *Kumamoto-Igakukae-Zasshi* **34**(8), 1735–6.

Izumova, N.A. (1958). [Seasonal dynamics in the parasitofauna (Mollusca, Annelida, Platyhelminthes, Ashelminthes) of fish in the R'binsk Reservoir.] *Tr. Biol. Sta. 'Borok' Akad. Nauk SSSR* **3**, 384–8. [In Russian.]

—— (1960). Sezonnaya dinamika parazitophaun' p'b R'binskogo vodokhranilischa (Soobschenie III. Schuka, Sinets, gustera). *Tr. Inst. Biol. Vodokhř.* **3**(6), 283–300. [In Russian.]

—— (1964). [The formation of the parasitofauna (Protozoa, Trematoda, Cestoda, Nematoda, Acanthocephala, Annelida, Crustacea) of fishes in the R'binsk Reservoir.] In *Proc. Symposium on Parasitic Worms and Aquatic Conditions*, Prague, Czechoslovakia, 29 October – 2 November 1962, pp. 49–55. Czechoslovak Academy of Sciences, Prague.

—— and Shilin, A.A. (1958). Parazitophauna p'b Volgi v paionakh

Gor'kovskogo i Kuib'shevskogo vodokhranililisch. *Tr. Biol. St. 'Borok'.* **3**, 364–83. [In Russian.]

Izvekova, E.I., Katchanova-Lvova, A.A., and Sokolova, N.Y. (1970). [The fauna of the lakes on Veliky Island and Kiudo Peninsula, in the bay of Kandalaksha, White Sea.] *Trudy belom. biol. Sta. M. G. U.* **3**, 113–48. [In Russian.] [Changes in leech abundance.]

Jacquet. E. (1885). Sur quelques parasites des Lymnées. *L. Exchange* **1**, 2. [Leeches.]

Jacoby, C. (1904). Über Hirudin. In Medizinische Gesellschaft in Göttingen. *Dt. Med. Wschr.* **30**(4), 1786–7.

Jakubski, A.W. (1910). See Autrum (1939c).

Jais, A.M.M., Kerkut, G.A., and Walker, R.J. (1983). The ionic mechanism associated with the biphasic glutamate response on leech (*Hirudo medicinalis*) Retzius cells. *Comp. Biochem. Physiol. C Comp. Pharmacol.* **74**(2), 425–32. [1983BA:65296.]

James, V.A., Sharma, R.P., Walker, R.J., and Wheal, H.V. (1980). Actions of glutamate, kainate, dihydrokainate and analogs on leech neurone amino acid receptors. *Eur. J. Pharmacol.* **62**(1), 35–40. [1980BA:9768.]

—— and Walker, R.J. (1978). Structure-activity studies on an excitatory glutamate receptor of leech neurones. (Proceedings.) *Br. J. Pharmacol.* **62**(3), 432P–433P.

—— —— (1979). The ionic mechanism responsible for L-glutamate excitation of leech Retzius cells. *Comp. Biochem. Physiol. C Comp. Pharmacol.* **64**(2), 261–6. [1980BA:58257.]

—— —— and Wheal, H.V. (1980). Structure–activity studies on an excitatory receptor for glutamate on leech Retzius neurones. *Br. J. Pharmacol.* **68**, 711–17.

Jamieson, B.G.M., and Daddow, L. (1979). An ultrastructural study of microtubules and the acrosome in spermiogenesis of Tubificidae (Oligochaeta). *J. ultrast. Res.* **67**(2), 209–24.

Janiszewska, J. (1963). Stan Badan Nad Fauna Pasozytow Bezkregowcow w Polsce. *Wiad. Parazyt.* **9**(4), 307–15.

—— (1963). [The state of studies of parasitic fauna of invertebrates in Poland.] *Wiad. Parazyt.* **9**(4), 307–15. [In Polish.]

—— and Zmijewska, M. (1968). *Chaetogaster limnaei* K. Baer and *Glossiphonia heteroclita* (L.) — parasites of water snails of the neighbourhood of Wroclaw. *Przegl. Zool.* **12**(1), 49–50. [In Polish; English summary.]

Jankowska-Siwinska, D. (1954). Badania nad trwaloścía cynnika swoistości we krwi probranej przez pijawke lekarska (*Hirudo medicinalis* L.). *Kosmos, Ser. Biol., Warsaw,* **3** 4(9), 465.

Jansen, J.K.S., Muller, K.J., and Nicholls, J.G. (1974). Persistent modification of synaptic interactions between sensory and motor nerve cells following discrete lesions in the central nervous system of the leech. *J. Physiol.* **242**(2), 289–305. [1975BA:49519.]

—— and Nicholls, J.G. (1972). Regeneration and changes in synaptic connections between individual nerve cells in the central nervous system of the leech. *Proc nat Acad. Sci. USA* **69**(3), 636–9.

—— —— (1973). Conductance changes, an electrogenic pump and the hyperpolarization of leech neurones following impulses. *J. Physiol.* **229**, 635–55.

Janzen, R. (1932). Der Farbwechsel von *Piscicola geometra*. 1. Beschreibung des Farbwechsels und seiner Elemente. *Z. Morphol. Ökol.* **24**, 327–41.

—— (1933). Über das Vorkommen eines physiologischen Farbwechsels bei einigen einheimischen Hirudineen. *Zool. Anz.* **101**, 35–40.

Jarcho, S. (1972). Cohnheim on inflammation. *Am. J. Cardiol.* **29**(4), 546–7.

Jarry, D. (1958). Les fluctuations dans l'espèce *Glossiphonia complanata*, Hirudinea. *Ann. Soc. hort. Hist. nat. Hérault*, 98e année, 4e trim., 211–17.

—— (1959*a-c*). Observations écologiques sur les Hirudinées. I. Á propos d'une maladie de captivité des Erpobdelles. II. Les systèmes de coactions chez les sangsues. III. Sur la présence de *Limnatis nilotica* (Savigny, 1820) dans un oued méditerranéen: la baillaurie a Banyuls-sur-Mer. *Bull. Soc. zool. Fr.* **84**, 62–7, 67–72, 73–6.

—— (1959*d*). Notes biologiques sur les Hirudinées. I. Les fluctuations dans l'espèce *Erpobdella octoculata* (L.). *Bull. Soc. zool. Fr.* **84**(5–6), 371–80.

—— (1959*e*). Notes biologiques sur les Hirudinées. II. Sur la présence d'*Hirudo michaelseni* Augener 1936 au Borkou. *Bull. Soc. zool. Fr.* **84**, 380–3.

—— (1959*f*). Notes biologiques sur les Hirudinées. III. A propos de Pontobdelles récoltées sur *Raja batis* L. à Sète. *Bull. Soc. zool. Fr.* **84**, 382–4.

—— (1959*g*). La place des Hirudinées dans quelques biocénoses dulcaquicoles de la région de Montpellier. *Vie Milieu* **10**(2), 147–59.

—— (1959*h*). Note préliminaire sur les hirudinées de la source du Lez. *Vie Milieu* **10**(3), 267–79.

—— (1960*a*). *Piscicola haranti* n. sp. (Hirudinea). *Ann. Parasitol. hum. Comp.* **35**, 305–15.

—— (1960*b*). Une curieuse coaction de parasitisme: l'association entre *Glossiphonia complanata* et *Erpobdella octoculata*. *Terre Vie* **107**(1), 51–5.

—— (1960*c*). Histoire des sangsues. *Bull. Soc. nat. Arch. Ain* **74**, 77–80.

Jaschke, W. (1933). Beiträge zur Kenntnis der symbiontischen Einrichtungen bei Hirudineen und Ixodiden. *Z. Parasitenk.* **5**, 515–41.

Jazdzewska, T. (1966). [Nouvelles données concernant la reproduction de la sangsue *Piscicola geometra* (L.).] *Zesz. Nauk. Univ. Lódz, Ser. 2, mat.-Przyr.* **21**, 57–61. [In Polish; French summary.] [Two generations within eight months; sexual maturity after four feedings in a period of 38 days.] [see also Pawlowska.]

—— (1967). [Les sangsues (Hirudinea) d'anciens étangs poissonneux de Zaborów dans les environs de Varsovie.] *Zesz. Nauk. Univ. Lódz, Ser. 2* **25**, 77–81. [In Polish; French summary.]

Jeal, F., and West, B. (1968–70)). *Oceanobdella blennii* (Knight-Jones) in Co. Mayo. *Ir. nat. J.* **16**(7), 209.

Jeannin and Destaing, (Incomplete reference) (1968). *Revue méd.* [Fibrinolysin — *Hirudo medicinalis*.]

Jennings, J.B., and Van Der Lande, V.M. (1967). Histochemical and bacteriological studies on digestion in nine species of leeches. (Annelida: Hirudinea). *Biol. Bull.* **133**(1), 166–83.

Jensen, A.S. (1895). En Tyroglyphide i Hesteiglens Aegkapsel. *Vidensk. Meddel. Dansk naturh. Foren.* (Incomplete reference)

Jensen, B. (1960). Laegeiglens (*Hirudo medicinalis* L.) forekomst i Danmark. *Flora Fauna* **66**, 25–32.

Jesus, Z. de (1934). Experiments on the control of the common water leech, *Hirudinaria manillensis*. *Philippine J. Sci.* **53**, 47–63.

Jeuniaux, C. (1963). *Chitine et chitinolyse*. Masson, Paris. [A chapter on the molecular biology. Doubts that chitin was found in leeches by Freytag (1953) and Ananichev (1961).]

Jijima, J. (1882). Origin and growth of the eggs and egg-strings in *Nephelis*. *Q. Jl microsc. Sci.* **22**, 189–211. [*Erpobdella* sp.]

Jiminiz, J.M., and Garcia-Más, I. (1980–81). Hirudineos de Espana: Catalogo provisional. *Boln. Sco. Port. Sciênc. Mat.* **20**, 119–25.

Jimenez, M.F. (1845). Sobre los accidentes a que de lugar en México, la applicación e sanguijuelas. *Soc. Filorat. Méx. Per.* **1**, 213–18.

Johansen, J., Hockfield, S., and McKay, R.D.G. (1984*a*). Axonal projections of mechanosensory neurons in the connectives and peripheral nerves of the leech, *Haemopis marmorata*. *J. comp. Neurol.* **226**, 255–62.

—— —— —— (1984*b*). Distribution and morphology of nociceptive cells in the central nervous system of three species of leeches. *J. comp. Neurol.* **226**, 255–62.

—— Yang, J., and Kleinhaus, A.L. (1984). Actions of procaine on specific nociceptive cells in leech central nervous system. *J. Neurosci.* **4**, 1253–61.

Johansson, L. (1896*a*). Bidrag till kännedomen om Sveriges ichthyobdellider. *Akad. Afhandl. Upsala Un.*

—— (1896*b*). Über den Blutumlauf bei *Piscicola* und *Callobdella*. *Festschr. Lilljeborg, Uppsala* 317–30.

—— (1898*a*). Einige systematisch wichtige Theile der inneren Organisation der Ichthyobdelliden. *Zool. Anz.* **21**, 581–95.

—— (1898*b*). Die ichthyobdelliden im Zool. Reichsmuseum in Stockholm. *Ofv. Ak. Förh.* **55**, 665–87.

—— (1909*a*). Über eine eigentümliche Öffnung des Darmes bei einem afrikanischen Egel (*Salifa perspicax*). *Zool. Anz.* **34**, 521–3.

—— (1909*b*). Über die Kiefer der Herpobdelliden. *Zool. Anz.* **35**, 1–5.

—— (1909*c*). Einige neue Arten Glossosiphoniden aus dem Sudan. *Zool. Anz.* **35**, 146–54.

—— (1909*d*). Hirudinea, Egel. In *Die Süsswasserfauna Deutschlands.* Hrsg. v. Brauer, H. 13, pp. 67–81. Fischer, Jena.

—— (1910*a,b*). Zur Kenntnis der Herpobdelliden Deutschlands. *Zool. Anz.* (*a*) **35**, 705–14; (*b*) **36**, 367–79.

—— (1910*c*). Überzählige Darmöffnungen bei Hirudineen. *Zool. Anz.* **36**, 405–8.

—— (1910*d*). *Hirudinea. Sjöstedt's Kilimandjaro Meru Expedition*, 3, Abt. 22, pp. 29–31.

—— (1911). *Hirudinea.* Fauna Südwest Austral., Hrsg. v. Michaelsen & Hartmeyer, Bd., 3, pp. 407–31.

—— (1913*a*). Über eine neu von Dr. K. Absolon in der Herzegowina entdeckte höhlenbewohnende Herpobdellide. *Zool. Anz.* **42**, 77–80.

—— (1913*b*). Hirudineen aus dem Sudan. *Res. Swed. zool. Exped. Egypt* Part V, Nr. 1.

—— (1914). See Autrum (1939*c*).

—— (1918). Hirudineen von Neu Caledonien und den Neuen Hebriden. Sarasin & Roux. *Nova Caledonia, Zool.* **2**, 373–96.

—— (1924). Ein neuer Landblutegel aus den Juan Fernandez-Inseln. *Nat. Hist. Juan Fernandez, Zool.* **3**, 439–60.

—— (1926*a*). See Autrum (1939*c*).

—— (1926*b*). Hirudineen aus dem nördlichen und östlichen Spanien, gesammelt von Dr. F. Haas in den Jahren 1914–1919. *Abh. Senckenberg nat. Ges.* **39**, 217–31.

—— (1928). Über den Bau von *Trematobdella perspicax* L. Joh. *Res. Swed. zool. Exped. Egypt.* Uppsala, Part 5, Nr. 2.

—— (1929). Hirudinea (Egel). *Tierwelt Dt.* **15**, 133–55.

—— (1935). Opredelitel pijavok s dopolnenijami dlja form, vstrečennych na territorii Sovetskogo Sojuza, E.A. Vasil'eva. *Trudy Otd. Gidrol. Leningr. Obl. Gidro-meteor, Upravl., Leningrad* **1**, 1–29. [In Russian.] [Key to leeches.] [Suppl. by E.A. Vasiliev.]

Johnson, G.M., and Klemm, D.J. (1977). A new species of leech, *Batra-cobdella cryptobranchii* n. sp. (Annelida: Hirudinea), parasitic on the Ozark hellbender. *Trans. Am. microsc. Soc.* **96**(3), 327–31.

Johnson, J.R. (1816). *A treatise on the medicinal Leech; including its medical and natural history, with a description of its anatomical structure; also, remarks upon the diseases, preservation and management of leeches,* I–XII, pp. 1–147. London.

Jones, P.G., Sawyer, R.T. and Berry, M.S. (1985). Novel preparation for studying excitation–secretion coupling in the isolated single cell. *Nature* **315**, 679–80.

Jordan H.J. (1939). *Het leven der dieren in het water,* pp. 21, 75, 93, 151, 180. Utrech.

Jörg, — (1931). [*Placobdella* sp. as intermediate host of gregarines; *Placobdella Gregarinophora,* nomen nudum.] *Actas Congr. Int. Biol. Montev.,* facs. V, *Parasitol.,* pp. 1155–6, 1163.

—— (1936). [*Placobdella* sp., from Argentina, as intermediate host of gregarine.] *Octava Reunion Soc. Arg. Patol. Reg. del Norte (S. del Estero)* Part 2, 1034.

Jörgensen, M. (1908). Untersuchungen über die Eibildung bei *Nephelis vulgaris* Moquin-Tandon (*Herpobdella atomaria* Carena). *Arch. Zellforsch.* **2**, 279–347.

—— (1913). Zellenstudien II. Die Ei- und Nährzellen von *Piscicola. Arch. Zellforsch.* **10**, 127–60. [See also pp. 1–126 for general review of nurse cells in oogenesis.]

Jourdane, J. (1977). [Ecology of the development and transmission of the platyhelminth parasites of Soricids of the Pyrenees] *Mem. Mus. Natl. Hist. nat. Ser. A Zool.* **103**, 1–174. [In French; English summary.] [1979BA:4515.] [*Erpobdella octoculata* as intermediate host.]

Joyeux, C. (1922). Recherches sur les *Taenia* des Ansériformes. Développement larvaire d'*Hymenolepis parvula* Kew. chez *Herpobdella octoculata* L. *Bull. Soc. Path. exot.* **15**, 46–51.

—— and Baer, J.G. (1936). See Autrum (1939*c*).

Joyner, D.E. (1980). Influence of invertebrates on pond selection by ducks in Ontario, Canada. *J. wildl. Manag.* **44**(3), 700–5.

Judd, W.W. (1969) Studies of the Byron Bog in Southwestern Ontario XXXVII. Leeches (Hirudinea) collected in the Bog. *Can. Field-Nat.* **83**(2), 168.

Julka, J.M., and Chandra, M. (1980). A small collection of leeches collected during the Daphabum and Subansiri Expeditions (Arunachal, Pradesh, India). *J. Bombay nat. Hist. Soc.* **77**(1), 160–1.

Jung, D. (1963). Bau und Feinstruktur der Augen auf dem vorderen und hinteren Saugnapf des Fischegels *Piscicola geometra* L. *Zool. Beitr.* **9**, 121–72.

Jung, T. (1955*a*). Zur Frage der Verbreitung der medizinish-pharmazeutisch nutzbaren Hirudineen in Niedersachsen. *Z. angew. Zool.* **4**(4), 457–60.

—— (1955*b*). Zur Kenntnis der Ernährungsbiologie der in dem Raum zwischen Harz und Heide vorkommenden Hirudineen. *Zool. Jb. (Allg. Zool.)* **66**(1), 79–123. [feeding habits of some leeches were more dependent on particle size.]

Jutisz, M., Charbonnel-Bérault, A., and Martinoli, G. (1963). Purification de l'Hirudine. *Bull. Soc. Chim. Biol.* **45**(1), 55–67. [English summary.]

—— Martinoli, G., and Tertrin, C. (1962). Micro-dosage d'une activité anti-thrombique: l'hirudine. *Bull. Soc. Chim. Biol.* **44**, 461–9.

Jyoti, M.K., Sehgal, H., and Sahi, D.N. (1977). On the occurrence of some rhynchobdellid leeches in Jammu, India. *Geobios* (Jodhpur) **4**(2), 71–2. [1977BA:8619.]

Kaburaki, T. (1921). See Kaburaki (1921*b*) in Autrum (1939*c*).

Kaczmarek, S. (1968). [Variability in the coloration and design of the dorsal part of the body in the leech *Hirudo medicinalis* L.] *Prz. Zool.* **12**(1), 50–2. [In Polish; English summary.]

Kadota, K., and Nagata, M. (1975). Bioassay of acetylcholine with a thin muscle strip of Japanese medical leech: a preliminary note on the reappraisal of the availability of this method. *Japan J. Pharmacol.* **25**(5), 602–5. [1976BA:14591.]

—— Inoue, H., and Egawa, T. (1978). Assay of rat brain acetylcholine with a thin muscle strip of the Japanese medical leech. *Med. J. Osaka Univ.* **28**(3/4), 215–17. [1978BA:40326.]

—— and Kadota, T. (1977). Fine structure of obliquely striated muscle of Japanese medical leech, *Hirudo nipponia. J. Electron Microsc.* **26**(3). 262–3. [1978BA:32629.]

—— and Nagata, M. (1977). Bioassay of acetylcholine with thin muscle strips of Japanese medical leech and frog. *Jap. J. Pharmacol.* **27**(Suppl.) pp. 26. [1977BA:50703.]

Kaestner, A. (1964). Hirudinea, Blutegel. In *Lehrbuch der Speciellen Zoologie*, pp. 433–50. Fischer, Jena.

Kahlson, G., and Uvnäs, B. (1935). Zur Theorie der Sensibilisierung für Azetylcholin, zugleich Bericht über eine erregbarkeitssteigernde Wirkung des Fluorids. *Skand. Arch. Physiol.* **72**, 215–39.

—— —— (1938). Die Bedeutung der Acetylcholinesterase sowie der spezifischen Rezeptoven für die Acetylcholinempfindlichkeit kontraktiler Substrate. *Skand. Arch. Physiol.* **78**, 40–58.

Kai-Kai, M.A. *et al.* (1981). The structure, distribution, and quantitative relationships of the glia in the abdominal ganglia of the horse leech, *Haemopis sanguisuga. J. comp. Neurol.* **202**(2), 193–210.

Kaiser, F. (1954). Beiträge zur Bewegungsphysiologie der Hirudineen. *Zool. Jahrb., Abt. Allg. Zool. Physiol.* **65**, 59–90.

Kakar and Sawheney (Incomplete reference) (1965). [Leech in nose.] *Ann. Otol. Rhino. Lar.* **74**, 158–9.

Kalbe, L. (1964). *Glossiphonia complanata nebulosa* nov. subspec., eine neue Hirudinee aus Fliessgewässern des Havelgebietes. *Mitt. Zool. Mus. Berl.* **40**(2), 141–4.

—— (1965). Die Verbreitung der Hirudineen in Fliessgewässern der Havelgebietes. Veröffentlichungen des Bezirksheimatmuseums Potsdam. *Beitr. Tierwelt Mark* **9**, 8–15.

—— (1966). Zur Ökologie und Saprobiewertung der Hirudineen im Havelgebiet. *Int. Rev. Ges. Hydrobiol.* **51**(2), 243–77.

Kalmbach, E.R., and Gunderson, M.F. (1934). Western duck sickness, a form of botulism. *US Dept Agric. Tech. Bull.* 411. [*Theromyzon* in nose; North America.]

Kaluzhskii, N.V. (1888). Nabludeniya nad piyavkami okr. *Moskv'. Izv. Obsch. Lubit. Estestv* **50**(2), 370–8. [In Russian.]

Kannangara, D.W.W. (1972). On some aspects of the parasites of Ceylonese fresh water crabs. *Ceylon J. Sci. biol. Sci.* **10**(1), 32–8.

Kano, T. (1930). [On the leeches parasitizing the throats of animals.] *Zool. Mag., Tokyo* **42**, 124–5. [In Japanese.]

Karassowska, K., and Mikulski, J.S. (1960). Studia nad zbiorowiskami zwierzecymi roslinnosci zanurzonej i plywajacej jeziora Druzno. *Ekol. Pol., A, Warsaw* **8**(16), 335–. [First and to date only report of *Cystobranchus fasciatus* in Poland, at Druzno.]

Kasprzak, K. (1972). *Haementeria costata* (Fr. Muller) in Wolin National Park.] *Prz. Zool.* **16**(1), 37. [In Polish.]

—— (1977). Hirudinea. In Bottom fauna of the heated Konin Lakes (ed. A. Wróblewski) pp. 147–50. *Monog. Fauny Polski, Kraków* 7. [1978BA:9380.]

Kate, C.G.B.T. (1941). De Avifauna van Schokland III. *Limosa* **14**, 89. [*Piscicola geometra* and *Theromyzon tessulatum.*]

Katsulov, A. (1968). [A leech in the vaginal wall of an elderly woman causing haemorrhage.] *Akush. Ginek, Sofiia* **7**, 379–80. [In Bulgarian.]

Kawaguti, S., and Ikemoto, N. (1958). Electron microscopy on the smooth muscle of the leech, *Hirudo nipponia. Biol. J. Okayama Un.* **4**, 79–91.

Kazanskii, V.V., Bazanova, I.S., Mashanskii, V.F., and Merkulova, O.S. (1980). [The ultrastructural modifications in the Retzius neuron of the medical leech due to microelectrode insertion and after a short synaptic stimulation.] *Tsitologiya* **21**(6), 662–6. [In Russian.] [1980BA:9765.]

Kazantsev, B.N. (1957). Material' K rasprostraneniu i ekologii piyavki *Limnatis nilotica* v Tadzhikistane. *Izv. Otd. Estestv. Nauk Akad. Nauk Tadzh. SSR* **18**, 195–204. [In Russian.]

Keegan, H.L., and Weaver, R.E. (1964). Studies of Taiwan leeches II. Field tests of effectiveness of insect repellents against aquatic leeches at Cha'o Chow, Pingtung, Taiwan. *Bull. Inst. Zool. Acad. Sinica* **3**(2), 83–92.

—— Radke, M.G., and Murphy, D.A. (1970). Nasal leech infestation in man. *Am. J. trop. Med. Hyg.* **19**(6), 1029–30.

—— Toshioka, S., and Suzuki, H. (1968). Blood-sucking Asian leeches of families Hirudidae and Haemadipsidae. 406th Med. Lab. Spec. Report, July 1968. US Army Medical Command, Japan.

—— Poore, C.M., Weaver, R.E., and Suzuki, H. (1964). Studies on

Taiwan leeches. I. Insecticide susceptibility-resistance tests. *Bull. Inst. Zool. Acad. Sinica* 3, 39–43.

—— —— Fleshman, P., and Zarem, M. (1964). Studies of Taiwan leeches III. Further tests of repellents against aquatic, blood-sucking leeches. *Bull. Inst. Zool. Acad. Sinica* 3(2), 93–106.

—— —— Toshioka, S., and Matsui, T. (1964). Some venomous and noxious animals of east and southeast Asia. 406th Med. Lab. Spec. Report, Dept. July 1964. US Army Medical Command, Japan. [Leeches, pp. 7–9.]

Keim, A. (1977). Electrophoretic analyses of the crop contents of *Helobdella stagnalis* (L.) (Hirudinea). *Z. Naturforsch. Sect. C Biosci.* 32(9/10), 739–42. [1978BA:46379.]

—— (1979). Zur Wirtsspezifität von Blutegeln. Untersuchungen zum Parasitismus des Entenegels *Theromyzon tessulatum* (O.F. Müller 1774) und des medizinischen Blutegels *Hirudo medicinalis* Linnaeus, 1758. Dissertation. Univ. Hohenheim, Stuttgart.

Keith, M.M. (1954). A survey of the leeches (Hirudinea) of the Duluth area. *Proc. Minn. Acad. Sci.* 22, 91–2.

—— (1955). Notes on some leeches (Hirudinea) from the Yukon Territory, Canada, and Alaska. *Proc. Minn. Acad. Sci. 23* , 103–4.

—— (1959). A simplified key to the leeches of Minnesota. *Proc. Minn. Acad. Sci.* 27, 190–9.

Kelen, E.M.A. (1972). Substância anticoagulante de sanguessuga brasileira. Tese Doutor em Ciências.) 105 pp. Instituto de Biociências, Sao Paulo, USP. ˙

—— (1982). Fibrinolysin from *Haementeria deptessa* (Abstr. of a presentation of the Sixth International Congress on Fibrinolysis). *Haemostasis* 11, Suppl. 1, p. 16. [Full paper in a chapter in *Progress in fibrinolysis* Vol. 6. Churchill Livingstone, Edinburgh.]

—— Hruby, V.J., de Tomy, S.C., and Pinheiro, M.E.P. (1983). Hementerin: a potent in vitro inactivator of the human fibrinolytic system. In: *Progress in fibrinolysis*, Vol. 6. (ed. J.F. Davidson, F. Backmann, C.A. Bouvier and E.K.O. Kruiltiof). Churchill Livingstone.

—— and Rosenfeld, G. (1975). Fibrinogenolytic substance (Hementerin) of Brazilian blood-sucking leeches (*Haementeria lutzi* Pinto, 1920). *Haemostasis* 4(1), 51–64.

Kender, J. (1944). Über die Hirudineen von Tata und Umgebung. *Fragment Faunistica hung.*, Budapest 71(1), 11–13.

—— (1948). Piocak.-Hirudinoidea. *Dudich, E. Az. Allatok Gyujtese.*, Budapest 123–5. [In Hungarian.] [Collection and preservation of leeches.]

Kennedy, C.R. (1974). A checklist of British and Irish freshwater fish parasites with notes on their distribution. *J. fish. Biol.* 6(5), 613–44. [1975BA:15851.]

Kennel, J. (1886). Über einige Landblutegel des tropischen America. (*Cylicobdella* Grube und *Lumbricobdella* n.g.). *Zool. Jb. Systemat.* **2**, 37–64.

Kephart, D.G. (1982). Microgeographic variation in the diets of garter snakes. *Oecologia* (Berlin) **52**(2), 287–91. [1982BA:16245.]

—— and Arnold, S.J. (1982). Garter snake diets in a fluctuating environment: a seven-year study. *Ecology* **63**(5), 1232–6.

Kerkut, G.A., Sedden, C.B., and Walker, R.J. (1967*a*). Cellular localisation of monoamines by fluorescence microscopy in *Hirudo medicinalis* and *Lumbricus terrestris*. *Comp. Biochem. Physiol.* **21**(3), 687–90.

—— —— —— (1967*b*). A fluorescence microscopic and electrophysiological study of the giant neurones of the ventral nerve cord of *Hirudo medicinalis*. *J. Physiol.* **189**, 83P–85P. (Abstr.)

—— and Walker, R.J. (1967). The action of acetylcholine, dopamine and 5-hydroxytryptamine on the spontaneous activity of the cells of Retzius of the leech, *Hirudo medicinalis*. *Br. J. Pharmacol.* **30**(3), 644–54.

Keve, — (1968). Über die Arealveränderungen von *Plegadis falcinellus* (L.). *Zool. Abhandl. Staat. Mus. Tierkunde Dresden* **29**(13), 169.

Keymer, I.F. (1969). Infestation of waterfowl with leeches (correspondence). *Vet. Rec.* **85**(22), 632–3. [*Theromyzon tessulatum.*]

Keyser, K.T. (1979). Leydig cells within the C.N.S. of the leech. Ph.D. thesis. State University of New York, Stony Brook.

—— and Lent, C.M. (1977*a*). On neuronal homologies within the central nervous system of leeches. *Comp. Biochem. Physiol.* **58A**, 285–97.

—— —— (1977*b*). Neuronal control of flattening in the leech. *Am. Zool.* **17**, 876. (Abstr.)

—— Frazer, B.M., and Lent, C.M. (1982). Physiological and anatomical properties of Leydig cells in the segmental nervous system of the leech. *J. comp. Physiol.* **146A**, 379–92. (see also 1980 *Neurosci. Abstr.* **6**, 214).

Keysselitz, C. (1906). See Autrum (1939*c*).

Khaibulaev, K. Kh. (1970). [The role of leeches in the life cycle of blood parasites of fishes.] *Parazitologiya, Leningrad* **4**(1), 13–17. [In Russian; English summary; English transl. available USNM.] [see also: (1979), *Uspekhi Protozoologii*. Nauka, Leningrad.]

Khan, H. (1944). Study in diseases of fish. Infestation of fish with leeches and fish lice. *Proc. Indian Acad. Sci.* **19B**, 171–5.

Khan, M.M. (1912). Notes on the rearing of leeches in Mavai, Bara Banki District, United Provinces. *Rec. Ind. Mus.* **7**, 206–7.

Khan, R.A. (1972). On a trypanosome from the Atlantic cod, *Gadus morhua* L. *Can. J. Zool.* **50**, 1051–4.

—— (1974). Transmission and development of the trypanosome of the Atlantic cod by a marine leech. (Abstr.) *3rd Int. Congr. Parasit.*, Munich, 25–31 August 1974. Proceedings, Vol. 3, Vienna, Austria.

FACTA Publication 1974:1608. [*Johanssonia arctica* nec Myzobdella.]

—— (1975). Development of *Trypanosoma murmanensis* in the Atlantic cod, *Gadus morhua. 50th Meeting Am. Soc. Parasitol.*, p. 80.

—— (1976). The life cycle of *Trypanosoma murmanensis* Nikitin. *Can. J. Zool.* **54**(11), 1840–9.

—— (1977*a*). Susceptibility of marine fish to trypanosomes. *Can. J. Zool.* **55**(8), 1235–41. [1978BA:2536.]

—— (1977*b*). Infectivity of *Trypanosoma murmanensis* to leeches (*Johanssonia sp.) Can. J. Zool.* **55**(10), 1698–700.

—— (1977*c*). Blood changes in Atlantic cod (*Gadus morhua*) infected with *Trypanosoma murmanensis. J. Fish. Res. Bd Can.* **34**(11), 2193–6.

—— (1978*a*). A redescription of *Trypanosoma cotti* Brumpt and Lebailly, 1904, and its development in the leech, *Calliobdella punctata. Ann. Parasitol. Hum. comp.*, *Paris* **53**(5), 461–6. [French summary.] [1979BA:16099.]

—— (1978*b*). Longevity of *Trypanosoma murmanensis* in the marine leech, *Johanssonia* sp. *Can. J. Zool.* **56**(9), 2061–3. [1979BA:23710.]

—— (1978*c*). A new haemogregarine from marine fishes. *J. Parasitol.* **64**, 35–44. [*Johanssonia arctica.*]

—— (1980). The leech as a vector of a fish piroplasm. *Can. J. Zool.* **58**(9), 1631–7.

—— (1982). Biology of the marine piscicolid leech, *Johanssonia arctica* (Johansson) from Newfoundland. *Proc. helminth. Soc. Wash.* **49**(2), 266–78. [1983BA:49904.]

—— and Cowan, G.I.M. (1976). A biochemical approach towards the taxonomy of leeches (Annelida: Hirudinea). *Can. J. Zool.* **54**: 1803–5.

—— and Emerson, C.J. (1981). Surface topography of marine leeches as revealed by scanning electron microscopy. *Trans. Am. microsc. Soc.* **100**(1), 51–5.

—— Barrett, M., and Murphy, J. (1980). Blood parasites of fish from the northwestern Atlantic Ocean. *Can. J. Zool.* **58**(5), 770–81. [1980BA:45741.]

—— Forrester, D.J., Goodwin, T.M., and Ross, C.A. (1980). A haemogregarine from the American alligator (*Alligator mississippiensis*). *J. Parasitol.* **66**(2), 324–8.

—— and Meyer, M.C. (1976). Taxonomy and biology of some Newfoundland marine leeches (Rhynchobdellae: Piscicolidae). *J. fish. Res. Bd Can.* **33**, 1699–714.

—— —— (1978). Evidence of a bi-annual life cycle in the marine leech *Oceanobdella sexoculata* (Hirudinea: Piscicolidae). *J. Parasitol.* **64**(4), 766–8.

Khlebovich, V.V. (1966). [Changes in optical density of tissue protein solutions obtained from some freshwater animals related to the medium salinity.](Incomplete reference)

Kiener, A., and Ollier, J. (1970). Contribution a l'étude écologique et biologique de la riviére le Gapeau (Var.) *Hydrobiologia* **36**, 189–251. [*Erpobdella octoculata, Helobdella stagnalis*, S. France.]

Kijek, H. (1978). Wystepowanie rzadko notowanych pijawek *Batracobdella paludosa* (Carena) i *Hirudo medicinalis* L. na Dolnum Slasku. *Przegl. Zool.* **22**, 251–3.

Kikuchi, S., Saito, K., Ohri, M., Oshima, T., Toshioka, S., Yamamoto, H., and Okuyama, Y. (1977). Scanning electronmicrography of some species of leech (Hirudidae and Haemadipsaidae). I. *Hirudo nipponia* from Japan. *Jap. J. Sanit. Zool.* **28**(4), 393–400. [In Japanese; English summary.] [1978BA:21706.]

Kimura, T., and Keegan, H.L. (1966). Toxicity of some insecticides and molluscides for the Asian blood sucking leech, *Hirudo nipponia* Whitman. *Am. J. trop. Med. Hyg.* **15**(1),113–15.

—— —— and Haberkorn, T. (1967). Dehydrochlorination of DDT by Asian blood-sucking leeches. *Am. J. trop. Med. Hyg.* **16**(5), 688–90.

King, D.K., McBain, A.E., Jais, M.M., Roberts, C.J., Collins, J.F., Wheal, H.V., and Walker, R.J. (1982). The excitatory actions of a series of piperidinedicarboxylates on central neurons of the rat, snail (*Helix aspersa*), leech (*Hirudo medicinalis*) and horseshoe crab (*Limulus polyphemus*) and crab (*Eupagurus bernhardus*). *Comp. Biochem. Physiol. C Comp. Pharmacol.* **73**(1), 71–8. [1983BA:74095.]

King-Wai Yau. see Yau, King-Wai.

Kinji, Y., Bern, H.A., and Hagadorn, I.R. (1963). Electrophysiology of neurosecretory neurons in the leech *Theromyzon rude. Proc. 16th Int. Congr. Zool.* **16**(2), 148.

Kinkelin, P. de, Gerard, J.P., and Tuffery, G. (1973). Pathologie des poissons d'eau douce. II. Causes infectieuses: les maladies — leur traitement. *Informat. Tech. Serv. Vétérin.* No. 43/44, 43–62.

Kirka, A., Nagy, S., Zahumensky, L., Libosvarsky, J., Penaz, M., and Krupa (Incomplete reference) (1978). Fish distribution community of diatoms and zoobenthos in the water system of the river Poprad and in the spring waters of rivers Hornad and Hnilec. *Biol. Pr.* **24**(3), 9–98.

Kirpichen, Ko, M.J. (1965). [The fauna of the filamentous algae coenosis in the Kuybyshev Reservoir.] *Trudy Inst. Biol. Vnutrennikh vod.* **8**(11), 137–9. [In Russian.]

Kirschner, L.B., Greenwald, L., and Kerstetter, T.H. (1973). Effect of amiloride on sodium transport across body surfaces of freshwater animals. *Am. J. Physiol.* **224**, 832–7. [*Haemopis marmorata.*]

Klabunde, E. (1939). Mein Pferdeegel. *Aquar., Berlin* **13**, 33–4.

Kleerekoper, H. (date undetermined). Limnological observations in Northeastern Rio Grande do Sul, Brazil I. *Arch. Hydrobiol.* **50**, 553–67.

Kleine, T.O., and Merten, B. (1981). A procedure for the simultaneous determination of small quantities of hyaluronate and isomeric chondroitin sulfates by chondroitinases. *Anal. Biochem.* **118**(1), 185–90. [1982BA:29523.]

Kleinhaus, A.L. (1975). Electrophysiological actions of convulsants and anticonvulsants on neurons of the leech suboesophageal ganglion. *Comp. Biochem. Physiol. C Comp. Pharmacol.* **52**(1), 27–34. [1976BA:43241.]

—— (1976). Divalent cations and the action potential of leech Retzius cells. *Pflueger's Arch. Eur. J. Physiol.* **363**(2), 97–104. [1976BA:60959.] [*Macrobdella decora.*]

—— (1980). Segregation of leech (*Macrobdella decora*) neurons by the effect of sparteine on action potential duration. *J. Physiol.* **299**, 309–22. [1980BA:2927.]

—— and Brand, S. (1981). A potassium-dependent inhibitory synaptic connection in leech segmental ganglia. *Comp. Biochem. Physiol.* **70A**, 37–44. (see also 1979, *Proc. Soc. Neurosci.* **5**, 250).

—— and Prichard, J.W. (1974*a*). Electrophysiological effects of tetraethylammonium chloride on leech Retzius cells. *Soc. Neurosci. Abstr.* **IV**, 286.

—— —— (1974*b*). Electrophysiological properties of the giant neurons of the leech suboesophageal ganglion. *Brain Res.* **72**, 332–6. [*Macrobdella.*] [1975BA:43439.]

—— —— (1975). Calcium dependent action potentials produced in leech Retzius cells by tetraethylammonium chloride. *J. Physiol.* **246**(2), 351–70. [1975BA:25735.]

—— —— (1976*a*). Sodium dependent tetrodotoxin-resistant action potentials in a leech neuron. *Brain Res.* **102**, 368–73.

—— —— (1976*b*). Magnesium-resistant excitatory synaptic potentials in the leech Retzius cell. *J. Neurobiol.* **7**(4), 371–6. [1976BA:66969.]

—— —— (1976*c*). Antagonistic action of barbituates and calcium on action potential repolarization in the leech Retzius cell. *Neurosc. Abstr.* **II**, 180.

—— —— (1977*a*). Pentobarbital actions on a leech neurone. *Comp. Biochem. Physiol.* **58C**, 61–5.

—— —— (1977*b*). Close relation between TEA responses and Ca-dependent membrane plenomena of four identified leech neurons. *J. Physiol.* **270**(1), 181–94. [1977BA:15182.]

—— —— (1977*c*). A calcium-reversible action of barbituates on the leech Retzius cell. *J. Pharmacol. exp. Ther.* **201**(2), 332–9. [1977BA:40643.]

—— —— (1979). Interaction of divalent cations and barbituates on 4 identified leech (*Macrobdella decora*) neurons. *Comp. Biochem. Physiol. C Comp. Pharmacol.* **63**(2), 351–8. [1980BA:17022.]

—— —— (1983). Differential action of tetrodotoxin on identified leech neurons. *Comp. Biochem. Physiol. C Comp. Pharmacol.* **74**(1), 211–18. [1983BA:34231.]

Klekowska, Z. (1950). [Research on the reproduction of leeches of the genus *Erpobdella* de Blainville.] *Spraw. Lodz. Tow. Nauk.* **5**, 1(9), 120–1.

—— (1951). [Recherches sur la reproduction des sangsues du genre *Erpobdella* de Blainville.] *Acta zool. oec. Un. Lodz.*, 5. Soc. Sc. Lodz., No. 5, 1–44. [In Polish; French summary.]

Klekowski, R. (1961). Die Resistenz gegen Austrocknung bein einigen Wirbellosen aus astatischen Gewässern. *Verh. Int. Ver. Limnol.* **14**, 1023–8.

Klemm, D.J. (1970). Ecology and taxonomy of the leeches (Hirudinea) of Washtenaw County, with the distribution and an illustrated key to the leeches of Michigan. M.S. thesis, Eastern Michigan University.

—— (1972*a*). Freshwater leeches (Annelida: Hirudinea) of North America. Indent. Manual No. 8, Biota of Freshwater Ecosystems. *Water Poll. Contr. Res. Ser.* Environmental Protection Agency, Washington, DC.

—— (1972*b*). The leeches (Annelida: Hirudinea) of Michigan. *Mich. Acad.* **4**(4), 405–44.

—— (1973). Incidence of a parasitic leech *Marvinmeyeria (= Oculobdella) lucida* Moore and two cercarial types in estivating *Stagnicola exilis* (Lea). *Malacol. Rev.* **6**(1), 66–7.

—— (1974). Studies on the feeding relationships of leeches (Annelida: Hirudinea) as natural associates of mollusks. Ph.D. Thesis, University of Michigan, Ann Arbor. [Available from University Microfilms, Ann Arbor, Mich.]

—— (1975). Studies on the feeding relationships of leeches (Annelida: Hirudinea) as natural associates of mollusks. *Sterkiana* **58**, 1–50; **59**, 1–20.

—— (1976). Leeches (Annelida: Hirudinea) found in North American mollusks. *Malacol. Rev.* **9**(1/2), 63–76. [1976BA:14789.]

—— (1977). A review of the leeches (Annelida: Hirudinea) in the Great Lakes region. *Mich. Acad.* **9**(4), 397–418. [1977BA:97067.]

—— (1982). *The leeches (Annelida: Hirudinea) of North America.* Aquatic Biology Section, Environmental Monitoring and Support Laboratory, US Environmental Protection Agency, Cincinatti, Ohio.

—— Huggins, D.G., and Wetzel, M.J. (1979). Kansas leeches (Annelida: Hirudinea) with notes on distribution and ecology. *Tech. Pub. State Biol. Surv. Kans.* **8**, 38–46. [1979BA:61184.]

Klie, W. (1951). Der Blutegel in Niedersachsen. *Beitr. Naturkunde Niedersach.* **3** Jahrg., 3. Heft, Hannover, 75–9.

Klimek, L. (1960). Studia nad Fauna denna jeziora Druzno Zesz. *Nauk. UMK Biol.* **5**(7), 29–69.

Knapp, M.F., and Mill, P.J. (1968). Chemoreception and efferent sensory impulses in *Lumbricus terrestris* Linn. *Comp. Biochem. Physiol.* **25**, 523–8. [See also (1971*a*)   J. Cell Sci. **8**, 413–25; and (1971*b*). Tissue cell **3**, 623–36.

Knight, D.P., and Hunt, S. (1974). Molecular and ultrastructural characterization of the egg capsule of the leech, *Erpobdella octoculata* L. *Comp. Biochem. Physiol.* **47**(3A), 871–80.

Knight-Jones, E.W. (1940). The occurrence of a marine leech *Abranchus blennii* n. sp. resembling *A. sexoculata* (Malm) in North Wales. *J. mar. Biol. Ass. UK* **24**(2), 533–41.

—— (1961). The systematics of marine leeches. In *Leeches (Hirudinea). Their structure, physiology, ecology and embryology* (ed. K.H. Mann) Pure and applied Biology, Div. Zool. II, Appendix B, pp. 169–86. Pergamon Press, Oxford.

—— and Llewellyn, L.C. (1964). British marine leeches. *Rep. Challenger Soc.* **3**, 25.

Kobakhidze, D.N. (1942*a*). [On explaining the habitats of the medicinal leech under natural conditions in Georgia.] *Soobsch. Akad. Nauk Gruz. SSR Zool.* **3** (1), 69–72.

—— (1942*b*). [Animals as food for the medicinal leech in certain parts of its range.] *Soobsch. Akad. Nauk Gruz. SSR Zool.* **3**(9), 923–7.

—— (1943*a*). Kolichestvo Krovi V'sosannoi meditsinskoi piyavkoi pri odnokratnom pitanii i ee dal'neishee izmenenie. *Soobsch. Akad. Nauk Gruz. SSR Zool.* **4**(2), 173–8.

—— (1934*b*). K eksperimental'nomu izuchniu ustoichivosti meditsinskoi pijavki k shizheniu vlazhnosti pochv'. *Soobsch. Akad. Nauk Gruz. SSR Zool.* **4**(5), 439–44.

——(1943*c*). K izucheniu Khobotka y *Haementeria costata* Müller. *Soobsch. Akad. Nauk Gruz. SSR Zool.* **4**(9), 917–20.

—— (1944*a*). K izucheniu gidrostoikosti meditsinskoi piyavki. *Soobsch. Akad. Nauk Gruz. SSR Zool.* **5**(4), 423–9.

—— (1944*b*). Pop'tka ustanovleniya termicheskogo poroga zhizni meditsinskoi piyavki v eksperimental'n'kh usloviyakh. *Soobsch. Akad. Nauk Gruz. SSR Zool.* **5**(5), 543–8.

—— (1944*c*). K izucheniu vliyaniya kolichestva pischi na populyatsiu *Haementeria costata* Müller. *Soobsch. Akad. Nauk Gruz. SSR Zool.* **5**(9), 921–26.

—— (1945). Material' o kolichestve pogloschennoi pischi pri odnokratnom pitanii egipetskoi piyavki (*Limnatis nilotica* Savigny). *Soobsch. Akad, Nauk Gruz. SSR Zool.* **6**(8), 641–5.

—— (1946). Material' k inventarizatsii gidrophaun' Gruzii. *Tr. Zool. Inst. Akad. Nauk Gruz. SSR* **6**, 291–7.

—— (1956). *Meditsinskaya Piyavka I Ee Primenenie*, pp. 1–51. Tbilisi.

—— (1958). [New subspecies of cave leech (Hirudinea, Herpobdellae) from Georgia, SSR.] *Soobsch. Akad. Nauk Gruz. SSR Zool.* **21**(5), 591–2. [In Russian.] [*Dina absoloni ratschaensis* in western Georgia.]

Kochhar, R.K., Dixit, R.S., and Somaya, C.I. (1974). A critical analysis of 'Deet' as a repellent against arthropods of public health importance and water leeches. *Indian J. med. Res.* **62**(1), 125–33. [1975BA:10306.] [70 per cent meta-Deet (diethyl-m-toluamide) was more effective than DMP (dimethyl-phthalate) as a repellent against *Hirudo medicinalis.*]

Koehler, K.A., and Magnusson, S. (1974). The binding of proflavin to thrombin. *Arch. Biochem. Biophys.* **160**(1), 175–84. [1974BA:53011.] [Hirudin binds tightly to thrombin and displaces proflavin from the enzyme.]

Koffler, B.R., Seigel, R.A., and Mendonca, M.T. (1978). The seasonal occurrence of leeches on the Wood Turtle, *Clemmys inscripta* (Reptilia, Testudines, Emydidae). *J. Herpetol.* **12**(4), 571–2. [BA:59699.]

Koidl, B. (1974). The GABA content of the central nervous system of Crustacea and Annelida: a comparison. *J. comp. Physiol.* **94**, 49–55.

Koli, L. (1960(1961)). Über die Hirudineen des Brackwassers in der Umgebung von Tvärminne, Sudwestfinnland. *Arch. Soc. Zool.-bot. Feninicae 'Vanamo'* **15**(1–2), 58–62.

—— (1961). [On the setiferous leech (*Acanthobdella peledina* Grube) and its distribution in Finland.] *Luonnon Tutkija* **65**, 139–41. [In Finnish.]

Kollar, V. (1862). *Piscicola fasciata.* In *Treitschke-Naturhistorische Bildersaal des Tierreiches.* **3**, p. 101. Pesth, Leipzig.

Komàrek, J. (1953). Herkunft der Süsswasser-Endemiten der Dinarischen Gebirge, Revision der Arten, Artenstehung bei Höhlentieren. *Arch. Hydrobiol.* **48**(3), 269–349.

Komarova, M.S. (1957). [Seasonal parasitofauna dynamics of the tench in the Donets River.] *Zool. Zh.* **36**, 654–7. [In Russian; English summary.]

Konietzko, B. (1952). Sur la présence en Belgique de l'hirudinée *Trocheta subviridis* Dutrochet, 1817. *Bull. Mus. Hist. nat. belg.* (*Inst. R. Soc. Nat. Belge*) **28**(66), 1–4.

Kopenski, M.L. (1969). Selected studies of Michigan leeches. Ph.D. thesis, Michigan State University, East Lansing, Michigan. [Diss. Abst. Int. 30B:1955. Available from University Microfilm, Inc., Ann Arbor, Mich.]

—— (1972). Leeches (Hirudinea) of Marquette County, Michigan, *Mich. Acad.* **4**(3), 377–83.

Kopp, R. (1945). Le problème du mode d'action thérapeutique des sangsues. Diss., Strasbourg.

Koppenheffer, T.L., and Ginsberg, S.M. (1978). A comparative immunochemical study of hemoglobins from the earthworm and horseleech. *Comp. Biochem. Physiol. A Comp. Physiol.* **61**(1), 97–100. [1979BA:15798.]

Korovaev, N.M. (1957). Sluchai parazitirovaniya piyavok *Glossiphonia tesselata* v nosov'kh polostyakh gusei. *Sb. Nauchn. Rabot Altaiskoi N.-issl. Vet. Stantsii* **1**, 242–4.

Korzhuev, P.A., and Khudaiberdiev, S.R. (1937a). [Characteristics of protein digestion in the leeches *Hirudo medicinalis and Haemopis sanguisuga*.] *Akademiya Nauk SSSR Isvestiya. Seriya Biologicheskaya* **2**, 907–11. [English trans. RTS–2691. Mar. 65. 8pp. National Lending Library for Science and Technology, Boston Spa, Yorkshire England.]

—— —— (1937*b*). Ob osobennostyakh perevarivaniya belkov u krovososuschikh i khischi'kh piyavok (*Hirudo medicinalis* i *Haemopis sanguisuga*). *Bull. Eksperiment. Biol. i Med.* **3**(5), 479–82. [In Russian.]

Kosareva, N.A. (1963). Parazitophauna prom'slov'kh p'b vodokhranilisch Volgo-Donskogo sudokhodnogo kanala im. V.I. Lenina. *Avtoreph. Diss., Leningrad* 1–23. [In Russian.]

—— and Epshtein, V.M., (1962). Dann'e po phaune, dinamike, chislennosti i patogennosti p'b'ikh piyavok vodokhranilisch Volgo-Donskogo Kanala. *Pervoe Nauchn. Sovesch. Zoologov Ped. Inst. P. S. Ph. S. R. Tez. Dokl., Moscow* 36–7.

Kosel, V. (1973). [*Batracobdella slovaca* sp. n., a new species of leech in southwest Slovakia (Hirudinoidea, Glossiphoniidae.] *Biológia, Bratislava* **28**(2), 87–90. *Slowacka Akademia Nauk Bratyslawa.* [In Czechoslovak.]

Kosheva, A. Ph. (1954). [Parasitic fauna of the most important economic fish in the Volga at Kuibiskev, in connection with the construction of the water-reservoir). *Trans. Spec. Probl. Conf.* (*7th Conf. on Problems of Parasitology*), Moscow. Vol. 4, pp. 66–9. [In Russian.]

—— and Lyakhov, S.M. (1954). Sluchai parazitirovaniya piyavki *Herpobdella octoculata* (L.) v nosovoi polosti. *Med. Parazitol. i Parazit. Bolezni* **4**, 355.

Koshi, T., and Varma, R.N. (1965). A preliminary study of the land leeches of Assam and NEFA. *Armed Forces med. J.* **21**(2), 99–104.

Kosswig, K. (1967). Sackwiesensee in den Ostalpen (Hochschwabgebiet). Zur Limnologie eines dystrophen Gipsgewässers. *Int. Rev. ges. Hydrobiol. Hydrogr.* **52**, 321–59.

Kostowski, W. (1965*a*). [Influence of acetylcholine and serotonin on ganglionic transmission in the leech (*Hirudo medicinalis*).] *Acta physiol. pol.* **16**(3), 392–7.

—— (1965*b*). [Studies on the effect of acetylcholine and serotonin on ganglion conduction in the leech (*Hirudo medicinalis*).] *Acta physiol. pol.* **16**, 457–63

Kotel'nikov, G.A. (1956). [On the parasitism of leeches on aquatic birds.] *Sb. Rabot Vologodskoi N.-i. Veterinarnoi Stantsii* **3**, 89–91.

Kotevski, D., Konjanovski, D., and Docovski, M. (1969). [Hirudiniasis of the eye.] *Acta ophthalmol. Iugosl.* **7**(4), 492–7. [In Serbo-Croatian.]

Kots, Y.L. (1951). [Occurrence of leeches in the upper respiratory tract and oesophagus and methods of extraction.] *Vestnik Oto-Rino-Laringol.* **3**, 19–24.

Koubková, B., and Vojtková, L. (1973). [The Czechoslovakian Hirudinean fauna.] *Folia Fac. Sci. Nat. Un. Purkynianae Brunensis Biol.* **14**(6), 103–18. [In Czechoslovak; Russian and German summaries.] [1975BA:9011.]

Kowalevski, A. (1871). Embryologische Studien an Würmern und Arthropoden. *Mém. Acad. Sci. St. Pétersb.* (7) 16, Nr. 12.

—— (1896*a*). Étude sur l'anatomie de l'*Acanthobdella peledina*. *Bull. Acad. St. Pétersb.* (5) 5, Nr. 4, 263–73.

—— (1896*b*). Étude sur l'anatomie de l'*Acanthobdella peledina* Grube et de l'*Archaeobdella esmontii*. *Bull. Ac. St Pétersb.* **5**(5), 227–31.

—— (1897). See Kowalevsky, A. (1897*a*) in Autrum (1939*c*).

—— (1900*a*). Phénomènes de la fécondation chez l'*Helobdella algira* (Moquin-Tandon). *Mém. Soc. Zool. France* **13**, 66–88.

—— (1900*b*). Étude biologique de l'*Haementeria costata* Müller. *Mém. Acad. Sci. St. Pétersb.* (8) 11, Nr. 1.

—— (1901). Phénomènes de la fécondation chez l'*Haementeria costata* de Müller. *Mém. Acad. Sci. St. Petersb.* (8) 11, Nr. 10, 1–19.

Kozarov, G.I., and Mikhailova, P. (1955). Polski i laboratorni izsledovaniya v'rkhu biologiyata na *Limnatis nilotica* Sav. *Izv. Na. Tsentr. Khelmintolog. Lab., Kn.* **1**, 73–90.

Kozhov, M.M. (1962). *Biologiya Ozera Baikal* 313 pp. Moscow.

—— (1963). Lake Baikal and its life. *Monogr. Biol.* **XI**, 1–344. [Hirudinea:88–92.]

Kozur, H. (1970). Fossile Hirudinea aus dem Oberjura von Bayern. *Lethala* **3**(3), 225–32.

Krahl, B., and Zerbst-Boroffka, I. (1983). Blood pressure in the leech *Hirudo medicinalis*. *J. exp. Biol.* **107**, 163–8.

Kramer, A.P. (1981). The nervous system of the glossiphoniid leech *Haementeria ghilianii*. II. Synaptic pathways controlling body wall shortening. *J. comp. Physiol.* **144**, 449–57.

—— and Goldman, J.R. (1981). The nervous system of the glossiphoniid leech *Haementeria ghilianii*. I. Identification of Neurons. *J. comp. Physiol.* **144**, 435–48.

—— —— and Stent, G.S. (1985). Developmental arborisation of sensory neurons in the leech *Haementeria ghilianii*. I. Origin of natural variations in the branching pattern. *J. Neurosci.* **5**(3), 759–67.

—— and Stent, G.S. (1985). Developmental arborisation of sensory neurons in the leech *Haementeria ghilianii*. II. Experimentally induced variations in the branching pattern. *J. Neurosci.* **5**(3), 768–75.

—— and Kuwada, J.Y. (1983). Formation of the receptive fields of leech mechanosensory neurons during embryonic development. *J. Neurosci.* **3**(12), 2474–86.

—— and Stuart, D.K. (1982). Muscle cells, glial cells and axonal guidance in leech embryos. *Neurosci. Abstr.*

—— and Weisblat, D.A. (1985). Developmental neural kinship groups in the leech. *J. Neurosci.* **5**, 388–407.

Kramers, P., Moen, J.E.T., and Roos, P.J. (1977). The macro-invertebrates of the rubble banks of the Abcoudermeer. *Bijdr. Dierkd.* **46**(2), 215–18. [8 leech species.]

Krasil'nikova, N.I. (1965). Piyavki, parasitiruuschie na p'bakh Verkhnego Dona. *Sb. Nauchn. Rab. Aspirantov Voronezhsk. Gos. Un.* **2**, 129–33.

Krause, M., and Wilke, B. (1931). Über die Fortpfanzung vor *Haementeria officinalis. Zool Anz.* **107**, 30–2.

Krejci, K., and Fritz, H. (1976). Structural homology of a trypsin-plasmin inhibitor from leeches (bdellin B–3) with secretory trypsin inhibitors from mammals. *Fed Eur. Biochem. Soc. Lett.* **64**(1), 152–5.

Krenz, W.D. (1975). Elektrophysiologische Untersuchungen zur Afferenzverarbeitung im Visuellen System des Blutegels (*Hirudo medicinalis* L.) Thesis, Math.-Naturn. Fakultät, Universität Göttingen.

Kretter, K. (1935). Therapeutic effect of leeches in thrombosis. *Polska Gaz. Lek.* **14**, 796.

Kretz, J.R., Stent, G.S., and Kristan Jr, W.B. (1976). Photosensory input pathways in the medicinal leech. *J. comp. Physiol.* **106**, 1–37.

Kristan, W.B. Jr. (1974a). Neural control of swimming in the leech. *Am. Zool.* **14**(3), 991–1001.

—— (1974b). Characterization of connectivity among invertebrate motor neurons by cross correlation of spike trains. In *Invertebrate neurons and behaviour* (ed. C.A.G. Wiersma) pp. 371–7. The neurosciences: a third study program. MIT Press, Cambridge, Mass.

—— (1977). Neural control of movement. In *Function and formation of neural systems* (ed. G.S. Stent) pp. 329–54. Dahlem Konferenzen, Berlin.

—— (1979). Neuronal changes related to behavioural changes in chronically isolated segments of the medicinal leech. *Brain Res.* **167**(1), 215–20. [1979BA:51972.]

—— (1980). The generation of rhythmic motor patterns. In: *Information processing in the nervous system* (eds. H. Pinsker and W.D. Willis) pp. 241–61. Raven Press, New York.

—— (1982). Sensory and motor neurons responsible for the local bending response in leeches. *J. Exp. Biol.* **96**, 161–80. [1983BA:18686.]

—— (1983). The neurobiology of swimming in the leech. *Trends Neurosci.* **6**(3), 84–8.

—— and Calabrese, R.L. (1976). Rhythmic swimming activity in neurones of the isolated nerve cord of the leech. *J. exp. Biol.* **65**(3), 643–68. [1977BA:51619.]

—— and Guthrie, P.B. (1977). Acquisition of swimming behaviour in chronically isolated single segments of the leech. *Brain Res.* **131**, 191–5.

—— and Nusbaum, M.P. (1983). The dual role of serotonin in leech swimming. *J. Physiol.* (Paris) **78**(8), 743–7.

—— and Stent, G.S. (1976). Peripheral feedback in the leech swimming rhythm. *Cold Spring Harbor Symp. quant. Biol.* **40**, 663–74.

—— and Weeks, J.C. (1983). Neurons controlling the initiation, generation and modulation of leech swimming. In: *Neural origin of rhythmic movements* (eds. A. Roberts and B. Roberts). Soc. Exp. Biol. Symposium No. 37. Cambridge University Press, Cambridge.

—— McGirr, S.J., and Simpson, G.V. (1982). Behavioural and mechanosensory neuron responses to skin stimulation in leeches. *J. exp. Biol.* **96**, 143–60. [1983BA:18685.]

—— Stent, G.S., and Ort, C.A. (1974*a*). Neuronal control of swimming in the medicinal leech. I. Dynamics of the swimming rhythm. *J. comp. Physiol.* **94**, 97–119.

—— —— —— (1974*b*). Neuronal control of swimming in the medicinal leech. III. Impulse pattern of the motor neurons. *J. comp. Physiol.* **94**, 155–76.

—— Weisblat, D.A., and Radojcic, T. (1984). Development and function of the leech nervous system. In: *Invertebrate models in aging research* (ed. D.H. Mitchell and T.E. Johnson). Chap. 5, pp. 95–119. CRC Press, Inc. Boca Raton, Fla.

Krogh, A. (1965). *Osmotic regulation in aquatic animals*, pp. 50–2. Dover, New York. [Based on 1939 edition.]

Krotas, P. (1968). [Fauna of piscine parasites in Lake Zuvintas.] In *Reservation of Zuvintas. Acad. Sci. Lithuanian SSR, Vilnius* 179–92. [In Russian; English summary.] [*Piscicola geometra; Hemiclepsis marginata; Haemopis sanguisuga.*]

Krotas, R.A. (date undetermined). [Complex ichthyological investigation of Kaunas water reservoir 5. Parasites of fishes] *Trud. Akad. Nauk Lith. SSR Ser. C* **2**(34), 87–96. [In Russian; Lithuanian summary.]

Kruglova, V.M. (1951). K izucheniu piyavok Tomskoi oblasti. *Tr. Tomsk. Goz. Un.* **155**, 273–8.

Kruglyi, I.M. *et al.* (1978). [Living foreign body (leech) in the nasopharynx.] *Zh. Ushn. Gorl. Bolezn.* **6**, 85–6. [In Russian.]

Krull, J.N. (1970). Aquatic plant-macroinvertebrate associations and waterfowl. *J. wildl. Management* **34**, 707–18.

Kudryashov, B.A., Baskova, I.P., Orlova, A.S., and Shapiro, F.B. (1981). [Effect of hirudin-thrombin and phenylethylsulfonl-thrombin on some blood coagulation characteristics.] *Byull. eksp. Biol. Med.* **91**(3). 308–9. [In Russian; English summary.]

Kufel, J. (1966). [Leeches (Hirudinea) found in the right bank tributaries of the Widawa River.] *Prz. Zool.* **10**(2), 174–9.

—— (1967). [Leech fauna (Hirudinea) of the Widawa River.] *Opolski Tow. Przyk. Natk, Zeszyty Przyrodnicze* **7**, 87–99.

—— (1969). [*Haementeria costata* (Fr. Mull.) in the southern part of the Voivodeship Olsztyn.] *Prz. Zool.* **13**, 184–6. [In Polish; English summary.]

—— (1970). [Fauna of leeches (Hirudinea) in streams of Karkonosze mountains.] *Zesz. Przy. Opol. Nauk* **10**, 59–90. [In Polish.]

—— (1975). Leeches (Hirudinea) of the reserve Stawy Milickie (Milicz Fishponds). *Zesz. Przy. Opol. Nauk* **15**, 219–28.

Kuffler, D.P. (1975). Neuromuscular transmission in the leech, *Hirudo medicinalis*. Dissertation, University of California, Los Angeles. [*Diss. Abst. Int.* B**36**(5), 2018–19.]

—— (1978). Neuromuscular transmission in longitudinal muscle of the leech, *Hirudo medicinalis. J. comp. Physiol. A* **124**, 333–8. [Same title, (1976). *Acta physiol. scand.* (Suppl.) **440**, 89.] [1977BA:34785.]

—— and Muller, K.J. (1974). The properties and connections of supernumary sensory and motor nerve cells in the central nervous system of an abnormal leech. *J. Neurobiol.* **5**, 331–48.

Kuffler, S.W. (1967). Neuroglial cells: physiological properties and a potassium mediated effect of neuronal activity on the glial membrane potential. The Ferrier Lecture. *Proc. R. Soc.* **B168**, 1–21.

—— and Nicholls, J.G. (1964). Glial cells in the central nervous system of the leech: their membrane potential and potassium content. *Arch. exp. Path. Pharmakol.* (*Naunyn-Schmiedebergs*) **248**(3), 216–27.

—— —— (1965). How do materials exchange between blood and nerve cells in the brain? *Perspect. Biol. Med.* **9**(1), 69–76.

—— —— (1966). The physiology of neuroglial cells. *Ergebn. Physiol. Biol. Chem. exp. Pharmakol.* **57**, 1–90.

—— —— (1976). A simple nervous system: the leech. In *From neuron to brain*. Sinauer Associates, Sunderland, Mass.

—— and Potter, D.D. (1964). Glia in the leech central nervous system: physiological properties, and neuron-glia relationship. *J. Neurophysiol.* **27**, 290–320.

—— Nicholls, J.G., and Potter, D.D. (1965). An approach to the study of neuroglia and of extracellular space based on recent work on the nervous system of the leech. In *Studies in physiology* (ed. D.R. Curtis and A.K. McIntyre). Springer, Berlin.

Kuflikowski, T. (1970). Fauna in vegetation in carp ponds at Goczalkowice. *Acta hydrobiol., Kraków* **12**, 439–56.

Kühn, G. (1940). Zur Ökologie und Biologie der Gewässer (Quellen und Abflusse) des Wassersprengs bei Wien. *Arch. Hydrobiol., Stuttgart* **36**(2), 157–262. [Vermes: 189–97.]

Kükenthal, W., and Matthes, E. (1960). *Leitfaden für das zoologische Praktikum*, 14. Aufl. Fischer, Stuttgart.

Kulayev, S.I. (1929). See Kulajew, S.I. (1929) in Autrum (1939c).

—— (1932a). [Structure et fonction du canal évacuateur du sperme de *Glossossiphonia complanata* L. et son évolution au cours du cycle annuel.] *Bull. Stat. Biol. Bolchevo* **5–6**, 17–30. [In Russian; German summary, pp. 28–30.]

—— (1932b). Hirudinea vodoemov Mescherskoi nizmennosti. *Zap. Bolshevsk. Biolog. St.* **5–6**, 67–74. [In Russian.]

—— (1932c). [Structure et genèse du conduit séminifère de la sangsue (*Glossossiphonia complanata*).] *TP. 1. Conférence d'Histologie de Moscou*, pp. 150–1. [In Russian.]

—— (1938). [Cyclomorphosis of the male reproductive apparatus and the spermatogenesis in *Glossossiphonia complanata*.] *Arch. Russ. Anat. Histo. Embryo., Leningrad* **18**(3), 353–88. [In Russian; English summary, pp. 478–84.]

Kulkarni, G.K. (1976). Studies on biochemical ecology of Indian leech, *Poecilobdella*. Ph.D. thesis. Marathwada University. Aurangabad, India.

—— and Hanumante, M.M. (1977a). A note on the ecology of the Indian cattle leech, *Poecilobdella viridis. Marathwada Un. J. Sci.* **16**(9), 221–3. [1979BA:40191.]

—— —— (1977b). The anatomical survey of the reproductive system of the Indian cattle leech, *Poecilobdella viridis. Marathwada Un. J. Sci.* **16**(9), 225–8. [1979BA:3105.]

—— and Nagabhushanam, R. (1977a). Circadian rhythm in the respiratory metabolism of Indian leech, *Poecilobdella viridis* (Blanchard). *J. anim. Morphol. Physiol.* **24**(2), 255–8. [1978BA:14992.]

—— —— (1977b). Annual changes in the biochemical constituents of Indian leech, *Poecilobdella viridis. Marathwada Un. J. Sci.* **16**(9), 241–7. [1979BA:3104.]

—— —— (1978a). Thermal acclimation induced compensation in biochemical constituents of the Indian freshwater leech, *Poecilobdella viridis* (Blanchard). *Hydrobiologia* **58**(1), 3–6. [BA:39969.]

—— —— (1978b). Hormonal control of osmoregulation in the Indian freshwater leech, *Poecilobdella viridis* (Blanchard). *Hydrobiologia* **59**(3), 197–201. [1979BA:28653.]

—— —— (1978c). Respiratory metabolism of Indian leech, *Poecilobdella viridis:* influence of body size. *Marathwada Un. J. Sci.* **17**(10), 97–100. [1980BA:64959.]

—— —— (1980). Role of brain hormone in oogenesis of the Indian freshwater leech, *Poecilobdella viridis* (Blanchard) during the annual reproductive cycle. *Hydrobiologia* **69**(3), 225–8. [1980BA:37504.]

—— —— (1981). Histochemical features of the brain neurosecretory profile of the Indian leech *Poecilobdella viridis* (Blanchard). *Biol. J.* **3**(1), 43–7.

—— Hanumante, M.M., and Nagabhushanam, R. (1977). Some aspects of osmotic biology of the freshwater leech, *Poecilobdella viridis* (Blanchard). *Hydrobiologia* **56**(2), 103–8. [1978BA:59071.]

—— —— —— (1979). Neuroscretory changes induced by salt loading in the freshwater leech. *Poecilobdella viridis* (Blanchard). *Hydrobiologia* **62**(1), 3–6. [1979BA:48055.]

—— Nagabhushanam, R., and Hanumante, M.M. (1978a). Implication of hormonal principle in thermal acclimation of the Indian freshwater leech, *Poecilobdella viridis*: (Blanchard) influence of brain homogenate on oxygen consumption. *Hydrogiologia* **57**(1), 53–5. [BA:39967.]

—— —— —— (1978*b*). Reproductive biology of the Indian freshwater leech *Poecilobdella viridus*. (Blanchard). *Hydrobiologia* **58**(2), 157–65. [1978BA:46433.]

—— —— —— (1979). Effect of thermal acclimation on the neurosecretory cells of the brain of Indian leech, *Poecilobdella viridis* (Blanchard). *Rev. Bio.* **72**(3/4), 219–28.

—— —— —— (1980). Impact of some neurohumors on the brain neurosecretory profile of the Indian leech, *Poecilobdella viridis* (Blanchard). *Biol. J.* **2**(3), 37–41.

—— —— Bhaskerrao, A., and Anand, C.S.K. (1982). Influence of neuroendocrinological and pharmacological stimulations on the oxygen consumption of a freshwater leech, *Poecilobdella viridis*. *Proc. Indian natn. Sci. Acad. B Biol. Sci.* **48**(6), 730–4. [1984BA:3547.]

Kunst, M. (1956). Faunistische Notizen über Oligochäten und Hirudineen aus Island und Färöern. *Hydrobiologia* **8**, 323–7.

Kuntz, R.E., and Myers, B.J. (1968). Helminths of vertebrates and leeches taken by the U.S. Naval Medical Mission to Yemen, Southwest Arabia. *Can. J. Zool.* **46**(5), 1071–5.

—— —— (1969). A checklist of parasites and commensals reported for the Taiwan Macaque (*Macaca cyclopsis* Swinhoe, 1862). *Primates* **10**(1), 71–80.

—— —— Bergner Jr, J.F., and Armstrong, D.E. (1968). Parasites and commensals of the Taiwan Macaque (*Macaca cyclopsis* Swinhoe, 1862). *Formosan Sci.* **22**(3/4), 120–36.

Kunze, W. (1907). Über *Orcheobius herpobdellae* Schuberg et Kunze, ein Coccidium aus *Herpobdella atomaria* Car. (*Nephelis vulgaris* Moq. Tand.) *Arch. Protistenk.* **9**, 382–429.

Kusnetsova, O.N. (1955). [Leech parasites of waterfowl.] *Pittsevodstvo* **5**, 32–4. [In Russian.]

Kussat, R.H. (1969). A comparison of aquatic communities in the Bow River above and below sources of domestic and industrial wastes from the city of Calgary. *Can. Fish. Cult.* **40**, 3–31. [*Marvinmeyeria lucida*.]

Kutakov, O.I. (1963). Nekotor'e vopros' mekhanizma deistviya piyavok. *Zdravookhr. Turkmenistana* **5**, 11–14. [In Russian.]

—— (1965). [Hirudization therapy in thrombophlebitis.] *Klin. Med., Moskva* **43**, 120–3. [In Russian.]

Kutschera, U. (1980). Bestandsregulation bei Egeln. *Mikrokosmos* **69**(3), 80–2. [1980BA:37630.]

—— (1983). Dichteregulation durch intraspezifische Kokonzerstörung und Untersuchungen zur Fortpflanzungsbiologie beim Egel *Erpobdella octoculata* L. (Hirudinea: Erpobdellidae). *Zool. Jb. Syst.* **110**, 17–29. [English summary.]

—— (1984). Untersuchungen zur Brudflege und Fortpflanzungsbiologie beim Egel *Glossiphonia complanata* L. (Hirudinea Glossiphoniidae). *Zool. Jb. Syst.* III, 427–38. [In German; English summary].

—— (1985). Beschreibung einer neuen Egelart, *Helobdella striata* nov. sp. (Hirudinea: Glossiphoniidae). *Zool. Jb. Syst.* **112**, 469–76.

Kuwada, J.Y. (1984). Normal and abnormal development of an identified leech motor neuron. *J. Embryol. exp. Morph.* **79**, 125–37.

—— and Kramer, A.P. (1983). Embryonic development of the leech nervous system: primary axon outgrowth of identified neurons. *J. Neurosci.* **3**(10), 2098–111. (see also 1981, *Neurosci. Abstr.*)

Kuz'mich, V.N. (1965). [Seasonal and age characteristics of the feeding of the roach (*Rutilus rutilus lacustris* Pall) in Shaksha Lake.] *Izv. Biol. Georgr. Nauch-Issled. Inst. Irkutsk. Un.* **18**, 1–2; 108–17. [In Russian.]

Kuzmina, L.V. (1968). Distribution of biogenic monoamines in the nervous system of the body segments of the leech, *Hirudo medicinalis*. *J. evol. Biochem. Physiol. Suppl. 4. Physiology and Biochemistry of Invertebrates*, pp. 50–6.

Kuznetsova, O.N. (1953). [Distribution of the medicinal leech.] *Zool. Zhurn.* **32**(5), 833–9. [In Russian.]

—— (1955). [Leech parasitism of aquatic birds.] *Ptitsevodstvo, Moscow* **5**, 32–4. [In Russian.] [*Theromyzon tessulatum.*]

—— (1961*a*). Nekotor'e biologicheskie nabludeniya nad vozbuditelyami protoklepsioza sel'skokhozyaistvennoi vodoplavauschei ptits'. *Tr. Moskovsk. Vet. Akad.* **33**, 146–9. [In Russian.]

—— (1961*b*). Nauchnoe piyavkovodstvo v Sovetskom Souze. *Tr. Moskovsk. Vet. Akad.* **33**, 149–52. [In Russian.]

Lagutenko, Yu. P. (1973). [Anatomy of the nervous system in *Hirudo medicinalis.*] *Zool. Zh.* **52**(4), 485–91. [In Russian.] [1974BA:20142.]

—— (1975*a*). [The morphology of dorsal and ventral associative part of ventral ganglion of the leech.] *Trudy Leningr. Obshch. Estest.* **77**(4), 49–63. [In Russian.]

—— (1975*b*). [The typical features of interneuronal connections in the posterior medial nucleus of ganglion of ventral chain of leech.] *Trudy Leningr. Obsch. Estest.* **77**(4), 63–9. [In Russian.]

—— (1975*c*). [Structure of sensory neuropile of the central nervous system of the leech, *Hirudo medicinalis.*] *Arkh. Anat. Gistol. Embriol.* **69**(8), 37–71. [In Russian; English summary.] [1976BA:48919.]

—— (1981). [Ultrastructural features of the medicinal leech interneuronal synapses.] *Arkh. Anat. Gistol. Embriol.* **80**(6), 27–33. [In Russian; English summary.] [1982BA:32518.]

—— and Sotnikov, O.S. (1982). (Ultrastructural changes in leech (*Hirudo medicinalis*) brain chain synapses on high frequency electrostimulation). *Tsitologiya* **24**(9), 1019–23. [In Russian; English summary.] [1983BA:10952.]

Lainson, R. (1981). On *Cyrilia gomesi* (Neiva and Pinto, 1926) gen. nov. (Haemogregarinidae) and *Trypanosoma bourouli* Neiva and Pinto, in the fish *Synbranchus marmoratus:* simultaneous transmission by the leech *Haementeria lutzi.* In: *Parasitological topics. A presentation volume to P.C.C. Garnham, F.R.S.* (ed. E.U. Canning) pp. 150–8. Society of Protozoologists.

Laird, M., and Bullock, W.L. (1969). Marine fish hematozoa from New Brunswick and New England. *J. Fish. Res. Bd Can.* 26, 1075–102.

Lal, M.B., and Chowdhury, N.K. (1950). Anticoagulant activity of the Indian cattle leech. *Nature, Lond.* **166**, 480.

—— —— and Kishor, K. (1951). Purification of the anticoagulant principle obtained for the Indian cattle leech, *Hirudinaria. Science, NY* **114**, 696–7.

Lameere, A. (1895). *Manuel de la Faune de Belgique*, Vol. 1, pp. 291–4. Bruxelles. (Publisher undetermined)

Lammert, H. (1974). Einige Beobachtungen zur Parasitologie der Seezunge, *Solea solea* (L.) *Ber. Dt. Wiss. Komm. Meeresforsch.* **23**(2), 149–52. [In German; English summary.] [1975BA:15854.] [*Calliobdella* occurrence correlates with length of host.]

Lampert, K. (1925). *Das leben der Binnengewässer*, pp. 314–20 and 779. Leipzig. (Publisher undetermined)

Lancaster, S. (1939). Nature of the chromaffin nerve cells in certain annulates and arthropods. *Trans. Am. microsc. Soc.* **58**, 90–6.

Lande, V.M. Van Der. See Van Der Lande, V.M.

Landis, B.H., Zabinski, M.P., Lafleur, G.J.M., Bing, D.H., and Fenton, J.W. (1978). Human alpha and gamma thrombin differential inhibition with hirudin. *Fedn Proc. Fedn Am. Socs exp. Biol.* [1978BA:2317.] (Incomplete reference)

Landmann, H. (1972). Hirudin und andere Thrombininnibitoren aus blutsaugenden Tieren. *Folia haem, Leipzig* **98**, 437–45. [In German.]

Landon, P. (1905). *The opening of Tibet.* New York. [Leeches, p. 49.]

Lang, A. (1890). Über die äussere Morphologie von *Haementeria ghilianii*, F. de Filippi. Zürich. Festschrift zum 50. Doktorjubiläum von Nägeli und Kölliker, pp. 199–211.

Lang, C. (1974). Macrofaune des fonds de cailloux du Léman. *Schweiz. Z. Hydrol.* **36**(2), 301–50. [In French; English and German summaries.]

Lang, D.C. (1969). Infestation of ducklings with leeches (correspondence). *Vet. Rec.* **85**(20), 566. [*Theromyzon tessulatum.*]

Lang, E. (1980). The leech in the laboratory. *Newsday* 24 April pt. II/3.

Lankester, E.R. (1880*a*). Observations of the microscopic anatomy of the medicinal leech (*Hirudo medicinalis*). *Zool. Anz.* **3**, 85–90.

—— (1880*b*). On intra-epithelial capillaries in the integument of the medicinal leech. *Q. Jl microsc. Sci.* (new series) **20**, 303–6.

—— (1880*c*). On the connective and vasifactive tissues of the medicinal leech. *Q. Jl microsc. Sci.* (new series) **20**, 307–17.

Lanza, B. (1970). Necrologio del Prof. Iginio Sciacchitano (Palermo, 10 Guigno 1897-Firenze, 12 Novembre 1968). *Monit. Zool. Ital.* (new séries) Suppl. III (14), 309–17.

Lanzavecchia, G. (1974). Morphofunctional correlations in the muscles of some Hirudinea. *Boll. Zool.* **41**(4), 498–9. [1977BA:31023.]

—— (1975). Variazoni morfologiche e funzionali nei muscoli degli invertebrati. *Riv. Istochim.* **19**, 50–9. [*Erpobdella* muscles.]

—— and Arcidiacono, G. (1981). Contraction mechanism of helical muscles experimental and theoretical analysis. *J. submicrosc. Cytol.* **13**(2), 253–66.

—— and de Eguileor, M. (1976). Studies on the helical and paramyosinic muscles: V. Ulstrastructural morphology and contraction speed of muscular fibers of *Erpobdella octoculata* and *Erpobdella testacea* (Annelida: Hirudinea). *J. submircrosc. Cytol.* **8**(1), 69–88. [Italian summary.] [1976BA:14594.]

—— —— and Valvassori, R. (1975). The junctions of the invertebrates; phylogenetic problems. *Boll. Zool.* **42**, 395–402.

—— —— Vailati, G., and Valvassori, R. (1977). Studies on the helical and paramyosinic muscles: VI. Submicroscopic organization and function of body wall muscle fibers in some leeches. *Boll. Zool.* **44**(4), 311–26. [1979BA:54513.]

Lapage, G. (1962). *Monnig's veterinary helminthology and entomology*, 5th edn, pp. 330–3. Williams and Wilkins, Baltimore. ['*Limnatis africana*' in nose of monkey in West Africa.]

Lapicque, L. (1936*a*,*b*). See Autrum (1939*c*).

Lapinskaite, J. (1968). [Benthic fauna of Lake Zuvintas.] *Reservation of Zuvintas. Acad. Sci. Lithuanian SSR, Vilnius,* 159–77. [In Russian.]

Lapitskii, V.P., and Rusinov, A.A. (1978). [Effect of cerebral ganglia on the bioelectrical activity of circular nerves in the leech.] *Vestn Leningr. Un. Biol.* **1**, 86–91. [In Russian; English summary.] [1979BA:2981.]

Lapkina, L.N., and Flerov, B.A. (1980). [Leeches in identification of pesticides in water.] *Gidrobiol. Zh.* **16**(3), 113–20. [In Russian.] [1982BA:6808.]

Larrey, D.J. (1803). See Autrum (1939*c*).

Larsen, A. (1957). Laegeiglens, *Hirudo medicinalis,* forekomst pa Bornholm. *Foren. Bornholms Naturhistoriske,* pp. 45–9.

Larsson, R. (1981). Description of *Nosema tractabile,* new species (Microspora, Nosematidae), a parasite of the leech *Helobdella stagnalis* (Hirudinea, Glossiphoniidae). *Protistologica* **17**(3), 407–22. [French summary.] [1982BA:32800.]

Lasansky, A., and Fuortes, M.G.F. (1969). The site of origin of electrical responses in visual cells of the leech, *Hirudo medicinalis. J. cell. Biol.* **42**(1), 241–52.

Lasserre, P. (1975). Clitellata. In A.C. Giese and J.S. Pearse *Reproduction of marine invertebrates* Vol. 3. Annelids and Echiurans, pp. 215–75. Academic Press, London.

Latzel, R. (1887). Beiträge zur Fauna Kärtens. V. Zur Kenntnis des europäischen Landegels. *Jb. Landesmus. Kärnten* **12**, 120–4.

Laupy, M. (1970). Zur Kenntnis der Bodenfauna Zweier Teiche im Blatná-Gebiet. *Vést. Csl. Spol. Zool.* **34**, 110–20. [Leeches in ponds, Blatná-Gebiet, Czechoslovakia.]

Lavelle, P. (1981). Un ver de terre carnivore des savanes de la moyenne Côte d'Ivoire: *Agastrodrilus dominicae* nov. sp. (Oligochetes: Megascolecidae). *Rev. Écol. Biol. Sol.* **18**(2), 253–8. [English summary.][1983BA:34479.]

Laverack, M.S. (1969). Mechanoreceptors, photoreceptors and rapid conduction pathways in the leech, *Hirudo medicinalis. J. exp. Biol.* **50**(1), 129–40.

Laveran, A., and Mesnil, F. (1912). *Trypanosomes et trypanosomiases.* 999 pp. Deuxiéme édition. Masson et Cie, Paris.

Lavier, G. (1951). Emile Brumpt (1877–1951). *Presse méd.* **59**(62), 1297–8.

Leake, L.D. (1975). *Comparative histology. An introduction to the microscopic structure of animals.* Academic Press, London. 738 pp.

—— (1977). The action of (S)-3-allyl-2-methyl-4-oxocyclopen-2-enyl (IR)-*trans*-chrysanthemate, S-Bioallethrin, on single neurones in the central nervous system of the leech, *Hirudo medicinalis. Pestic. Sci.* **8**, 713–21.

—— (1982). Do pyrethroids activate neurotransmitter receptors? *Comp. Biochem. Physiol.* **72C**, 317–23.

—— (1983). The leech as a scientific tool. *Endeavour* (N.S.) **7**(2), 88–93.

—— and Sunderland, A.J. (1981). Pharmacological aspects of the control of leech body wall tension by Retzius cells. *Br. J. Pharmacol.* **74**(4), 983P–984P.

—— and Walker, R.J. (1980). Pesticides. In: *Invertebrate neuropharmacology.* pp. 272–7. Blackie, Glasgow.

—— Lauckner, S.M., and Ford, M.G. (1979). Relationship between neurophysiological effects of selected pyrethroids and toxicity to the leech *Haemopis sanguisuga* and the locust *Schistocerca gregaria. Proc. Symp. Insect Neurobiology and Pesticide Action.* London Soc. of Chemical Industry. 423–30.

—— Mason, A.J.R., and Sunderland, A.J. (1981). Is 5-hydroxytryptamine a neurotransmitter in the leech? [*Neurotransmitters in invertebrates.* (ed. K.S. Rozsa. Veszprem, Hungary.) Academic Press.] *Adv. Physiol. Sci.* **22**, 391–406.

—— Sunderland, A.J., and Walker, R.J. (1977). Different ionic pump components in nueronal resting potentials of two leech species. *J. Physiol.* **272**(1), 46P–47P.

—— —— —— (1980). Is there a common receptor mediating the inhibitory actions of 5-hydroxytryptamine (5-HT) and dopamine on leech Retzius cells? *Br. J. Pharmacol.* **70**(1), 138–9.

—— Griffiths, S.G., Crowe, R., and Burnstock, G. (1985). 5-Hydroxytryptamine-like immunoreactivity in the peripheral and central nervous systems of the leech *Hirudo medicinalis*. *Cell Tissue Res.* **239**. 123–30.

—— *et al.* (1973). Studies on neurones from the segmental ganglion of the leech, *Hirudo medicinalis*. *J. Physiol.* **232**, 63P–64P.

Learner, M.A., and Potter, D.W.B. (1974). Life-history and production of the leech *Helobdella stagnalis* (L.) (Hirudinea) in a shallow eutrophic reservoir in South Wales. *J. anim. Ecol.* **43**(1), 199–208.

Lebedev, J.V. (1967). [Spontaneous impulse potentials of single neurons in nonisolated nerve ganglia of *Hirudo medicinalis*] *Fiziol. Zh. SSSR Sechenov* **53**, 915–21. [In Russian.]

Lebels, J. (1838). [The use of leeches.] *Pam. Towarz. Lek. Warzaw* 1838–9, **2**, 375–91. [In Polish.]

Leblanc, J.M., and McClung, R.P. (1979). Leech (Hirudinoidea) records for the Province of Nova Scotia. (MS.) Curatorial Report No. 41, pp. 1–27. Nova Scotia Museum, Halifax.

Lechenault, H., and Pastisson, C. (1973). Analyse cytochimique des protéines nucléaires du spermatozoïde d'*Hirudo medicinalis* (L). *Ann. Histochim.* **18**(2), 141–7.

Leconti, —, and Faivre, E. (1857). Études sur la constitution chimique éléments et des tissus nerveux chez la sangsue médicinale. *Arch. gén. méd., Paris.* **2**, 666–83. [Also *C. r. Soc. Biol. Paris*. **2**, x, IV, pt 2m, 163–81.]

Lee, K.S. (1936). Beiträge zur blutgerinnungsaufhebenden Wirkung des Hirudins. *Fol. Pharmacol. Jap.* **22**, 36–48.

Legendre, R. (1959). Sur la présence de cellules neurosécrétrices dans les ganglions sous-oesophagiens de la sangsue médicinale (*Hirudo medicinalis* L.), suivre de quelques considérations sur la neurosécrétion. *Bull. Biol.* **93**, 462–71.

Léger, L. (1905). See Autrum (1939*c*).

Le Gore, R.S., and Sparks, A.K. (1971). Repair of body wall incision on the rhynchobdellid leech *Piscicola salmositica*. *J. invert. Path.* **18**, 40–5.

—— —— (1973). Repair of body wall burns in the rhynchobdellid leech *Piscicola salmositica*. *J. invert. Path.* **22**(2), 298–9. [1974BA:26099.]

Le Gros, A.E. (1970). Survey of Bookham Common — Hirudinea. *Lond. Nat.* **49**: 98.

—— (1974). Survey of Bookham Common. Vermes: Hirudinea: the medicinal leech. *Lond. Nat.* **53**: 74–5.

Lehmann, D.L. (1952). Notes on the life cycle and infectivity of *Trypanosoma barboni*. *J. Parasitol.* **38**, 550–3.

—— (1958). Notes on the biology of *Trypanosoma ambystomae* Lehmann, 1954. II. The life cycle in the invertebrate host. *J. Protozool.* **5**(1), 96–8. [Probably *Placobdella picta* not *Erpobdella* sp.]

Lehmensick, R. (1941). Über einen neuen bakteriellen Symbionten im

Darm von *Hirudo officinalis* L. *Zbl. Bakteriol.* **147**. (Pages undetermined) (1942).

—— Weitere Untersuchungen über den bakteriellen Darmbewohner des medizinischen Blutegels *Zbl. Bakteriol.* **149**. (Pages undetermined)

Leigh-Sharpe, W.H. (1914). *Calliobdella lophii. Parasitology* **7**, 204–18.

—— (1916). *Platybdella anarrhichae.* With a note, erratum, and an appendix. *Parasitology* **8**, 274–93.

—— (1933). See Leigh-Sharpe, W.H. (1933*a*) in Autrum (1939*c*).

Leiper, R.T. (1909). Check-list of the generic names of leeches, with their type species. *Zoologist* **13**, 422–6.

Lenaers, W. (1948). Bloedzuigerkwekerijin te Stamproy Natuurhistorisch Maandblad. *Org. v. h. Natuurhist. Genootschap in Limburg* **37**, 83–5.

Lenggenhager, K. (1936). Das Rätsel des Blutegelbisses. *Schweiz. med. Wschr.* **9**, 227–8. [Leech anaesthetic.]

Lent, C.M. (1971). A comparative study of the neuronal geometry of Retzius' cells in leeches. *Am. Zool.* **11**, 675.

—— (1972). Electrophysiology of Retzius' cells of segmental ganglia in the horse leech, *Haemopis marmorata* (Say). *Comp. Biochem. Physiol.* **42A**, 857–62.

—— (1973*a*). Retzius cells: neuroeffectors controlling mucus release by the leech. *Science, NY* **179**, 693–6.

—— (1973*b*). Retzius' cells from segmental ganglia of four species of leeches: comparative neuronal geometry. *Comp. Biochem. Physiol. A Comp. Physiol.* **44**(1), 35–40.

—— (1973*c*). Physiological characterization of motoneurons involved in swimming behaviour of leeches. *Am. Zool.* **13**, 1299–300. (Abstr.)

—— (1974). Neuronal control of mucus secretion by leeches: toward a general theory for serotonin. *Am. Zool.* **14**, 931–41.

—— (1976). Endogenous muscles within the central nervous system of the leech. *Am. Zool.* **16**, 177. (Abstr.)

—— (1977). Retzius cells within the central nervous systems of leeches. *Prog. Neurobiol.* **8**, 81–117.

—— (1981). Morphology of neurons containing monoamines within leech segmental ganglia. *J. exp. Zool.* **216**(2), 311–16.

—— (1982*a*). Serotonin-containing neurones within the segmental nervous system of the leech. In: *The biology of serotonergic transmission* (ed. N.N. Osborne). Ch. 17, pp. 431–456. John Wiley & Sons, Ltd.

—— (1982*b*). Fluorescent properties of monoamine neurons following glyoxylic acid treatment of intact leech ganglia. *Histochemistry* **75**, 77–89.

—— (1985). Serotonergic modulation of the feeding behavior of the medicinal leech. *Brain Res. Bull.* **14**, 643–55.

—— and Dickinson, M.H. (1984*a*). Serotonin integrates the feeding behavior of the medicinal leech. *J. Comp. Physiol. A.* **154**, 457–71.

—— —— (1985*b*). Retzius cells retain functional membrane properties following "ablation" of the neurotoxin, 5, 7-DHT. *Brain Res.* **300**, 167–171.

—— and Frazer, B.M. (1977). Connectivity of the monoamine-containing neurones in the central nervous system of leech. *Nature* **266**, 844–7.

—— Moore, G., Frazer, B., and Boles, J. (1977). White noise analysis of electrotonic junctions in the leech nervous system. *Neurosci. Abstr.* **3**, 186.

—— Mueller, R.L., and Haycock, D.A. (1983). Chromatographic and histochemical identification of dopamine within an identified neuron in the leech nervous system (*Hirudo medicinalis*). *J. Neurochem.* **41**(2), 481–90. [1984BA:11677.]

—— Ono, J., Keyser, K.T., and Karten, H.J. (1979). Identification of serotonin within vital-stained neurons from leech ganglia. *J. Neurochem.* **32**(5), 1559–63. (see also 1978. *Neurosci. Abstr.* **4**, 199). [1979BA:54126.]

Leong, T.S., and Holmes, J.C. (1981). Communities of metazoan parasites in open water fishes of Cold Lake, Alberta. *J. Fish. Biol.* **18**, 693–713.

Lepage, P., Serufilira, A. and Bossuyt, M. (1981). Severe anaemia due to leech in the vagina. *Ann. Trop. Paediatr.* **1**(3), 189–90.

Leslie, C.J. (1951). Mating behaviour of leeches. *J. Bombay nat. hist. Soc.* **50**, 422–3.

Lestage, J.A. (1936). La présence dans les eaux belges de l'ichtyoparasites *Cystobranchus respirans* Trosch. (Hirudinea). *Ann. Soc. zool. belg.* **66**, 127–32.

Lester, R.J.G., and Daniels, B.A. (1976). The eosinophilic cell of the White sucker, *Catostomus commersoni*. *J. Fish. Res. Board Can.* **33**(1), 139–44.

Letch, C.A. (1977). Studies on trypanosomes of small fishes from the River Lee. Ph.D. thesis, Council for National Academic Awards.

—— (1979). Host restriction, morphology and isoenzymes among trypanosomes of some British freshwater fishes. *Parasitology* **79**(1), 107–17.

—— (1980). The life-cycle of *Trypanosoma cobitis*. *Parasitology* **80**(1), 163–70.

—— and Ball, S.J. (1979). Prevalence of *Trypanosoma cobitis* Mitrophanow 1883 in fishes from the River Lee. *Parasitology* **79**, 119–24.

Leuckart, R., and Brandes, G. (1886–1901). *Die Parasiten des Menschen und die von ihnen herrührenden Krankheiten. Ein Hand- und Lehrbuch für Naturforscher und Ärzte.* Leipzig. [Hirudinea: pp. 535–897.]

Levanidova, I.M. (1968). [The benthos of tributaries of the Amur river: an ecological-zoogeographic survey). *Izv. Tikhookean Nauchissled. Inst. Ryb. Khoz. Okeanograf.* **64**, 181–289. (Transl. from *Ref.Zh. Biol.* 1969 7U108) [1970BA:70587.]

Levi, J.U., Cowden, R.R., and Collins, G.H. (1966). The microscopic anatomy and ultrastructure of the nervous system in the earthworm (*Lumbricus*) with emphasis on the relationship between glial cells and neurones. *J. comp. Neurol.* **127**, 489–507.

Levinsen, G.M.R. (1882*a*, 1882*b*, 1883). See Levinsen, G.M.R. (1881*a*, 1881*b*, 1883) in Autrum (1939*c*).

Lewis, J.W., and Ball, S.J. (1979). Attachment of the epimastigotes of *Trypanosoma cobitis* (Mitrophanow, 1883) to the crop wall of the leech vector *Hemiclepsis marginata*. *Z. Parasitenkd.* **60**(1), 29–36. [1980BA:58256.]

Leydig, F. (1849*a*). Zur Anatomie von *Piscicola geometrica* mit theilweiser Vergleichung anderer einheimischer Hirudineen. *Z. wiss. Zool.* **1**, 103–34.

—— (1849*b*). Zum Cirkulations und Respirations-System von *Nephelis* und *Clepsine*. *Ber. zootom. Anst., Würzburg* **2**, 14–20.

—— (1861). See Autrum (1939*c*).

—— (1862). Uber das Nervensystem der Anneliden. *Arch. Anat. Physiol., Lpz.* 90–124.

—— (1885). [Leeches.] In *Zelle und Gewebe*, p. 91, pl. II, fig. 31. Bonn.

Licht, L.E. (1969). Palatability of *Rana* and *Hyla* eggs. *Am. Midl. Nat.* **82**(1), 296–8.

Lichtenberg, R. (1972). Hydrobiologische Untersuchungen an einem suedlich von Wien gelegen Zieglteich (Hallateich). *Sitzungsber Oesterr. Akad. Wiss. Math.-Naturwiss. Kl.* **180**(8–10), 279–316. [1974BA:30445.] [Leeches in brick pool in Vienna.]

Lilienthal, H. (1943). Coronary thrombosis: proposed treatment by hirudin. *J. Mt. Sinai Hosp.* **10**(1). 135–7.

Li Mai Qun, and Yang Tong. (1982). Studies of internal hirudiniasis in human bodies in western Algeria. *Acta zool. sinica.* **28**(4), 377–82.

Lindegaard, C. (1979*a*). The invertebrate fauna of Lake Mývatn, Iceland. *Oikos* **32**, 151–61. [Russian summary.] [*Theromyzon; Glossiphonia complanata; Helobdella stagnalis.*]

—— (1979*b*). A survey of the macroinvertebrate fauna, with special reference to Chrinomidae (Diptera) in the rivers Laxá and Kráká, Northern Iceland. *Oikos* **32**, 281–8. [Russian summary.] [*Glossiphonia complanata.*]

—— and Jonasson, P.M. (1979). Abundance, population dynamics and production of zoobenthos in Lake Mývatn, Iceland. *Oikos* **32**(1/2), 202–27. [Russian summary.] [1980BA:8663.] [*Helobdella stagnalis.*]

Lindeman, E. (1954). That curious worm — the leech. *Nature Mag.* **47**(7), 355–7.

Lindeman, V.F. (1935). The relation of temperature to respiratory regulation in the leech (*Hirudo medicinalis*). *Physiol. Zool.* **8**, 311. [See also: **5**: 560–5 (1932).]

Lindemann, B.A. (1937). *Das Verhalten der Kapillaren in der Umgebung des Blutegelbisses*. Preisaufgabe der Medizinischen Fakultät der Friedrich-Wilhelms-Universität für 1937.

—— (1939). Das Verhalten der Kapillaren in der Umgebung des Blutegelbisses. *Archiv. Fuer exp. Path. Pharmak.* **193**, 490–502. [Histamine in salivary glands.]

Linfield, R.S.J. (1980). Ecological changes in a lake fishery and their effects on a stunted roach, *Rutilus rutilus*, population. *J. fish. Biol.* **16**, 123–44.

Ling, Han (1981). The bloodsuckers of Borneo's jungles. *Borneo Bull.* **30**(13), 14–15.

Linker, A., Hoffman, P., and Meyer, K. (1957). The hyaluronidase of the leech: an endoglucuronidase. *Nature, Lond.* **180**, 810–11.

—— Meyer, K., and Hoffman, P. (1960). The production of hyaluronate oligosaccharides by leech hyaluronidase and alkali. *J. biol. Chem.* **235**(4), 924–7.

Linnaeus, C. (1758). *Systema naturae.* Lipsiae. 10th edn. [Hirudinea: 648–51.]

Linton, L.R., Davies, R.W., and Wrona, F.J. (1982). Osmotic and respirometric responses of two species of Hirudinoidea to changes in water chemistry. *Comp. Biochem. Physiol.* **71A**(2), 243–8. [1982BA:10466.]

—— —— —— (1983*a*). The effects of water temperature, ionic content and total dissolved solids on *Nephelopsis obscura* and *Erpobdella punctata* (Hirudinoidea, Erpobdellidae): 1. Mortality. *Holarct. Ecol.* **6**(1), 59–63. [1983BA:47931.]

—— —— —— (1983*b*). The effects of water temperature, ionic content and total dissolved solids on *Nephelopsis obscura* and *Erpobdella punctata* (Hirudinoidea, Erpobdellidae): 2. Reproduction. *Holarct. Ecol.* **6**(1), 64–8. [1983BA:47932.]

Liskewich, S. (1922). Material' k poznaniu Hirudinea Kazanskoi gubernii. *Tr. Stud. Kruzhka Lubit Pripod' Pri Kazansk. Un.* **2**, 27–35. [In Russian.]

Litton, R.A. (date undetermined). Leeches attacking common newt. *Br. J. Herpetol.* **3**, 61–2.

Livanow, N.A. (1902). Die Hirudineen-Gattung *Hemiclepsis* Vejd. *Zool. Jb. Syst.* **17**, 339–62.

—— (1903). Untersuchungen zur Morphologie der Hirudineen. I. Das Neuro- und Myosomit der Hirudineen. *Zool. J. Anat.* **19**, 29–90.

—— (1904). Untersuchungen zur Morphologie der Hirudineen. II. Das Nervensystem des vorderen Körperendes und seine Metamerie. *Zool. J. Anat.* **20**, 153–226.

—— (1905). *Acanthobdella peledina* Grube, 1851. Morphologicheskoe issledovanie. *Uch. Zap. Kazansk. Un.* **22**,(5-ya i 6-ya knigi), 1–271.

—— (1906). *Acanthobdella peledina* Grube, 1851. *Zool Jb. Anat.* **22**, 637–866.

—— (1907). Untersuchungen zur Morphologie der Hirudineen. III. Das Nervensystem und die Metamerie des vorderen Körperendes von *Herpobdella atomaria* Carena. *Zool. Jb. Anat.* **23**, 683–702.

—— (1910). Untersuchungen zur Morphologie der Hirudineen. IV. Zur Atatomie des Blutgefässsystems. *Biol. Z.* **1**, 46–63. [Russian summary: pp. 64–7.]

—— (1931). Die Organisation der Hirudineen und die Beziehungen dieser Gruppe zu den Oligochaeten. *Ergbn. Fortschr. Zool.* **7**, 378–484.

—— (1937). Pijavki — Hirudinea. In *Zhivatn'i mir SSSR* (ed. S.A. Ziernov and N.J. Kuzniecov.) pp. 558–61.

—— (1940). *Klass Piyavok* (*Hirudinea*). *Rukovodstvo Po Zoologii*, [Vol. 2, pp. 203–57.] Izdatel'stvo Akad. Nauk SSSR, Moscow. [In Russian.]

—— (1955). *Puti Evolutsii Zhivotnogo Mira*, pp. 1–398. Analiz Organizatsii glavneishikh tipov mnogokletochn'kh zhivotn'kh. Moscow.

Llewellyn, L.C. (1965). Some aspects of the biology of the marine leech *Hemibdella soleae*. *Proc. zool. Soc. Lond.* **145**(4), 509–28.

—— (1966). Pontobdellinae (Piscicolidae: Hirudinea) in the British Museum (Natural History) with a review of the subfamily. *Bull. Br. Mus.* (*Nat. Hist.*) *Zool.* **14**(7), 389–439.

—— and Knight-Jones, E.W. (1984). A new genus and species of marine leech from British coastal waters. *J. mar. Biol. Ass. UK* **64**, 919–34.

Llosa, P. de la, Tertrin, C., and Jutisz, M. (1963). Composition en acides amines de l'hirudine. Identification du résidu N-terminal. *Bull. Soc. chim. biol.* **45**(1), 69–74. [English summary.]

—— —— —— (1964). L'enchaînement C-terminal de l'hirudine. *Biochim. biophys. Acta* **93**, 40–4.

Lochner, L. (1916). Uber geschmacksphysiologische Versuche mit Blutegeln. *Pflug. Arch.* **163**, 239–46.

Lockery, S.R., Rawlins, J.N.P., and Gray, J.A. (1985). Habituation of the shortening reflex in the medicinal leech. *Behavioural Neuroscience* **99**(2), 333–41.

Loeb, J. (1906). See Loeb, L. (1906) in Autrum (1939*c*).

Loeb, L, and Fleisher, M.S. (1913). Intravenous injections of various substances in animal cancer. *J. Am. med. Ass.* **60**(3), 1857–8. [Treatment of mouse carcinoma with crude leech extract.]

Löffler, H. (1974). Die Kleintierfauna des Schilfgürtels. In: *Der Neusiedlersee*, Chapter 12, Vienna: Verlag Fritz Molden. [*Hirudo* in Austria.]

Lom, J. (1970). Protozoa causing diseases in marine fishes. In *A Symposium on diseases of Fishes and Shellfishes* (ed. S. Snieszko) pp. 101–23.

—— (1973). Experimental infections of freshwater fishes with blood flagellates. *J. Protozool.* **20**, 537.

—— (1979). Biology of the trypanosomes and trypanoplasms of fishes. In: *Biology of the kinetoplastida*, Vol. 2 (eds. W.H.R. Lumsden and D.A. Evans) pp. 270–330. Academic Press, London.

Lönneberg, E. (1936). Borstigeln, Acanthobdella, funnen i Sverige, en intressant nyhet i var fauna. *Fauna Flora* **3**, 249–54.

Lora Lamia Donin, C., and Lanzavecchia, G. (1974). Morphogenetic effects of microtubules: III. Spermiogenesis in Annelida Hirudinea. *J. submicrosc. Cytol* **6**(2), 245–59. [1975BA:25539.]

Löser, R. (1909). See Loeser, R. (1909) in Autrum (1939*c*).

Lothian, A. (1959). English leeches and leech jars. *Chem. Drugg.* 153–9.

Lotz, R.G.A. (1968*a*). Biologische Experimente im interplanetaren Raum. *Sci. Indust. spat.* **7/8**, 3–12.

—— (1968*b*). Die Entwicklung eines extraterrestrischen biologischen Experimentes bis zur Flugreife. *Sonderdruck Forsch. w 68–30 Bundesminister. wiss. Forsch.* 101–13.

—— (1968*c*). Exposition extra-terrestre de *Hirudo medicinalis* pendant un temps supérieur à une année grâce a une biosonde automatique. In *European Colloquium on Space Radiation Doses Computation,* Toulouse.

—— (1969). The problem of sterilizing a life-support system in biosatellite experiments over one year. *Life Sci. Space Res.* **VII**, 130–3.

—— and Bowman, G.H. (1970). Basic studies on *Hirudo medicinalis* for a space experiment: II. Behaviour of *Hirudo medicinalis* in unnatural environments. *Space Life Sci.* **2**(1), 45–7.

—— and Fuchs, M.E.A. (1969). Oxygen supply and carbon dioxide absorption in long-term life-support systems. *Life Sci. Space Res.* **VII**, 134–7.

—— —— and Moyat, P.E.A. (1970). Basic studies on *Hirudo medicinalis* for a space experiment: I. Behavior of *Hirudo medicinalis* in natural environments. *Space Life Sci.* **2**(1), 40–4.

Love, A. (1969). *The zodiac.* Lomac Productions, Chicago.

Low, J.B. (1945). Ecology and management of the Redhead, *Nyroca americana,* in Iowa. *Ecol. Monog.* **15**, 35–69. [*Theromyzon.*]

Lubinsky, G.A., and Loch, J.S. (1979). Ichthyoparasites of Manitoba, Canada: literature review and bibliography. *Can. Fish. Mar. Serv. Manuscr. Rep.* **I–IV**, 1–20.

Lubits'kii, M.M. (1931). Materiali do vivchennya p'yavok stochischa pivdennogo Bura. *Tr. Prirodno-Tekhn. Vid. Vseukr. Akad. Nauk* **13**, 161–70. [In Russian.]

Lucky, Z., Dyk, V., and Bartik, M. (1955). Parasitofauna ryb v. oblasti reservace velky tisy y Lomnia nad Luznia. *Cas. nar. Mus.* **124**, 55–64.

Luferov, V.P. (1963) [Observations on the biology of leeches of the genus *Herpobdella.*] In *Material' po biologii gidrobiologii Volzhskikh vodokhranilits,* pp. 61–6. Akad. Nauk SSSR, Moscow. [Transl. *Ref. Zhur., Biol.* (1963), no. 1OD37.] [In Russian.]

Lukin, E.I. (1929). [Biological remarks on the leeches of the Donetz basin.] *Tr. Kharkivs'k Tov. Doslidn. Prirodi* **52**, 33–76. [In Russian.]

—— (1936*a*). [Questions concerning the evolution of freshwater fauna.] *Pratsi Nauk.-Dosl. Zool.-Biol. Inst. Khar'kovsk. Gos. Un.* **1**, 130–43. [In Russian.]

—— (1936*b*). [The biological peculiarities of the fish leech *Piscicola geometra.*] *Pratsi Nauk.-Dosl. Zool.-Biol. Inst. Khar'kovsk. Gos. Un.* **1**, 144–61. [In Russian; English summary.]

—— (1940). [Darwinism and the geographical laws of evolved organisms.] pp. 1–311. Moscow.

—— (1953). [On the composition of the leech fauna of Lake Sevan.] *Tr. Sevanks Hydribiol. Stantsii* **13**, 213–25. [In Russian.]

—— (1954*a*). [On the hydrobiological characteristics of a pond.] *Sb. Trudov. Kharkovsk Zootech. Inst.* **7**, 83–102. [In Russian.]

—— (1954*b*). [On the leech fauna of the Komi ASSR.] *Izv. Komi. Fil. Vsesouznogo Geograph. Obshch.* **2**, 61–8. [In Russian.]

—— (1955*a*). [Leech fauna of the Amur basin.] *Zool. Zhurn.* **34**(2), 279–85. [In Russian.]

—— (1955*b*). [Materials on the leech fauna of Siberia.] *Tr. Tomsk. Gos. Un.* **131**, 83–96. [In Russian.]

—— (1955*c*). [The leeches of Western Siberia.] *Zametki Po Faune I Flore Siberi*, Vol. 18. pp. 43–9. Tomsk.

—— (1956*a*). [Parasitological observations on leeches, distribution in the Ukraine.] *Tr. II Nauknoi Konferentsii Parazitologov, USSR*, pp. 82–4.

—— (1965*b*). [On the fauna of the White Russian Forest-Preserve and of the northwestern part of White Russia.] *Tr. Kompleksnoi Ekspeditsii Po Izucheniu Vodoemov Polesya, Minsk* 200–4. [In Russian.]

—— (1956*c*). [On the occurrence in the USSR of an interesting species of leeches, *Boreobdella verrucata* (Fr. Müller.] *Zool. Zhurn.* **35**(9), 1417–19. [In Russian; English summary.]

—— (1957*a*). [New data on the structure of the leech fauna of the Komi ASSR and on the importance of worms in the diet of fish.] *Izv. Komi Filiala Vsesouzn. Geogr. Obshch.* **4**, 111–18.

—— (1957*b*). [On the distribution of the medicinal leech in the USSR.] *Zool. Zhurn.* **36**(5), 658–69. [In Russian; English summary.]

—— (1958*a*). [On the classification of leeches.] *Zool. Zhurn.* **37**(11), 1740–1. [In Russian; English summary.]

—— (1958*b*). [On the characteristics of the leech fauna of the streams of the Carpathian Mountains.] *Nauch. Zap. Uzhgorodskogo Gos. Un.* **30**, 63–7. [In Russian.]

—— (1958*c*). [On the leech fauna of the Mongolian People's Republic.] *Izv. Biologo-geogr. Nauchno-issled. Inst. Pri Irkutskom Gos. Un.* **17**(1–4), 271–8. [In Russian.]

—— (compiler) (1958*d*). [Geographical ranges of freshwater leeches in the territories of the SSSR.] [*Problems in terrestrial zoogeography*,] pp. 144–9. L'vov. [In Russian.] [Erroneously reports *Oligobdella orientalis* from the Amur basin.]

—— (1958*e*). [On the fauna of leeches in Latvia.] *Tr. Inst. Biol., Akad. Nauk Latv. SSR Riga.* **5**, 219–23. [In Russian.]

—— (1959*a*). [On leech fauna of the Volga River.] *Uchen. Zap. Kuibishevskogo Gos. Ped. Inst. Im. V.V. Kuibisheva, Estestvoznanie* **22**, 32–8. [In Russian.]

—— (1959*b*). [On the leech fauna of the northern Caucasia Mountains.] *Tr. Zool. Inst. Akad. Nauk SSSR* **26**, 354–9. [In Russian.]

—— (1959*c*). [Zoogeographical features of the freshwater leech fauna of the Ukraine.] *Rep. Acad. Sci. Ukr. SSSR (Dopov. Akad. Nauk Ukrainskoi RSR)* **5**, 554–6. [In Russian; English summary.]

—— (1960*a*). [Elements of the leech fauna of China and Japan in the fauna of the Amur basin within the ranges of the USSR.] *Zool. Zhurn.* **39**(1), 40–4. [In Russian.]

—— (1960*b*). Non-miscibility of the Baikal and common Palearctic leech faunas. *Dokl. Akad. Nauk SSSR* [English translation] *Biol. Sci.* **135**, 987–9. *Dokl. Akad. Nauk SSSR* **135**(2), 489–92. [In Russian.]

—— (1960*c*). [On the Amur transitional district (geographical extensions of leeches.] *Material' Konpherentsii po Voprosam Zoogeographii Sishi 15–21 Avgsta 1960 p. Tez. Dokl. Alma-Ata*, pp. 91–2.

—— (1962*a*). [Fauna of the Ukraine. Leeches.] *Fauna Ukraini. Pijavki. Inst. Zool. Akad. Nauk Ukr. RSR, Kiev* **30**, 1–196. [In Ukrainian.]

—— (1962*b*). [On fauna of leeches of the Khram and Sangor Reservoirs and the Lake Taparavamy (Georgian S. S. R.).] *Byull. Inst. Biol. Vodokhran.* **13**, 19–21. [In Russian.]

—— (1962*c*). [Leeches in the basin of the River Usa and their importance in the feeding of fishes.] In [*Fishes of the basin of the River Usa and their food resources.*] (ed.-in-chief O.S. Zvereva) pp. 225–30. Akad. Nauk SSSR Komi Filia, Leningrad. [In Russian.]

—— (1962*d*). [On the study of the fauna of leeches of the Kuibisher water reservoir.] *Byull. Inst. Biol. Vodokhran.* **12**, 30–1. [In Russian.]

—— (1962*e*). [Characteristics of the leech fauna of the Amur basin.] *Izv. Tikhookean. Nauchno Issled. Inst. Ribn. Khoz. i. Okeanogr., Vladivostok* **48**, 195–202. [In Russian.]

—— (1962*f*). [On the internal ecological differentiation of organisms.] *In Problem' Vnutrividov'ch, Otnosheniy Organizmov. Material' K Soveshchaniu po problemam vnutividov'ch Otnosheniy Organizmov 10–14 Sentyabrya. 1962 g., Tomsk*, pp. 105–8. [In Russian.]

—— (1962*g*). [Ecological changes of organisms.] *Sb. 'Voshros' Ekoligii'. Po Materialam Chetvertoy ekologicheskoy Konpherentsii, Kiev* **4**, 49–53. [In Russian.]

—— (1963*a*) [Comparative analysis of freshwater leeches (Hirudinea) of the Palearctic and other zoogeographical regions.] In *Zoogeograficheskie Sushi* [*Zoogeography of the land.*] pp. 183–5. Tashkent From: *Referat. Zhurn., Biol.* (1964), 3 D54 (transl.).

—— (1963*b*). [The leech fauna of the Irkutsk reservoir in relation to the problem of the immiscibility of the Baikal and the common Palearctic fauna.] *Dokl. Akad. Nauk SSSR* **151**(5), 1225–7. [In Russian; English transl. *Biol. Sci.* **151**, 1090–1.]

—— (1963*c*). [Leeches.] *Radyans'ka Entsiklopediya, Kiev.* **12**, 43. [In Russian.]

—— (1964). [Leeches (Hirudinea) in the Oka River based on collections made in 1959.] *Tr. Zool. Inst. Akad. Nauk SSSR* **32**, 123–6. [In Russian; from *Referat. Zhurn., Biol.* (1964), 8 D119 (tranl.).]

—— (1966*a*). [New data on the distribution of leeches in the Pechora River basin.] In *A hydrobiological study and reclamation for fisheries of the lakes of the Far North of the USSR.*] pp. 71–5. Nauka, Moscow.

—— (1966*b*). Die endemischen Egel des Baikalsees und das Problem der Entstehung der Baikalfauna. *Verh. Int. Vereir. Theor. Angew. Limnol., Warszawa* **16**(3), 347–50.

—— (1968*a*). [Evolution of Acanthobdellidae-parasites of primitive bony fishes.] (Abstr.) [*All Union Meeting (Vth) on the diseases and parasites of fish and aquatic invertebrates, 29 Nov.–4 Dec, 1968.*] *Leningrad*, 70–1. Nauka, Izdat.

—— (1968*b*). [Reconciliation of some structural questions common to animal systems.] *Problemi Evolutsii*, I. *Novosibirsk* 71–81. [In Russian.]

—— (1968*c*). Hirudinea. In *Izdatel'stvo Prosveshchenia Moskva* (ed. L.A. Zenkevich) Vol. 1, pp. 509–26. [In Russian.]

—— (1969). [Peculiarities of the palaearctic leech fauna presented in parasitological terms.] *Sb. Problemi Parazitologii*, II, *Kiev*, pp. 253–5. [In Russian.]

—— (1970*a*). [On the phylogeny of primitive leeches.] *Sb. Vopros' Evolutsionnoy Morphologii i Biotsenolii, Kazan'*, pp. 61–7. [In Russian.]

—— (1970*b*). [Ecology of palaearctic leeches.] *Sb. Biologicheskie Protsess' v Morskih i Konginental'n'ch Vodoemah, Kishinev*, pp. 224–5. [In Russian.]

—— 1970*c*). [Zoogeographical comparison of animals which inhabit continental reservoirs.] *Sb. Voprosi P'bohozyaystvennogo Osvoeiya i Sanitarno-biologicheskogo. Rejima Vodoemov Ukrain'*, I, *Kiev*, pp. 28–30. [In Russian.]

—— (1973). [Results of a study and problems for further studies of freshwater leeches in Siberia.] In *Vodoemy Sibiri i perspektivy ikh rybokhozyaistvennogo ispol'zovaniya. Tomsk, USSR; Izdatel'stvo Tomskogo Universiteta*, pp. 251–4. [In Russian.]

—— (1976). [Leeches.] *Fauna USSR*, Vol. 1. Academy of Science of the USSR. [In Russian.] [Excellent review of leeches of Soviet Union, including extensive Russian bibliography.[[See: Lyakhov (1977).]

—— and Epshtein, V.M. (1959). [Baikal leeches.] [The 10th conference on the problems of parasitology and diseases with natural reservoirs,] Vol. 2, pp. 189–90, Leningrad. [In Russian.]

—— —— (1960*a*). [New data on the leech fauna of fresh water of the Crimea.] *Zool. Zhurn.* **39**(9), 1429–32. [In Russian; English summary.]

—— —— (1960*b*). [Endemic Baikalian leeches from the family Glossiphoniidae.] *Dokl. Akad. Nauk SSSR* **131**(2), 457–60. [English transl. available.]

—— —— (1960*c*). Leeches of the subfamily Toricinae subfam. n. and their geographical distribution. *Dokl. Akad. Nauk SSSR Biol. Sci.* **131**(5), 478–81. [English version **131**, 812–14.]

—— —— (1964). [On the geographical distribution of two southern

Palaearctic species of leeches: *Batracobdella algira* (Moq-Tand.) and *Herpbodella stschegolewi* Lukin and Epshtein.] *Zool. Zhurn.* **43**(4), 607–9. [In Russian; English summary.]

Lumsden, W.H.R., and Evans, D.A. eds (1979). *Biology of the Kinetoplastida.* Vol. 2. Academic Press, London.

Lux, H.D., and Müller-Mohnssen, H. (1965). Stationare Strom-Spannungs-kennlinien von Blutegel-Ganglien-Zellen. *Pflüger Arch. ges. Physiol. Mens. Tiere* **285**(4), 287–95.

Lyakhov, S.M. (1977). Review: *Leeches of fresh and brackish water bodies* by E.I. Lukin, Vol. 1 of *Fauna of the USSR. Zool. Zhurnal.* **56**(8), 1263–4.

Lynch, D.L., Fiordelisi, R., and Feyerherm, H.A. (1968). Characteristics of leech blood protein (*Haemopis marmoratis*, Say). *Trans. Ill. State Acad. Sci.* **61**(3), 310–12.

Lyskova, V.N. (1966). [The feeding and the feeding relationships of fishes acclimatized in Lake Gusinoe (Selenga River system).] In *Biologicheskie osnovy rybnogo klozyaistva na vodoemakh Srednei Azii i Kazakhstana,* pp. 203–5. Nauk, Alma-Ata. [In Russian.] [Sculpins and leeches play a major role in the diet of sheatfish in summer.]

Lyubarskaya, O.D. (1970). [Seasonal dynamics of the parasite fauna of *Abramis brama* in the Kuibyshev reservoir, at the inlet of the Volga.] In *Voprosy evolyutsionnoi morfologii i biogeografii,* pp. 40–9. Izdatel'stvo Kazanskogo Universiteta. [In Russian.]

McAdoo, D.J. (1977). The Retzius cell of the leech *Hirudo medicinalis.* In *Biochemistry of characterised neurones* (ed. N.N. Osborne) pp. 19–45. Pergamon, Oxford.

—— and Coggeshall, R.E. (1976). Gas chromatographic-mass spectrometric analysis of biogenic amines in identified neurones and tissues of *Hirudo medicinalis. J. Neurochem.* **26**(1), 163–7. [1976BA:48940.]

Macagno, E.R. (1978). Mapping synaptic sites between identified neurons in the leech CNS by means of 3-D computer reconstructions from serial sections. *Brain Theor. Newsl.* **3**, 186–9.

—— (1980). Number and distribution of neurons in leech segmental ganglia. *J. comp. Neurol.* **190**(2), 283–302.

—— Stewart, R.R., and Zipser, B. (1983). The expression of antigens by embryonic neurons and glia in segmental ganglia of the leech *Haemopis marmorata. J. Neurosci.* **3**(9), 1746–59.

—— Muller, K.J., Kristan, W.B., DeRiemer, S.A., Stewart, R., and Granzow, B. (1981). Mapping of neuronal contacts with intracellular injection of horseradish peroxidase and Lucifer Yellow in combination. *Brain Res.* **217**, 143–9.

Macan, T.T. (1977). A twenty-year study of the fauna in the vegetation of a moorland fishpond. *Arch. Hydrobiol.* **81**(1), 1–24. [1978BA:1438.]

—— and Kitching, A. (1976). The colonization of squares of plastic suspended in midwater. *Freshwater Biol.* **6**(1), 33–40. [*Erpobdella.*] [1976BA:7425.]

—— and Maudsley, R. (1969). Fauna of the stoney substratum in lakes in the English Lake District. *Verh. Int. Ver. Limnol.* **17**, 173–80.

McAnnally, R.D., and Moore, D.V. (1965). Relationship of *Helobdella* sp. to laboratory-reared *Australorbis glabratus*. *J. Parasitol.* **51**(2), sec. 2, 32.

—— —— (1966). Predation by the leech *Helobdella punctatolineata* upon *Australorbis glabratus* under laboratory conditions. *J. Parasitol.* **52**(1), 196–7.

MacCallum, W.G., and MacCallum, G.A. (1918). On the anatomy of *Ozobranchus branchiatus* (Menzies). *Bull. Am. Mus. nat. Hist.* **38**(2), 295–408.

McCaman, M.W., Weinreich, D., and McCaman, R.E. (1973). The determination of picomole levels of 5-hydroxytryptamine and dopamine in *Aplysia, Tritonia* and leech nervous tissues. *Brain Res.* **53**(1), 129–37.

McCarthy, P.H. (1962). Infestation of the bovine udder and teat-canal by leeches. *Aust. Vet. J.* **38**, 355–6.

McCarthy, T.K. (1975). Observations on the distribution of the freshwater leeches (Hirudinea) of Ireland. *Proc. R. Ir. Acad. Sci.* **75** (Section B) (21), 401–51.

McClung, R.P. (1974). Leeches of the St. Mary's River Watershed and leech records of the Nova Scotia Museum. MS. Nova Scotia Museum (Halivax), Curatorial Report No. 11, 11 pp.

MacCulloch, R.D. (1981). Leech parasitism on the western painted turtle *Chrysemys picta belli* in Saskatchewan, Canada. *J. Parasitol.* **67**(1), 128–9.

MacCurdy, E. (transl.) (1938). *Leonardo da Vinci. The notebooks,* Vol. 1. Folio 67. Reynal and Hitchcock, New York. [Illustrates swimming leech.]

McDonald, M.E. (1969). Catalogue of helminths of waterfowl (Anatidae). US Bureau of Sport Fish. and Wildlife, Special Scientific Report — Wildlife, Number 126. Washington, DC [Hirudinea: 689–92.] [*Placobdella ornata* feeding on various aquatic birds.]

Mace, T.F., and Davis, C.C. (1972). Energetics of a host–parasite relationship as illustrated by the leech *Malmiana nuda*, and the shorthorn sculpin *Myoxocephalus scorpius*. *Oikos* **23**(3), 331–43. [Russian summary.]

MacGinitie, G.E. (1955). Distribution and ecology of the marine invertebrates of Point Barrow, Alaska. *Smithsonian Misc. Coll.* **128**(9), 1–201. [Hirudinea: 144.]

McGregor, E.A. (1963). Publication of fish parasites and diseases 330 B.C.– A.D. 1923. *US Fish Wildlife Serv. Spec. Sci. Rept. Fish.* **474**, 1–84.

McIntosh, F.C., and Perry, W.L.M. (1950). Biological estimation of acetylcholine. *Meth. med. Res.* **3**, 78–92.

McKay, R.D. *et al.* (1983). Surface molecules identify groups of growing axons. *Science* **222**(4625), 788–94.

Mackay-Dick, J. (1970). Leech out of reach. *Lancet* **i**, 192.

McKee, P.M., and Mackie, G.L. (1979). Incidence of *Marvinmeyeria lucida* (Hirudinea: Glossiphoniidae) in the fingernail clam, *Sphaerium occidentale. Can. J. Zool.* **57**(3), 499–503. [French summary.] [1979BA:28981.]

McKinney, F., and Derrickson, S.R. (1979). Aerial scratching, leeches and nasal saddles in green-winged teal. *Wildfowl* **30**, 151–3.

MacLeod, K.I.E. (1950). Leech in the nasopharynx. *Br. med. J.* **ii**, 1058.

MacPhee, F.M. (1971*a*). The distribution of leeches (Hirudinea) in Scotland, with special reference to the lower valley of the Aberdeenshire Dee. *Glasgow Nat.* **18**, 401–6.

—— (1971*b*). The distribution of leeches (Hirudinea) in the River Clyde, Lanarkshire, with notes on their ecology. *Glasgow Nat.* **18**, 535–8.

Madanmohanrao, G. (1960). Salinity tolerance and oxygen consumption of the cattle leech, *Hirudinaria granulosa. Proc. Indian Acad. Sci.* **11**(5), 211–18.

Madill, J. (1983). The preparation of leech specimens: relaxation, the key to preservation. In: Proceedings of 1981 Workshop on Care and Maintenance of Natural History Collections (ed. D.J. Faber). *Syllogeus* No. 44, pp. 37–41.

Madsen, B.L. (1963). [Ecological investigations on some streams in East Jutland (Denmark). 2. Planarians and leeches.] *Flora Fauna* **69**(4), 113–25. [In Danish; English summary.]

Magazanik, L.G. (1976). Functional properties of postjunctional membrane. *Ann. Rev. Pharmacol.* **16**, 161–75.

—— and Potapieva, N.N. (1975). [Blocking action of snake venom polypeptides on cholinergic mechanisms of the leech dorsal muscle.] *Bull. Eksp. Biol. Med.* **79**(1), 38–40. [In Russian; transl.]

Magnetti, P.C., Dotti, M., Di Chiara, A.P., Smedile, E., and Tibaldi, E. (1971). [Investigation into the periphytic populations of an area alongside the river Po.] *Atti. Accad. Naz. Lincei Rc. (Sci. Fiz. Mat. Nat.)* **51**(5), 414–21. [In Italian.]

Magni, F., and Pellegrino, M. (1975). Nerve cord shortening induced by activation of the fast conducting system in the leech. *Brain Res.* **90**, 169–74.

—— —— (1978*a*). Patterns of activity and the effects of activation of the fast conducting system on the behaviour of unrestrained leeches. *J. exp. Biol.* **76**, 123–35. [1979BA:28654.]

—— —— (1978*b*). Neural mechanism underlying the segmental and generalized cord shortening reflexes in the leech. *J. comp. Physiol.* **124**(4), 339–51.

Magnus, W. (1928). Der Blutegelbiss und seine Wirkung auf menschliche Gefässe. *Zentralbl. Chir.* **55**, 2355–7.

Magnusson, S. (1971). (Hirudin possibly involving multiple binding sites). In: *The enzymes*. (ed. P.D. Bayer) 3rd edn. Vol. 3, pp. 277–321.

—— Sottrup-Jensen, L., Claeys, H., Zajdel, M., and Petersen, T.E. (1975). Proceedings: complete primary structure of prothrombin. Partial primary structures of plasminogen and hirudin. *Thromb. Diath. Haemorrh.* **34**(2), 562–3.

Maier, B.L. (1892). Beiträge zur Kenntniss des Hirudineen-Augen. *Zool. Jb. Anat.* **5**, 552–80.

Maitland, P.S. (1963). *Hemiclepsis marginata* and *Batracobdella paludosa* in Stirlingshire with notes on the ecology and morphology of the latter species. *Glasg. Nat.* **18**(5), 219–27.

—— (1966*a*). New vice-county records of leeches (Hirudinea) in Scotland, with special reference to the lower valley of the Aberdeenshire Dee. *Glasg. Nat.* **18**(8), 401–6. [*Hirudo medicinalis.*]

—— (1966*b*). The fauna of the River Endrick. *Glasg. Un. Publ. Stud. Loch Lomond* **2**, 1–194.

—— (1973). The freshwater and terrestrial fauna of the Clyde Sea Area. II. Freshwater leeches (Hirudinea). *Glasg. Nat.* **19**, 39–43.

—— and Hudspith, P.M.G. (1974). The zoobenthos of Loch Leven, Kinross, and estimates of its production in the sandy littoral area during 1970 and 1971. *R. Soc. Edinb.* **74**, 220–39. [Leeches: 222.]

—— and Kellock, E. (1971). The freshwater leeches (Hirudinea) of Orkney. *Glasg. Nat.* **18**(10), 558–64.

Maitland, R.F. (1897). *Prodrome de la Faune des Pays Bas et de la Belgique flamande*, 44 p. Leiden.

Major, E., and Fuhrmann, O. (1913). Voyage d'exploration scientifique en Colombie, Vol. V In: *Mémories de la Société neuchateloises des sciences naturelles, Attinger Freres.* (Incomplete reference)

Makhotin, Yu.M. (1964). [The food of burbot in Kuibyshev Reservoir.] *Tr. Talarskogo Otd. Gos. Nauch-Issled. Inst. Ozern Rechn Rybn Khoz.* **10**, 163–5. [Leeches were observed as food for burbot.]

Malecha, J. (1965). Culture organotypique d'Hirudinées. *C.r. Séanc. Soc. Biol., Paris* **159**(8/9), 1674–5.

—— (1967). Étude en culture organotypique de l'influence endocrine de la masse nerveuse peripharyngienne sur la maturation testiculaire chez *Hirudo medicinalis* L. *C.r. hebd. Séanc. Acad. Sci. Ser D Sci.Natur., Paris* **265**(23), 1806–8.

—— (1970*a*). Étude, en culture organotypique, du contrôle hormonal de la spermatogenèse chez *Hirudo medicinalis* (Hirudinée: Gnathobdelliforme). *Gen. comp. Endocr.* **14**(2), 313–20.

—— (1970*b*). Influence de la température sur la spermatogenèse et l'activité neurosécrétrice d'*Hirudo medicinalis* L. (Hirudinée: Gnathobdelliforme). *Gen. comp. Endocr.* **14**(2), 368–80.

—— (1970*c*). Étude expérimentale de la maturation sexuelle chez les Hirudinées. *Bull. Soc. zool. Fr.* **95**, 517–28.

—— (1975). Étude ultrastructurale de la spermiogenèse de *Piscicola geometra* L. (Hirudinée, Rhynchobdelle). *J. ultrastr. Res.* **51**(2), 188–203. [English summary.] [1975BA:43712.]

—— (1976*a*). Influence de la decérébration et de l'ovariectomie sur l'évolution des glandes clitelliennes de *Piscicola geometra* L. (Annélide, Hirudinée, Rhynchobdelle). *C.r. hebd Séanc. Acad. Sci. Paris*, **D283**, 655–7. [1977BA:30669.]

—— (1976*b*). Film: "Biologie des Sangues". Réalisation: M. Guillon. distribution: Service du film de Recherche Scientifique (16mm, 25min). English version available.

—— (1979*a*). Contribution à l'étude de la biologie de l'Hirudinée Rhynchobdelle, *Piscicola geometra* (L.). Thèse. Université des Sciences et Techniques de Lille, no. 453. 160 pp., 242 figs.

—— (1979*b*). Mise en évidence d'une action du système nerveux central sur les échanges d'eau chez l'Hirudinée Rhynchobdelle *Theromyzon tessulatum* (O.F.M.). *C.r. hebd Séanc. Sci. Paris* **288**(7), 693–9. [1980BA:4560.]

—— (1980). Controle de la maturation sexuelle chez l'Hirudinée Rhynchobdelle *Piscicola geometra* L. *Bull. Soc. zool. Fr.* **105**(1), 133–40.

—— (1983). L'osmorégulation chez l'Hirudinée Rhynchobdelle *Theromyzon tessulatum* (O.F.M.). Localisation expérimentale de la zone sécrétice d'un facteur de régulation de la balance hydrique. *Gen. Comp. Endocrinol.* **49**, 344–51.

—— (1984*a*). Cycle biologique de l'hirudinée rhynchobdelle *Piscicola geometra* L. *Hydrobiologia* **118**, 237–43.

—— (1984*b*). Influence des facteurs externes sur l'activité reproductrice de l'Hirudinée rhynchobdelle *Piscicola geometra* L. *Hydrobiologia* **118**, 245–54.

—— and Prensier, G. (1974). Les glandes clitelliennes de *Piscicola geometra* L. : structure et cycle annuel. *Bull. Soc. zool. Fr.* **99**(3), 433–40. [English Summary.] [1975BA:8772.]

—— and Vinckier, D. (1983). Formation du cocoon chez l'hirudinée Rhynchobdelle *Piscicola geometra* L. *Arch. Biol.* **94**(2), 183–205.

Malek, E.A. (1958). Factors conditioning the habitat of Bilharziasis intermediate hosts of the family Planorbidae. *Bull. Wld Hlth Org.* **18**, 785–818. [Pages 809–10, predatory feeding habits of *Helobdella*.]

Malinconico, S.M., Katz, J.B., and Budzynski, A.Z. (1983). Unique fibrinogen cleavage by a protease from the leech *Haementeria ghilianii*. IX Inter. Congress on Thrombosis and Haemostasis **50**, 354.

—— —— —— (1984). Fibrinogen degradation by hementin, a fibrinogenolytic anticoagulant from the salivary glands of the leech *Haementeria ghilianii. J. Lab. Clin. Med.* **104**(5), 842–54.

Malm, C.W. (1863). Svenska Iglar, Disciferae. *Götheborg. Vetensk. n. Handl.* **8**, 153–263.

—— (1865). Ichthyologiska Bidrag till Skandinaviens Fauna. *Forh. Skand. Naturf.*, 9te Möte, 404–14. [*Platybdella:* 413–14.]

Maloney, S.D. (1974). Leeches of the Stone River drainage area in middle Tennessee, USA. *J. Tenn. Acad. Sci.* **49**(2), 57.

—— and Chandler, C.M. (1976). Leeches (Hirudinea) in the upper Stones River drainage of Middle Tennessee. *Am. Midl. Nat.* **95**(1), 42–8.

Mancino, G., and Puccinelli, I. (1964). Corredo cromosomico a meiosi di *Pontobdella muricata* (L.). (Hirudinea, Piscicolidae). *Boll. Zool.* **31**(2, Pt.2), 1311–21.

Mandal, B.K., and Moitra, S.K. (1975). Studies on the bottom fauna of a fresh water fish pond at Burdwan, India. *J. Inl. Fish Soc. India* **7**, 43–8.

Mané-Garzón, F. (1973). Un nuevo tipo de Hirudinea *Colombobdella ringueletin.* gen., n. sp. parasito, de una tortuga de Colombia. *Trabajos V Congreso Latinoamericano de Zoologia.* Montevideo, December 1971, Vol. 1, pp. 129–37. [In Spanish.]

—— and Montero, R. (1977). *Myzobdella uruguayensis* n.sp. (Hirudinea Piscicolidae) parasita de las branquias del bagre amarillo *Rhambdia sapo* (Vall.). *Rev. Biol. Uruguay* **5**(2), 59–65. [English summary.]

Mann, K.H. (1949). Wildfowl and leeches. *Scott. Nat.* **61**, 192.

—— (1951). On the bionomics and distribution of *Theromyzon tessulatum* (O.F. Müller, 1774) (= *Protoclepsis tesselata*). *Ann. Mag. nat. Hist.* **4**(12), 956–61.

—— (1952). A revision of the British leeches of the family Erpobdellidae with an account of *Dina lineata* (O.F. Müller, 1774), a leech new to the British fauna. *Proc. zool. Soc. Lond.* **122**, 395–405.

—— (1953a). A revision of the British leeches of the family Glossiphoniidae, with a description of *Batracobdella paludosa* (Carena, 1824), a leech new to the British fauna. *Proc. zool. Soc. Lond.* **123**, 377–91.

—— (1953b). The life history of *Erpobdella octoculata* (L.). *J. anim. Ecol.* **22**, 197–207.

—— (1953c). The segmentation of leeches. *Biol. Rev.* **28**(1), 1–15.

—— (1954a). The anatomy of the horse leech, *Haemopis sanguisuga* (L.) with particular reference to the excretory system. *Proc. zool. Soc. Lond.* **124**, 69–88.

—— (1954b). A key to the British freshwater leeches with notes on their ecology. *Sci Publ. freshwat. Biol. Ass. Br. Emp.* **14**, 1–21.

—— (1955a). The ecology of the British freshwater leeches. *J. anim. Ecol.* **24**(1), 98–119.

—— (1955b). Some factors influencing the distribution of freshwater leeches in Britain. *Verh. int. Ver. Limnol.* **12**, 582–7.

—— (1956). A study of the oxygen consumption of five species of leech. *J. exp. Biol.* **33**, 615–26.

—— (1957*a*). The breeding, growth and age structure of a population of the leech *Helobdella stagnalis* (L.). *J. anim. Ecol.* **26**, 171–7.

—— (1957*b*). A study of a population of the leech *Glossiphonia complanata* (L.). *J. anim. Ecol.* **26**, 99–111.

—— (1958). Seasonal variation in the respiratory acclimatisation of the leech *Erpobdella testacea* (Sav.). *J. exp. Biol.* **35**(2), 314–23.

—— (1959). On *Trocheta bykowskii* Gedroyć, 1913, a leech new to the British fauna, with notes on the taxonomy and ecology of other Erpobdellidae. *Proc. zool. Soc. Lond.* **132**, 369–79.

—— (1961*a*). *Leeches* (*Hirudinea*): *their structure, physiology, ecology and embryology*. Pergamon, New York. [Oxford edition published 1962.]

—— (1961*b*). The life history of the leech *Erpobdella testacea* (Sav.) and its adaptive significance. *Oikos* **12**(1), 164–9.

—— (1961*c*). The oxygen requirements of leeches considered in relation to their habitats. *Verh. int. Verein. Limnol.* **15**, 1009–13.

—— (1964). *A key to the British freshwater leeches with notes on their ecology*, 2nd edn. Sci. Publ. Freshwater Biol. Ass. No. 14, pp. 1–50.

—— (1967). Hirudinea. In *Limnofauna Europaea* (ed. Illes) pp. 118–20. Gustav Fischer Verlag, Stuttgart.

—— (1971). In: *A manual on methods for the assessment of secondary productivity in fresh water* (IBP Handbook No. 17). (eds. W.T. Edmondson, and G.G. Winberg). Blackwell Scientific, Oxford.

—— and Tyler, M.J. (1967). Leeches as endoparasites of frogs. *Nature, Lond.* **197**, 1224–5.

Mannsfeld, W. (1928, 1934). See Autrum (1939*c*).

Manoleli, D. (1972*a*). Contributions concernant la systématique et la distribution du genre *Erpobdella* de Blainville 1818 (Hirudinoidea, Erpobdellidae) en Roumania. *Trav. Mus. Hist. nat. 'Gr. Antipa'* **12**, 39–44. [English, Romanian, and Russian summaries.]

—— (1972*b*). A new species of leech *Limnatis bacescui* sp. nov. (Hirudinoidea: Hirudinidae). *Rev. Roum. Biol. Ser. Zool.* **17**(4), 237–9. [1974BA:31816.]

—— (1974). Contributions to the knowledge of the Hirudinea from the eastern Romanian Plain and from the Mounts of Vrancea. *Trav. Mus. Hist. nat. 'Gr. Antipa'* **14**, 79–83.

—— (1976). [Ecological and biological observations on the species *Erpobdella monostriata* (Godroyć, 1916) Pawlowski, 1948 (Hirudinea, Erpobdellidae).] *Trav. Mus. Hist. nat. 'Gr. Antipa'* **17**, 1–8. [In French; English, Romanian, and Russian summaries.] [1977BA:39053.]

Manson-Barh, (1960). [Leeches in nose.] In: *Manson's tropical diseases*, 15th edn, p. 840.

Mansouri, M.D. *et al.* (1966). [The leech as a live foreign body.] *Bratisl. Lek. Listy* **46**, 495–7. [In Czechoslovak.]

Manton, S.M. (1964). Mandibular mechanisms and the evolution of the arthropods. *Phil. Trans. R. Soc.* **B247**, 1–183.

—— (1972). The evolution of arthropod locomotory mechanisms, Part 10. Locomotory habits, morphology and evolution of the hexapod classes. *J. Linn. Soc. Zool.* **51**, 203–400.

—— (1977). *The Arthropoda*. Oxford University Press, Oxford, 527 pp.

Manumante, M.M. See Hanumante, M.M.

Maranto, A. (1982). Neuronal mapping: A photo-oxidation reaction makes Lucifer yellow useful for electron microscopy. *Science* **127** (4563), 953–5. [1983BA:49832.]

Marchbanks, R.M. (1967). Compartmentation of acetylcholine in synaptosomes. *Biochem. Pharmacol.* **16**(5), 921–3.

Marcinkowski, T. *et al.*(1973). [Leeches in the digestive tract of a drowned Man.] *Wiad. Lek.* **26**, 655–7. [In Polish; English summary.]

Marcus, H. (1913). See Autrum (1939*c*).

Margalef, R. (1950). Datos para la hidrobiologia de la cordillera cantabrica, especialmente del macizo de los Picos de Europe. *Publ. Inst. Biol. apl., Barcelona* **7**, 37–76. [English summary.]

—— (1952). Màteriales para la hidrobiologia de la isla de Menorca. *Publ. Inst. Biol. apl., Barcelona* **11**, 5–112. [English summary.]

—— (1955). Comunidades bioticas de las aguas dulces del noroeste de España. *Publnes Inst. Biol. Apl., Barcelona* **21**, 5–85. [*Hirudo* in Spain.]

Margolis, L., and Arthur, J.R. (1979). Synopsis of parasites of fishes of Canada. *Fish. Res. Bd. Can. Bull.* **199**, 1–269. [1980BA:11477.] [21 leech taxa.]

Marine Biological Association (1957). *Plymouth marine fauna*, 3rd edn. Plymouth. [Hirudinea: 149.]

Markov, G.S., Reshetnikova, A.V., and Ivanov, V.P. (1965). [Factors affecting the degree of infection of the Russian sturgeon by Hirudinea.] *Mater. Nauch. Konph. Vsesouzn. Obshh. Gel'mintologov, Moscow* **3**, 162–6. [In Russian.]

—— —— and Trusov, V.Z. (1965). [New data on the biology and parasitology of the Volga-Caspian herd of giant sturgeon, *Huso huso* L.] In *Parazity i parazitozy cheloveka i zhivotnykl*, pp. 208–19. Nauk Dumka, Kiev. [In Russian.]

Markwardt, F. (1955). Untersuchungen über Hirudin. *Naturwissenschaften* **42**, 537–8.

—— (1956*a*). Untersuchungen über Hirudin. *Arch. exp. Path. Pharmakol.* **228**, 220–1.

—— (1956*b*). Untersuchungen über den Mechanismus der blutgerinnungschemmenden Wirkung des Hirudins. *Arch. exp. Path. Pharmakol.* **229**, 389–99.

—— (1956*c*). Die antagonistische Wirkung des Hirudins gegen Thrombin *in vivo*. *Naturwissenschaften* **43**, 111.

—— (1957*a*). Die Bestimmung des Thrombins durch Titration mit Hirudin. *Arch. Pharmaz.* **290**(6), 280–5.

—— (1957*b*). Die Isolierung und chemische Charakterisierung des Hirudins. *Z. Physiol. Chem.* **308**, 147–56.

—— (1958*a*). Hirudin, der blutgerinnungshemmende Wirkstoff des medizinischen Blutegels. *Blut* **4**, 161–70.

—— (1958*b*). Die quantitative Bestimmung des Prothrombins durch Titration mit Hirudin. *Arch. exp. Path. Pharmakol.* **232**, 487–98.

—— (1959). Der Hirudintoleranztest. *Klin. Wschr.* **37**, 1142–3.

—— (1963). *Blutgerinnungsehemmende Wirkstoffe aus blutsaugenden Tieren.* Fischer, Jena.

—— (1970). Hirudin as an inhibitor of Thrombin. In *Proteolytic enzymes* (ed. G.E. Perlmann and L. Lorand). *Meth. Enzymol.* **19**, 924–32.

—— and Walsmann, P. (1958). Die Reaktion zwischen Hirudin und Thrombin. *Z. Physiol. Chem.* **312**, 85–98.

—— —— (1967). Reindarstellung und Analyse des Thrombininhibitor Hirudin. *Z. Physiol. Chem.* **348**, 1381–6.

—— and Wenzel, G. (1961). Die Anwendung der Prothrombingestimmung mit Hirudin zur Überwachung der Behandlung mit Dicumarolderivaten. *Blut* **7**, 37–42.

—— Nowak, G., and Hoffmann, J. (1977). The influence of drugs on disseminated intravascular coagulation (DIC). II. Effects of naturally occurring and synthetic thrombin inhibitors. *Thromb. Res.* **11**, 275–83.

—— Schäfer, G., Töpper, H., and Walsmann, P. (1967). Die Isolierung des Hirudins aus medizinischen Blutegeln. *Pharmazie* **22**(5), 239–41.

—— Hauptmann, J., Nowak, G., Klessen, C., and Walsmann, P. (1982). Pharmacological studies on the antithrombotic action of hirudin in experimental animals. *Thromb. Haemostasis* **47**(3), 226–9.

—— *et al.* (1977). The influence of drugs on disseminated intravascular coagulation (DIC). II. Effects of naturally occurring and synthetic thrombin inhibitors. *Thromb. Res.* **11**(3), 275–83.

Marliave, J.B., and Elderton, V.J. (1979). Cleaning symbioses for temperate marine fishes in display aquaria. *Drum Croaker* **19**(2), 1–8.

Marsden, C.A., and Kerkut, G.A. (1969). Fluorescence microscopy of the 5-HT and catecholamine containing cells in the central nervous system of the leech *Hirudo medicinalis. Comp. Biochem. Physiol.* **31**, 851–62.

—— —— (1970). Quantitative studies on the neurons of the leech, *Hirudo medicinalis. Comp. gen. Pharmacol.* **1**(3), 293–8.

Marsden, P.D., and Pettitt, L.E. (1969). The survival of *Trypanosoma cruzi* in the medicinal leech (*Hirudo medicinalis.*) (Correspondence.) *Trans. R. Soc. trop. Med. Hyg.* **63**(3). 413–14.

Marshall, B.E. (1972). Some effects of organic pollution of benthic fauna. *Rhodesia Sci. News.* **6**(5), 142–5.

Marshall, C.G. (1983). Regenerative electrical activity in the salivary gland cells of leeches. *Am. Zool.* **23**, 902.

—— and Lent, C.M. (1983). Functional properties of salivary gland cells in the leech. East Coast Nerve Net. Ninth Annual Meeting. M.B.L., Wood's Hole, Mass., 22–24 April 1983.

—— —— (1984). Calcium dependent action potentials in leech giant salivary cells. *J. exp. Biol.* **113**, 367–80.

Marshall and Gilbert (Incomplete reference) (1904). Notes on the food and parasites of some freshwater fishes from the lakes at Madison, Wisconsin. In *Appendix to the Report of the Commissioner of Fisheries,* pp. 513–32.

Marshall, E.K. (1915–1916). The toxicity of certain hirudin preparations. *J. Pharmacol. exp. Ther.* **7**, 517.

Martin, D.R. (1972). Distribution of helminth parasites in turtles native to southern Illinois. *Trans. Ill. St. Acad. Sci.* **65**(3/4), 61–7. [1974BA:56261.]

Marx, M. (1979). Qualitative structure of the phytophilous fauna in Lake Marica, Dolj County, Romania. *Stud. Cercet Biol.* **31**(2), 101–4. [In Romanian; English summary.] [1980BA:31133.] [*Erpobdella nigricollis,* new to Romania.]

—— Paun, E., Sorescu, C., and Dragu, C. (1967). Contributü la studiul Hidrobiologie al Lacului Victoria (Reg. Oltenia). *Hidrobiologia, Bucurest* **8**, 99–119. [In Romanian; English summary.] [*Erpobdella octoculata* from Lake Victoria, Romania.]

Marx, R., and Martin du Theil, A.-M. (1964). Quelques cas curieux et inhabituels de parasitisme. *Bull. Soc. Path. exot.* **57**(5), 955–7. [English summary.]

Marzinowsky, E. (1927). Du développement de l'*Haemogregarina stepanovi. Ann. Parasitol.* **5**(2), 140–2.

Mashanskii, V.F., Bazanova, I.S., Kazanskii, V.V., and Merkulova, O.S. (1982). [Functional rearrangements of ultrastructure in the giant (Retzius) neuron of the medical leech (*Hirudo medicinalis*) and a possible role of calcium in these processes.] *Arkh. Anat. Gistol. Embriol.* **82**(3), 36–42. [In Russian; English summary.] [1983BA:26674.]

Mason, A.J.R. (1980). Morphology and physiology of leech Retzius cells. Ph.D. thesis, Portsmouth Polytechnic, England.

—— and Kristan Jr., W.B. (1982). Neuronal excitation, inhibition and modulation of leech longitudinal muscle. *J. comp. Physiol.* **146A**, 527–36.

—— and Leake, L.D. (1978). Morphology of leech Retzius cells demonstrated by intracellular injection of horseradish peroxidase. *Comp. Biochem. Physiol. A Comp. Physiol.* **61**(2), 213–16. [1979BA:35392.]

—— and Muller, K.J. (1982). Axon segments sprout at both ends: tracking growth with fluorescent D-peptides. *Nature* **296**, 655–7.

—— —— (1983). Regeneration and plasticity of neuronal connections in the leech. *Trends in Neurosciences* **6**(5), 172–6.

Sunderland, A.J., (1979). Effects of leech Retzius cells on body wall muscles. *Comp. Biochem. Physiol.* **63C**, 359–61.

Mason, J. (1974). Studies on the freshwater and terrestrial leeches of New Zealand. 1. Family Glossiphoniidae. *Jl R. Soc. NZ* **4**(3), 327–43.

—— (1976). Studies on the freshwater and terrestrial leeches of New Zealand: 2. Orders Gnathobdelliformes and Pharyngobdelliformes. *Jl R. Soc. NZ* **6**(3), 255–76. [1977BA:45198.]

Mastermann, E.W.G. (1908). Hirudinea as human parasites in Palestine. *Parasitology* **1**, 182–5.

Mathers, C.K. (1948). The leeches of the Okobojii region. *Proc. Iowa Acad. Sci.* **55**, 397–425.

—— (1954). *Haemopis kingi,* new species (Annelida, Hirudinea). *Am. Midl. Nat.* **52**(2), 460–8.

—— (1961). A study of North American species of the genus *Haemopis* (Annelida, Hirudinea, Hirudidae) with a description of a new species, *Haemopis latero-maculatum*. Ph.D. dissertation, State University of Iowa. [University Microfilms, Ann Arbor, Mich., Order No. 61–5593. *Diss. Abstr. Int.* **22**(8), 2920.]

—— (1963). *Haemopis latero-maculatum,* new species (Annelida: Hirudinea). *Am. Midl. Nat.* **70**(1), 168–74.

Mathis, B.J., Cummings, T.F., Gower, M., Taylor, M., and King, C. (1977). Dynamics of manganese, cadmium and lead in experimental power plant ponds. Univ. Ill. Urbana-Champaign Water Resour. Cent. Res. Rep. (125), 1–62. [1978BA:17781.] [These metals in leeches.]

—— —— —— —— —— (1979). Dynamics of manganese, cadmium and lead in experimental power plant ponds. *Hydrobiologia* **67**(3), 197–206. [1980BA:6381.] [*Glossiphonia heteroclita.*]

Matthai, G. (1921). Preliminary observations on cocoon-formation by the common Lahore leech, *Limnatis (Poecilobdella) granulosa* (Sav.). *J. Asiat. Soc. Bengal* (new series) **16**, 341–6.

Matthews, R.S. (1954). Land leeches. *J. Bombay nat. Hist. Soc.* **52**, 655–6.

Matysiak, K. (1964). [Leeches (Hirudinea) of the Bialtowieza forest.] *Przegl. Zool.* **8**(2), 154–6. [In Polish.]

—— (1965). Nowe stanowisko *Hirudo medicinalis* L. na Podlasiu wschodnim. *Przegl. Zool.* **9**, 366–8.

—— (1967). [Remarks on the leech *Piscicola geometra* L.). *Przegl. Zool.* **11**(3), 286–8. [In Polish.]

—— (1976*a*). Structure of leech groups (Hirudinea) in polluted parts of the catchment area of the Rivers Bzura and Ner. I. Field investigations. *Acta hydrobiol.* **18**(3), 259–76. [Polish summary.] [1977BA:62212.]

—— (1976*b*). [Contribution to the knowledge of the influence of water pollution of the occurrence of leeches in the Ner River, Poland.] *Przegl. Zool.* **20**(3), 326–8. [In Polish.] [1978BA:32586.]

—— (1977). An attempt to determine the settlement order of leech species (Hirudinea) in open bathing pools in the city of Lodz. *Acta hydrobiol.* **19**(3), 75–82. [1977BA:55738.]

—— (1978). [Structure of leech groups (Hirudinea) in polluted parts of the catchment area of the Bzura and Ner rivers: 2. Experimental-laboratory investigations.] *Acta hydrobiol.* **20**(2), 187–94. [1979BA:22416.]

Mayer, W. (1906). Beiträge zue Kenntnis der Hautsinnesorgane bei Rhynchobdelliden. *Z. wiss. Zool.* **81**, 599–631.

Mbahizireki, G.B. (1980). Observations on some common parasites of *Bagrus docmac* (Pisces: Siluroidea) of Lake Victoria, East Africa. *Hydrobiologia* **75**(3), 273–80.

Meeraus, A. (1929). Der Wasserschlinger von buisia. *Mitt. Hohlen-n. Karstf., Berlin* 124–9.

Mellanby, J. (1909). The coagulation of blood. Part II. The actions of snake venoms, peptone and leech extract. *J. Physiol.* **38**, 441–503.

Mendoza, G., and Herrara, A. (1865). Observation on the leech which is used in this capital. Imprenta de Inclan. Cerca de Sto. Domingo No. 12. Mexico. (from Pamphlets Vol. 4768, Army Medical Library, Washington, D.C.)

Menge, A. (1866). Ueber ein Rhipidopteron und einige andere im Bernstein eingeschlossene Thiere. Schriften der Naturforschenden Gesellschaft in Danzig, neue Folge, **1**, 1–8.

Merkulova, O.S., Bazanova, I.S., and Sergeeva, S.S. (1982). [The membrane potential, impulse activity and oxygen tension at the Retzius neuron surface during its synaptic activation.] *Fiziol. Zh. SSSR IM.I.M. Sechenova* **68**(3), 350–4. [In Russian; English summary.] [1983BA:18791.]

Mertens, R. (1929). *Glossiphonia algira* Moquin-Tandon als Parasit von *Hydromantes genei* Schlegel. *Blätt. Aquarien Terrarienk.* 40 Jahrg., Heft 12,1.

Meshkova, A.M. (1956). [Leeches of Lake Sevan, Armenia.] *Avtoreph. Diss. Baku* 1–13. [In Russian.]

—— (1957). [Leeches of Lake Sevan, Armenia.] *Tr. Sevansk. Gidrobiol. St. Akad. Nauk Armyanskoi SSR* **15**, 48–87. [In Russian.]

—— (1958). K izucheniu piyavok Armenii. *Izv. Akad. Nauk Armyanskoi SSR* **11**(5), 81–6. [In Russian.]

Meštrov, M., Krkač, N., Kerovec, M., Lui, A., Lattinger-Penko, R., Tavčar, V., Žnidarič, D. (1978). Effects of temperature on dominant macrozoobenthos species of the River Sava under laboratory conditions. *Verhandlungen int. Verein. Theor. angew. Limnol.* **20**(3), 1910–14. [Sudden rise of temp. on mortality of *Erpobdella octoculata*.]

Metcalfe, J.L., Fox, M.E., and Carey, J.H.. (1984). Aquatic leeches (Hirudinea) as bioindicators of organic chemical contaminants in freshwater ecosystems. *Chemosphere* **13**, 143–50.

Metuzals, J. (1966). [*Macrobdella* ganglion ultrastucture.] In *Proc. 6th Int.*

Cong. Electron Microscopy, Kyoto (ed. R. Uyeda) Vol. 2, p. 459. Maruzen, Tokyo.

—— (1967*a*). Helical arrangement of the subunits of the neurofibrillar bundles isolated from leech nervous system. *J. cell. Biol.* **34**(2), 690–6.

—— (1967*b*). Fine structural organisation of the neurofibrillar bundles and glia fibrils, isolated from leech nervous system. *Anat. Rec.* **157**, 287–400.

Meuche, A. (1937). Nahrungsuntersuchungen an den Schlundegeln *Herpobdella octoculata* and *H. testacea. Arch. Hydrobiol. Plankt.* **31**, 501–7.

Meyer, F.P. (1968*a*). Dylox as a control for ectoparasites of fish. *Proc. 22nd Ann. Conf. Southeast Assoc. Game and Fish Commissioners*, pp. 1–5. [Control of leech infestation.]

—— (1968*b*). A review of the parasites and diseases of fishes in warm-water ponds in North America. *FAO Fish. Rep.* (1966). **44**(5), 290–318.

—— (1969). A potential control for leeches. *Progressive Fish-Culturist* **31**(3), 160–3. [Reprinted in *The Aquarium* 3(2), 18–19, 52.]

Meyer, G.F. (1955). Vergleichende Untersuchungen mit der supravitalen Methylenblaufarbung am Nervensystem Wirbelloser Tiere. *Zool. Jb.* **74**. (Incomplete reference)

——(1957). Elektronenmikroskopische Untersuchungen an den Apathyschen neurofibrillen von *Hirudo medicinalis* L. *Z. Zellforschung* **45**, 538–42.

Meyer, K., Hobby, G.L., Chaffe, E., and Dawson, M.H. (1940). Relationship between 'spreading factor' and hyaluronidase. *Proc. Soc. exp. Biol. Med.* **44**, 294–6.

Meyer, M. (1904). Beiträge zur Kenntnis der Reparationsprozesse bei Hirudineen. Diss. Leipzig, Halle a. S. (Druck v. E. Karras), 1–33.

Meyer, M.C. (1937*a*). Leeches of southeastern Missouri. *Ohio J. Sci.* **37**, 248–51.

—— (1937*b*). Notes on some leeches from Ontario and Quebec. *Can. Field-Nat.* **51**, 117–119.

—— (1939). Demonstration of a species of marine Piscicolidae from Florida. *J. Parasitol.* **25**(6 supplement), 22.

—— (1940). A revision of the leeches (Piscicolidae) living on freshwater fishes of North America. *Trans. Am. microsc. Soc.* **59**, 354–76.

—— (1941). The rediscovery together with the morphology of the leech, *Branchellion ravenellii* (Girard, 1850). *J. Parasitol.* **27**(4), 289–98.

—— (1946*a*). A new leech *Piscicola salmositica*, n. sp. (Piscicolidae), from Steelhead trout (*Salmo gairdneri gairdneri* Richardson, 1838). *J. Parasitol.* **32**(5), 467–76.

—— (1946*b*). Further notes on the leeches (Piscicolidae) living on freshwater fishes of North America. *Trans. Am. microsc. Soc.* **65**, 237–49.

—— (1949a). Research notes: further note on leech infestation in man. *J. Parasitol.* **35**, 215.

—— (1949b). On the parasitism of the leech, *Piscicola salmositica* Meyer 1946. *J. Parasitol.* **35**, 215.

—— (1951). Hirudinea. *Exploration du Parc national Albert Mission G.F. de Witte* (1933–1935), Brussels. No. 76, pp. 1–29.

—— (1954). The larger animal parasites of freshwater fishes of Maine. *Bull. Maine Dept. Inland Fish Game Res. Mgt. Bull.* **1**, 36–7.

—— (1959). Another unusual case of erratic hirudiniasis. *J. Parasitol.* **45**(4), sect. 2, 39.

—— (1963). Hirudinea. *Encyclopaedia Britannica.* Chicago.

——(1965). Fish leeches (Hirudinea) from tropical West Africa. *Atlantide Rep. Sci. Results Danish Exped. Coast Trop. West Afr.* **8**, 237–45.

—— (1968). Moore on the Hirudinea with emphasis on his type-specimens. *Proc. US Natl. Mus.* **125**(3664), 1–32.

—— (1975). A new leech, *Macrobdella diplotertia* sp. n. (Hirudinea: Hirudinidae), from Missouri. *Proc. helminthol. Soc. Wash.* **42**(2), 83–5.

—— and Bangham, R.V. (1950). Erratic hirudiniasis in a lake trout (*Cristivomer namaycush*). *J. Parasitol.* **36**(6), 2–20.

—— and Barden Jr, A.A. (1955). Leeches symbiotic on Arthropoda, especially decapod Crustacea. *Wasmann J. Biol.* **13**(2), 297–311.

—— and Burreson, E.M. (1983). Redescription of the piscicolid leech *Trulliobdella capitis* Brinkmann. *Proc. helminth. Soc. Wash.* **50**(1), 138–42. [1983BA:3654.]

—— and Khan, R.A. (1979). Taxonomy, biology and occurrence of some marine leeches in Newfoundland waters. *Proc. helminthol. Soc. Wash.* **46**(2), 254–64.

—— and Moore, J.P. (1954). Notes on Canadian leeches (Hirudinea), with the description of a new species. *Wasmann J. Biol.* **12**(1), 63–96.

—— and Olsen, O.W. (1971). *Essentials of parasitology.* Brown, Dubuque, Iowa.

—— and Roberts, L.S. (1977). *Cystobranchus mammillatus* (Malm), a Piscicolidae leech new to North America. In *Excerta Parasitológica en memoria del doctor Eduardo Caballero y Caballero. Instituto de Biologia Publicaciones Expeciales. Universidad Nacional de Mexico,* No. 4, pp. 513–17.

Meyer, O. (1944). New principle in treatment of poliomyelitis. *Int. Bull. ed. Res. publ. Hyg.* (A 44-P), 5–9. [Application of leeches over inflamed jugular vein.]

Meyers, T.R. (1978). Prevalence of fish parasitism in Raritan Bay, New Jersey. *Proc. helminthol. Soc. Wash.* **45**(1). 120–8. [1978BA:10639.]

Michaelsen, W. (1905). Die Oligochaeten des Baikal-Sees. *Wiss. Ergebnisse Zool. Exped. Baikal-See Prof. Alexis Korotneff J. 1900–1902.* **1**,1ff. Kiev. [*Agriodrilus vermivorus*: 2.]

—— (1919). Über die Beziehungen der Hirudineen zu den Oligochaeten. *Mitt. Naturh. Mus. Hamburg.* **36**, 131–53.

—— (1926). *Agriodilus vermivorus* aus dem Baikal-See, ein Mittelglied zwischen typischen Oligochäten und Hirudineen. *Mitt. zool. Inst. Mus. Hamburg* **42**, 1–20.

—— (1928). *Clitellata*-Gürtelwürmer. 3. Klasse der *Vermes polymera (Annellida)*. Kükenthal and Krumbach, *Handb. Zool. Berlin* **2** (8), 1–118.

Michelson, E.H. (1957). Studies on the biological control of schistosome-bearing snails. Predators and parasites of freshwater mollusca. A review of the literature. *Parasitology* **47**, 413–26. [Page 22, predatory feeding habits of *Helobdella*.]

Michler, A. (1974). Anordnung, Gestalt und Häufigkeit der Kanäle zwischen der Vakuole der Sehzelle des Blutegels, *Hirudo medicinalis* L. und dem Intersititium. Thesis, Math.-Naturwiss. Fakultät Universität Göttingen.

Migala, K. (1971). [Observations on the infection by protozoa from the genus *Cryptobia (Trypanoplasma)* in the blood-vascular system of grass carp (*Ctenopharyngodon idella* Val.) bred in carp ponds.] *Rocz. Nauk Roln. Ser. H Rybactwo* **93**(3), 65–73. [In Polish; English summary.] [1972BA:21384.]

Mill, P.J., and Knapp, M.F. (1967). Efferent sensory impulses and the innervation of tactile receptors in *Allolobophora longa* Ude and *Lumbricus terrestris* Linn. *Comp. Biochem. Physiol.* **23**, 263–76.

—— —— (1970). Neuromuscular junctions in the body wall muscles of the earthworm, *Lumbricus terrestris* Linn. *J. cell. Sci.* **7**, 263–71.

Miller, C.C. (Incomplete reference). *Parasites of freshwater fishes in North Carolina.*

Miller, J.A. (1929). The leeches of Ohio. *Ohio St. Un. Franz T. Stone Lab., Columbus* **2**, 1–38.

—— (1933–45). Studies in the biology of the leech. I–IX. (Title varies.) *Ohio J. Sci.* **33**(6), 460–3; **34**(1), 57–61; **34**(5), 318–22; **36**(6), 343–8; **42**(1), 45–52; **43**(5), 198–200; **44**(1), 31–5; **44**(4), 177–87; **45**(6), 233–46.

—— (1937). A study of the leeches of Michigan with key to orders, suborders and species. *Ohio J. Sci.* **37**, 85–90.

Miller, J.B. (1974). Mechanical properties of a leech muscle. Ph.D. thesis, University of East Anglia, England.

—— (1975). The length–tension relationships of the dorsal longitudinal muscle of a leech. *J. exp. Biol.* **62**(1), 43–53. [1975BA:67054.]

—— and Aidley, D.J. (1973). Two rates of relaxation in the dorsal longitudinal muscle of a leech. *J. exp. Biol.* **58**, 91–103.

Miller, R.L., Olson, A.C., and Miller, L.W. (1973). Fish parasites occurring in thirteen southern Californian reservoirs. *Calif. Fish Game* **59**(3), 196–206.

Mills, D.H. (1967). The occurrence of the fish leech, *Piscicola geometra* L., on salmonid in the river Tweed and its tributaries. *Salm. Trout Mag.* **181**, 234–5. [First record of this species in Scotland.]

Minelli, A. (1971). Una nuova *Xerobdella* della Prealpi venete (Hirudinea). *Mem. Mus. Civico Storia Nat., Verona* **19**, 355–62. [In Italian; English summary.] [1974BA:43702.]

—— (1977). Irudinei (Hirudinea). *Guide per il Riconoscimento delle specie Animali delle Acque Interne Italiane.* Consiglio Nazionale Delle Ricerche. Verona, Stamp. Valdonega.

—— (1978). '*Dina vignai*' n. sp., a new cave leech from Turkey (Hirudinea, Erpobdellidae). *Fauna Ipogea di Turchia. Quaderni di Speleologia, Circolo Speleologico Romano* **3** (1978–79), 9–14.

—— (1979*a*). Hirudinea. *Fauna d'Italia,* Vol. XV, pp. 1–152. Edizioni Calderini, Bologna.

—— (1979*b*). Sanguisughe d'Italia. Catalogo orientativo e considerazioni biogeografiche. *Lavori Soc. ital. Biogeogr.* (nuova serie) **IV**, 1–35.

—— (In press). La sanguisughe italiane del genere *Dina. (Erpobdellidae).* (Incomplete reference)

Minz, B. (1932). See Autrum (1939*c*).

Mironov, P. *et al.* (1966). [Treatment of myocardial infarct by leeches.] *Folia med., Plovdiv* **8**, 273–7. [In French]

Mishra, G.C., and Dev, B. (1971). Histochemical localization of alkaline phosphatase and physiological significance in the body wall and receptor organs of the common Indian leech, *Poecilobdella granulosa* (Savigny, 1822). *Acta morphol., neerl.-scand.* **8**, 357–61.

—— —— (1972*a*). On the histoenzymology and functional significance of alkaline and acid phosphatase in the ventral nerve cord of Indian medicinal leech, *Poecilobdella granulosa* (Savigny, 1822). *Ann. Histochem.* **17**, 71–8.

—— —— (1972*b*). Histochemical localization of acid and alkaline phosphatase in the ventral nerve cord of the common Indian cattle leech, *Poecilobdella granulosa* (Savigny, 1822). *41st Ann. Sess. Nat. Acad. Sci., India, Varanasi,* p. 95, 24–26 Feb.

—— —— (1973*a*). On the presence of adenosine triphosphatase and its biochemical behaviour in the salivary complex of the common Indian leech, *Poecilobdella granulosa* (Savigny, 1822). 60th (Diamond Jubilee) Session of the *Indian Sci. Congress,* Pt III, pp. 437–8.

—— —— (1973*b*). Biochemical observations on the behaviour of phosphomonoesterase in the intestine and stomach of Indian cattle leech *P. granulosa* (Savigny, 1822), *43rd Ann. Sess. Nat. Acad. Sci. India, Jodphur,* p. 25, 20–22 Oct.

—— —— (1973*c*). Distribution of adenosine triphosphatase in various layers of stomach and intestine of Indian cattle leech, *P. granulosa,* (Savigny, 1822). *43rd Ann. Sess. Nat. Sci. India, Jodphur,* p. 27, 20–22 Oct.

—— —— (1974). Distribution of alkaline phosphatase in the gut wall of

Indian cattle leech *Poecilobdella granulosa* (Savigny, 1822). *61st Sess. Indian Sci. Cong., Nagpur*, p. 104.

—— —— (1976). Studies on the host-parasite interaction and role of esterases during biting of the Indian cattle leech, *Poecilobdella granulosa. Z. Parasitenkd.* **50**(1), 43–51. [1977BA:16177.]

—— —— (1977*a*). Adenosine triphosphatase in the nervous system of *Poecilobdella granulosa. Cell molec. Biol.* **22**(2), 169–76. [French summary.][1979BA:41813.]

—— —— (1977*b*). Studies on the distribution of certain esterases in the central nervous system of Indian cattle leech, *Poecilobdella granulosa. Cell. molec. Biol.* **22**(2), 177–90. [French summary.][1979BA:41814.]

—— —— (1977*c*). Effect of different amino acids on the phosphatase system of an ectoparasite, *Poecilobdella granulosa. Z. Parasitenkd.* **52**(1), 103–12. [1977BA:62941.]

Mishra, G.O., and Sharma, P.N. (1978). Effect of cobra venom on the phosphatase system of Indian cattle leech, *Poecilobdella granulosa. Riv. Parassitol.* **39**(2–3), 97–102.

Mishra, G.S., and Gonzalez, J.P. (1978). Les parasites des tortues d'eau douce en Tunisie. *Arch. Inst. Pasteur, Tunis* **55**(3), 303–26. [English summary.] [1980BA:18450.] [*Placobdella costata; Haemogregarina stapanovi;* turtles (*Clemmys caspica leprosa; Emys orbicularis*).]

Mishra, N.K. (1967). Neurosecretory cells in the ventral nerve cord of the leech (*Hirudinaria granulosa*). *Experientia, Suisse* **23**(12), 1055–6.

—— and Das, S.C. (1978). Functional correlates of the neurosecretory cells of ventral nerve cord of leech *Poecilobdella* (= *Hirudinaria*) *granulosa. Indian J. exp. Biol.* **16**(12), 1267–70. [1979BA:60929.] [Gameto-genesis.]

Mishra, T.N., and Chubb, J.C. (1969). The parasite fauna of the fish of the Shropshire Union Canal, Cheshire. *J. Zool., Lond.* **157** (Part 2), 213–24.

Misra, K.K., Haldar, D.P., and Chakravary, M.M. (1972). Observations on *Mesnilium malariae* gen. nov., spec. nov. (Haemosporidia, Sporozoa) from the freshwater teleost, *Ophicephalus punctatus.* (Incomplete reference)

Mistick, D.C. (1974). Rohde's fiber; a septate axon in the leech. *Brain Res.* **74**(2), 342–8. [1975BA:43440.]

—— (1978). Neurons in the leech that facilitate an avoidance behaviour following nearfield water disturbances. *J. exp. Biol.* **75**, 1–24. [1979BA:15815.]

Mitchell, J.F. (1966). Acetylcholine release from the brain (sheep, cat, rabbit, leech). In *Proceedings of an International Wenner-Gren Center Symposium on mechanisms of release of biogenic amines, Feb. 1965, Stockholm*, Vol. 5, pp. 425–37. Pergamon Press, New York.

Mitchell, J.F.O. (1951). Leeches as endoparasite (in upper respiratory tract). *J. Laryngol. Otol.* **65**, 370.

Mitenev, V.K., and Zubchenko, A.V. (1975). [Parasitic fauna of the whitefish *Coregonus lavaretus* (L.) of some waters of the Kola Peninsula.] *Vopr. Ikhtiol.* **15**(2), 356–60. [In Russian; English summary.] [1976BA:44521.]

Mitrophanow, P. (1883). Beiträge zur Kenntnisse der Hämatozoen. *Biol. Zentralb.* **3**, 35–44.

Miyadi, D. (1933). Studies on the bottom fauna of Japanese Lakes. XI. Lakes of Etorohu-Sima surveyed at the expense of the Keimeikwai fund. *Jap. J. Zool.* **5**(2), 171–208. [Hirudinea: 182.]

—— (1935). Limnological reconnaissance of Southern Sakhalin. I. General features of the fauna. *Bull. Jap. Soc. Sci. Fish.* **4**(2), 113–21.

—— (1937). Limnological survey of the North Kurile Islands. *Arch. Hydrobiol.* **31**(3), 433–83.

—— (1938). Bottom fauna of the lakes in Kunasiri-Sima of the South Kurile Islands. *Int. Revue ges. Hydrobiol. Hydrograph.* **37**(1/3), 125–63.

—— (1940). Leeches from Manchonkuo. In *Report of the Limnological Survey of Kwantung and Manchonkuo*, pp. 395–6.

Miyao, I. (1932). Experimental inquiry into the possibility of transmission of yaws by leeches. *Philipp. J. Sci.* **47**, 463–6.

Miyata, A. (1976). Anuran hemoprotozoa found in the vicinity of Nagasaki City, Japan. Part 1. *Trypanosoma rotatorium. Trop. Med.* **18**(3), 125–34.

—— (1977*a*). *Haemogregarina shirikenimori* n. sp. (Protozoa: Haemogregarinidae) detected from *Triturus pyrrhogaster ensicauda* (Hallowell, 1860) (Amphibia: Salamandridae) in Okinawa Island. *Trop. Med.* **19**(2), 105–12. [Japanese summary.] [1978BA:9896.] [*Haemadipsa zeylanica japonica* possible vector.]

—— (1977*b*). *Trypanosoma ogawai* n. sp. (Protozoa: Trypanosomatidae) detected from *Triturus pyrrhogaster ensicauda* (Hallowell, 1860) (Amphibia: Salamandridae) in Okinawa Island. *Trop. Med.* **19**(2), 113–22. [1978BA:9901.] [*Haemadipsa zeylanica japonica* possible vector.]

Miyazaki, S., and Nicholls, J.G. (1976). The properties and connections of nerve cells in leech ganglia maintained in culture. *Proc. R. Soc. Lond. B Biol. Sci.* **194**(1116), 295–311. [1977BA:21200.]

—— —— and Wallace, B.W. (1976). Modification and regeneration of synaptic connections in cultured leech ganglia. In *The synapse. Cold Spring Harbor Symp. quant. Biol.* **40**, 483–93.

Modi, N.J. (1972). *Modi's Textbook of medical jurisprudence and toxicology* (18th ed.). N.M. Tripathi, Bombay. [Leeches used for criminal abortions.]

Modig, M. (1975). On therica (antidotes) and leeches. *Nord. med.-hist. Arsb.* 71–86. [In Swedish.] [1977BA:16784.] [English abstract: 195.]

Moeller, H. see Möller, H.

Moffett, S. (1977). Neuronal events underlying rhythmic behaviors in invertebrates. *Comp. Biochem. Physiol. A Comp. Physiol.* **57**(2), 187–95. [1977BA:39028.]

Mohan, R.N. (1968). Diseases and parasites of buffaloes, part III. Parasitic and miscellaneous diseases. *Vet. Bull., Weybridge* **38**, 735–56.

Möller, H. (1974). Untersuchungen ueber die Parasiten der Flunder (*Platichthys flesus* L.) in der Kieler Foerde. *Ber. Dt. wiss. Komm. Meeresforsch* **23**(2), 136–49. [1974BA:56578.] [*"Piscicola geometra"* in small numbers.]

Moller Pillot, H.K.M. (1971). Faunistische beoordeling van de fauna in laaglandbeken. Pillot-Standaardboekhandel Tilburg 1971. [Leeches in Holland.]

Molnar, K. (1970). Beitrage zur Kenntnis der Fischparasitenfauna Ungarns: VI. Cestoda, Nematoda, Acanthocephala, Hirudinea. *Parasitol. Hung.* **3**(3), 51–76.

—— Hanek, G., and Fernando, C.H. (1974). Parasites of fishes from Laurel Creek, Ontario. *J. Fish. Biol.* **6**, 717–28.

Molner, B., and Szabó, S. (1967). Caracteristicile sistemului neurosecretor la lipitoarea de cal (*Haemopis sanguisuga*). *Studii Cercet. Biol. Cluj. Acad. RPR. (Ser. Zool.)* **19**(4), 317–22. [In Romanian; French summary.] [Seasonal variation of neurosecretory system.]

Moment, G., and Johnson, J. (1979). *J. Morph.* **159**, 1–15. (Incomplete reference)

Monakov, A.V. (1972). Review of studies of feeding of aquatic invertebrates conducted at the Institute of Biology of Inland Waters, Academy of Science, USSR. *J. Fish. Res. Bd. Can.* **29**, 363–83.

Moon, H.P. (1940). An investigation of the movement of freshwater invertebrate fauna. *J. anim. Ecol.* **9**, 76–83.

Moore, G., and Myers, B.J. (1973). Parasites of non-human primates. In *Ann. Proc. two conferences of the Am. Ass. Zoo. Veter.*, 1972 Houston, Texas; 1973 Colombus, Ohio. Hill's Division Riviana Foods, Topeka, Kansas.

Moore, G., Frazer, B., Lent, C., and Boles. J. (1977). White noise analysis of electrotonic junctions in the leech nervous system. *Neurosci. Soc. Abstr.* **3**, 186.

Moore, J.E. (1964). Notes on the leeches (Hirudinea) of Alberta. *Natl. Mus. Can. nat. Hist. Pap.* **27**, 1–15.

—— (1966*a*). Further notes on Alberta leeches (Hirudinea). *Natl. Mus. Can nat. Hist. Pap.* **32**, 1–11.

—— (1966*b*). New records of leeches (Hirudinea) for Saskatchewan. *Can. Field-Nat.* **80**, 59–60.

Moore, J.P. (1894, 1895*a,b*, 1897). See separate bibliography to Branchiobdellida, H. pp. 805–6.

—— (1898). Leeches of the US National Museum. *Proc. US nat. Mus.* **21**(1160), 543–63.

—— (1900*a*). A description of *Microbdella biannulata* with especial regard to the constitution of the leech somite. *Proc. Acad. nat. Sci. Philadelphia* **52**, 50–73.

—— (1900*b*). Note on Oka's biannulate leech. *Zool. Anz.* **23**, 474–7.

—— (1901*a*). The Hirudinea of Illinois. *Bull. Ill. St. Lab. nat. Hist.* **5**, 479–547.

—— (1901*b*). Descriptions of two new leeches from Porto Rico. *Bull. US Fish Comm.* **2**, 211–22.

—— (1906). Hirudinea and Oligochaeta collected in the Great Lakes region. *Bull. Bur. Fish.* 25, 153–71.

—— (1908). The leeches of Lake Amatitlan. In *The zoology of Lakes Amatitlan and Atitlan, Guatemala, with special reference to ichthyology. Field Columbian Mus., Zool. Ser.* **7**, 199–201. [*Erpobdella triannulata.*]

—— (1910). *Platybdella chilensis*, sp. nov. *Rev. Chil. Hist. Nat.* **14**, 29–30.

—— (1911). Hirudinea of southern Patagonia. *Rep. Princeton Un. Exped. Patagonia, 1896–1899* **3**, 669–89.

—— (1912). Classification of the leeches of Minnesota. In *The leeches of Minnesota. Geol. nat. Hist. Surv. Minnesota, Zool. Ser.* no. 5, pt. 3, 63–150.

——(1918). The leeches (Hirudinea). In *Fresh-water biology* (ed. H.B. Ward and G.C. Whipple), pp. 646–60. Wiley, New York.

—— (1920). The leeches of Lake Maxinkuckee. In *Lake Maxinkuckee: a physical and biological survey* (ed. B.W. Evermann, and H.W. Clark), Vol. 2, pp. 87–95. Dept. Conserv., State of Indiana.

—— (1921). Hirudinea: annelids, parasitic worms, protozoans, etc. *Rep. Can. Arctic Exped. 1913–18, Ottawa* **9**, pt. C.

—— (1922). The fresh-water leeches (Hirudinea) of southern Canada. *Can. Field-Nat.* **36**, 6–11, 37–9.

—— (1923). The control of blood-sucking leeches, with an account of the leeches of Palisades Interstate Park. *Roosevelt Wild Life Bull.* **2**(1), 7–53.

—— (1924*a*). The leeches (Hirudinea) of Lake Nipigon. *Un. Toronto Stud., Publ. Ontario Fish. Res. Lab.* No. 23, 17–31.

—— (1924*b*). The anatomy and systematic position of the Chilean terrestrial leech, *Cardea valdiviana* (Philippi). *Proc. Acad. nat. Sci. Philadelphia* **76**, 29–48.

—— (1924*c*). Notes on some Asiatic leeches (Hirudinea) principally from China, Kashmir, and British India. *Proc. Acad. nat. Sci. Philadelphia* **76**, 343–88.

—— (1927). The segmentation (metamerism and annulation) of the Hirudinea; Arhynchobdellae, In *The fauna of British India* (ed. W.A. Harding and J.P. Moore) Hirudinea, pp. 1–12, 97–302. London: Taylor and Francis.

—— (1929). Leeches from Borneo with descriptions of new species. *Proc. Acad. nat. Sci. Philadelphia.* **81**, 267–95.

—— (1930*a*). The leeches (Hirudinea) of China. *Peking Soc. nat. Hist. Bull.* 1929–30, **4** (pt. 3), 39–43.

—— (1930*b*). Leeches (Hirudinea) from China with descriptions of new species. *Proc. Acad. nat. Sci. Philadelphia* **82**, 169–92.

—— (1931*a*). The life of the Darjeeling land leech. *J. Darjeeling nat. Hist. Soc.* **5**, 106–17.

—— (1931*b*). A remarkable South American leech. *Actas Congr. Arch. Soc. Biol. Montevideo* Suppl., **5**, 1220–5.

—— (1932*a*). How abundant are land leeches? *J. Bombay nat. Hist. Soc.* **35**, 701–2.

—— (1932*b*). Land leeches in the *Fauna of British India* — Some corrections. *Rec. Indian Mus.* **34**(pt. 1), 1–6.

—— (1932*c*). Leeches and planarians compared. *J. Darjeeling nat. Hist. Soc.* **7**, 61–7.

—— (1933). Leeches. In *Scientific results of the Cambridge Expedition to the East African Lakes*, 1930–31. *J. Linn. Soc.* **38**, 297–9.

—— (1935*a*). More about land leeches: how abundant are they? *J. Darjeeling nat. Hist. Soc.* **9**, 116–24.

—— (1935*b*). A description of *Aetheobdella hirudoides* gen. et sp. n., from New South Wales, with notes on leeches collected by the Oxford University Sarawak Expedition. *Ann. Mag. nat. Hist.*, Ser. 10, **16**, 296–304.

—— (1935*c*). Leeches from Borneo and the Malay peninsula. *Bull. Raffles Mus.* No. 10, 67–79.

—— (1936*a*). Hirudinea from Yucatan. *Carnegie Inst. Washington* No. 457, 41–3.

—— (1936*b*). Report on Hirudinea. In *Yale North Indian Expedition. Mem. Connecticut Acad. Arts Sci.* **10**, Part. 11, pp. 191–2.

—— (1936*c*). The leeches of Lake Nipissing. *Can. Field-Nat.* **50**, 112–14.

—— (1936*d*). *Macrobdella ditetra.* In *Parasites of certain North Carolina salientia* (ed. B.B. Brandt) *Ecol. Monogr.* **6**, 491–532.

—— (1937). Laboratory care of leeches. In *Culture methods for invertebrate animals* (ed. Galtsoff, Lutz, Welch, and Needham) pp. 201–4. Ithaca, New York.

—— (1938*a*). Leeches. In *Australasian Antarctic Expedition, 1911–1914. Sci. Rept.*, Ser. C, **10**(pt. 3). 1–15.

—— (1938*b*). Leeches (Hirudinea) from Yucatan caves. *Carnegie Inst. Washington* No. 491, 67–70.

—— (1938*c*). Leeches (Hirudinea) principally from the Malay Peninsula, with descriptions of new species. *Bull. Raffles Mus.* No. 14, 64–80.

—— (1939*a*). Leeches (Hirudinea) from the Atlas Mountains of Morocco. *Ann. Mag. nat. Hist.* Ser. 11, **3**, 80–7.

—— (1939*b*). *Helobdella punctato-lineata,* a new leech from Puerto Rico. *Puerto Rico J. Hyg. trop. Med.* **14**, 422–9.

—— (1939*c*). Additions to our knowledge of African leeches (Hirudinea). *Proc. Acad. nat. Sci. Philadelphia* **90**, 297–360.

—— (1939*d*). In memoriam. Edwin Linton (1855–1939). *J. parasitol. Urbana* **25**, 450–3.

—— (1940). *Austrobdella anoculata,* a new species of fish leech from Greenland. *J. Wash. Acad. Sci.* **30**(12), 520–4.

—— (1944*a*). The leeches (Hirudinea) of Lake Huleh, Palestine. *Ann. Mag. nat. Hist.* Ser II, **11**, 182–90.

—— (1944*b*). Leeches in the British Museum, mostly Haemadipsinae from the South Pacific with descriptions of new species. *Ann. Mag. nat. Hist.* Ser. II, **11**, 383–409.

—— (1945*a*). A water-squirting Indian leech. *J. Bengal nat. Hist. Soc.* **20**, 16–19.

—— (1945*b*). Two new leeches (Hirudinea) in the collection of the United States National Museum. *J. Washington Acad. Sci.* **35**, 261–5.

—— (1946*a*). Leeches (Hirudinea) from the Hawaiian Islands, and two new species from the Pacific region in the Bishop Museum collection. *Occ. Pap. Bernice P. Bishop Mus.* **18**(11), 171–91.

—— (1946*b*). The anatomy and systematic position of *Myzobdella lugubris* Leidy (Hirudinea). *Notulae Nat. Acad. nat. Sci. Philadelphia* No. 184, 1–12.

—— (1949). Hirudinea. In *The animal life of temporary and permanent ponds in southern Michigan* (ed. R. Kenk), pp. 38–9. *Misc. Publ. Mus. Zool., Univ. Michigan* No. 71, 1–66.

—— (1952*a*). New Piscicolidae (leeches) from the Pacific and their anatomy. *Occ. Pap. Bernice P. Bishop Mus.* **21**(2), 17–44.

—— (1952*b*). Professor A.E. Verrill's fresh-water leeches — a tribute and a critique. *Notulae Nat. Acad. nat. Sci. Philadelphia* No. 245, 1–15.

—— (1953). Three undescribed North American leeches (Hirudinea). *Notulae Nat. Acad. nat. Sci. Philadelphia* No. 250, 1–13.

—— (1957). Hirudinea. *BANZ Antarctic Research Expedition, 1929–1931* Ser. B (pt 6), 99–105.

—— (1958). The leeches (Hirudinea) in the collection of the Natal Museum. *Ann. Natal Mus.* **14**(pt 2) 303–40.

—— (1959*a*). Hirudinea. In *Ward and Whipple's Fresh-water biology,* 2nd edn. (ed. W.T. Edmondson) pp. 542–57. [1963:646–60.] John Wiley & Sons, Inc.: New York.

—— (1959*b*). Leeches. In *Encyclopaedia Americana,* Vol. 17, pp. 196–7. Chicago.

—— (1960). On the contributions of Doctor Eduardo Cabellero y C. to Mexican hirudinology. In *Libro Homenaje al Dr. Eduardo Caballero y Caballero, Jubileo 1930–1960,* pp. 29–31. Instituto Politecnico Nacional, Mexico.

—— (1963). Leeches. In *Encyclopaedia Britannica*, Vol. 13, pp. 890–1. Chicago.

—— and Meyer, M.C. (1951). Leeches (Hirudinea) from Alaskan and adjacent waters. *Wassmann J. Biol.* **9**, 11–77.

—— —— (eds.) (1955). *Études morphologiques et systematiques sur les Hirudinées, I: L'Organisation des Ichthyobdellides* (Russian edition by W.D. Selensky, 1915) [English trans.] Published by editors. [Copy in US National Museum of Natural History. Worms Division.]
[See also Caballero (1969); Meyer (1968); Wenrich (1965).]

Moosbrugger, G., and Reisinger, E. (1971). Zur Kenntnis des europäischen Landblutegels *Xerobdella lecomtei* (Frauenfeld) (mit besonderer Berucksichtigung der postembryonalen Entwicklung des Kopulationsapparates und seiner Funktion). *Z. wiss. Zool.* **183**(1/2), 1–50. [English summary.]

Mootz, R. (1936). [Endosymbiotic bacteria in *Hirudo medicinalis*.] *Med. Welt* **33**, 1552.

Moquin-Tandon, A. (1846). *Monographie de la famille des Hirudinées.* Paris.

Morawitz, P. (1903). See Autrum (1939c).

Moretti, G. (1931). See Autrum (1939c).

Morhardt, P.E. (1949). Le renouveau des sangsues. *Presse méd.* **57**, 882.

Moriarty, C. (1973). Distribution of freshwater macroinvertebrates in Ireland, 1967–1972. *Ir. nat. J.* **17**, 409–12. [Leeches found in stomachs of eel *Anguilla anguilla*.]

Morris, D.L., and Brooker, M.P. (1979). The vertical distribution of macroinvertebrates in the substratum of the upper reaches of the River Wye, Wales. *Freshwater Biol.* **9**, 573–83.

Moser, M., and Anderson, S. (1977). An intrauterine leech infection: *Branchellion lobata* Moore, 1952 (Piscicolidae) in the Pacific angel shark (*Squatina californica*) from California. *Can. J. Zool.* **55**(4), 759–60. [1977BA:28542.]

Moszynski, A. (1930). *Dzdzownica i pijawka.* Warsaw. (Publisher undetermined)

Mothes, G. (1967). Einige Tiergruppen mit geringer Artendichte innerhalb der makroskopischen Bodenfauna des Stechlinsee. *Limnologica* **5**(1), 11–21.

Mrázek, A. (1908, 1913). See Autrum (1939c).

Mühling, — (See Autrum) (1939c).

Mukherjee, G. (1974). Unusual foreign body causing haematuria. *J. Indian med. Ass.* **63**(9), 284–5.

Müller, H.E., Pinus, M., and Schmidt, U. (1980). [*Aeromonas hydrophila* as a normal intestinal bacterium of the vampire bat, *Desmodus rotundus*]. *Zentralbl. Veterinaermed. Reihe B.***27**(5), 419–24.

Muller, K.J. (1932). Über normale Entwicklung, inverse Asymmetrie und Doppelbildungen bei *Clepsine sexoculata*. *Z. wiss. Zool.* **142**, 425–90.

Muller, K.J. (1979). Synapses between neurones in the central nervous system of the leech. *Biol. Rev.* **54**, 99–134.

—— (1981). Synapses and synaptic transmission. In: *The neurobiology of the leech.* (ed. K.J. Muller, J.G. Nicholls and G.S. Stent) Chap. 6, pp. 79–111. Cold Spring Harbor, New York.

—— and Carbonetto, S.T. (1976). Signalling in the nervous system synapses of specific neurons in the leech. *Carnegie. Inst. Washington Yearb.* **75**, 1975–76. [1977BA:91444.]

—— —— (1977). Two ways that an electrical connection is re-established in the leech. *Neurosci. Abstr.* **3**, 353.

—— —— (1979). The morphological and physiological properties of a regenerating synapse in the C.N.S. of the leech. *J. comp. Neurol.* **185**(3), 485–516.

—— and McMahan, U.J. (1975*a*). The arrangement and structure of synapses formed by specific sensory and motor neurons in segmental ganglia of the leech. *Anat. Rec.* **181**, 432. (Abstr.)

—— —— (1975*b*). Synapses of specific sensory and motor neurons in the leech. *Neurosci. Abstr.* **1**(1024), 662.

—— —— (1976). The shapes of sensory and motor neurons and the distribution of synapses in ganglia of the leech: a study using intracellular injection of horseradish peroxidase. *Proc. R. Soc. Lond.* **B194**, 481–99.

—— and Nicholls, J.G. (1974). Different properties of synapses between a single sensory neurone and two different motor cells in the leech C.N.S. *J. Physiol.* **238**(2), 357–69. [1974BA:43654.]

—— —— (1981). Regeneration and plasticity. In: *The neurobiology of the leech.* (eds. K.J. Muller, J.G. Nicholls and G.S. Stent) Chap. 10, pp. 197–234.

—— and Scott, S.A. (1979*a*). Correct axonal regeneration after target cell removal in the central nervous system of the leech. *Science* **206**(4414), 87–9. [1980BA:15202.]

—— —— (1979*b*). Direct electrical synaptic connection is mediated by an interneuron. *Neurosci. Abstr.* **5**, 744.

—— —— (1980). Removal of the synaptic target permits sprouting of a mature intact axon. *Nature, Lond.* **283**, 89–90.

—— —— (1981). Transmission of a 'direct' electric connexion mediated by an interneurone in the leech. *J. Physiol.* **311**, 565–83.

—— and Thompson, B. (1976). An electrical synapse in the leech. *Carnegie Inst. Wash. Yearb.* **75**, 95–7.

—— Carbonetto, S.T., and Thomas, B. (1976). Synapses of specific neurons in the leech: sprouting of sensory neuron terminals. *Carnegie Inst. Wash. Yearb.* **75**, 91–5.

—— Nicholls, J.G., and Stent, G.S. eds. (1981). *The neurobiology of the leech.* Cold Spring Harbor Laboratory Publications, Cold Spring Harbor, pp. 288.

—— Scott, S.A., and Thomas, B.E. (1978). Specific associations between sensory cells. *Carnegie Inst. Wash. Yearb.* **77**, 69–70.

—— Carbonetto, S.T., Scott, S.A., and Thomas, B.E. (1977). Regeneration in the nervous system formation of specific synapses in the leech. *Carnegie Inst. Wash. Yearb.* **77**, 62–71.

Müller, M. (1971). Anatomische Studien an Blutegein. *Mikroskosmos* **60**(12), 353–7.

Müller, M. (1971). Zur Hirudineenfauna des Naturschutz-gebietes 'Heiliges Meer' bei Hopsten, Kreis Tecklenburg. *Abh. Landesmus. Naturk. Münster Westafalen* **33**(1), 3–15.

Mulloney, B. (1970). Structure of the giant fibers of earthworms. *Science* **168**, 994–6.

Mumford, E.P. (1936). Terrestrial and fresh-water fauna of the Marquesas Islands. *Ecology* **17**(1), 143–57. [No leeches from the Marquesas and residents agree they are absent, but Dordillon (1904) gives a native name *toke omo toto* for 'sangsue'.]

Münchberg, P. (1936). Über das Vorkommen des medizinische Blutegels (*Hirudo medicinalis* L.) in der Grenzmark Posen-Westpreussen. *Abh. Ber. Grenzmärk. Ges. Naturwiss. Abt., Schneidemühl.* **11**, 82–4.

Muntz, W.R.A. (1952). Fresh water leeches. *Rep. Oundle Sch. nat. Hist. Soc.* 24–7.

Murer, E.H., James, H., Budzynski, A.Z., Malinconico, S.M. and Gasic G.J. (1984). Protease inhibitors in *Haementeria* leech species. *Thromb. Haemostas.* (Stuttgart). **51**(1), 24–6.

Murie, J. (1865). See Autrum (1939c).

Murnaghan, M.F. (1958). The morphinized-eserinized leech muscle for the assay of acetylocholine. *Nature, Lond.* **182**, 317.

Murphy, P.M. (1980). An ecological study of the macroinvertebrate fauna of the River Ely. Ph.D. thesis. University of Wales.

—— and Learner, M.A. (1982a). The life history and production of the leech *Erpobdella octoculata* (Hirudinea: Erpobdellidae) in the River Ely, South Wales. *J. Anim. Ecol.* **51**, 57–67.

—— —— (1982b). The life history and production of the leech *Helobdella stagnalis* (Hirudinea: Glossiphoniidae) in the River Ely, South Wales. *Freshwat. Biol.* **12**, 321–9.

Murray, A.B., Nance, W.S., and Tarter, D.C. (1977). The occurrence of the leech *Piscicolaria reducta* on *Etheostoma blennoides* from West Virginia. *Trans. Am. microsc. Soc.* **96**(3), 412. [1978BA:33997.]

Musacchio, M., and Macagno, E.R. (1982). Quantitative aspects of growth of an identified neuron in the leech *Hirudo medicinalis. Biol. Bull.* **163**(2), 388 (Abstr.)

Mustafa, S., and Ahmad, Z. (1982). Diversity and productivity of benthic microfauna of Baigul and Nanaksagar Reservoirs (India) (Naintal District). *Z. angew. Zool.* **69**(3), 337–44. [German summary.] [1983BA:63602]

Myers, R.J. (1935). Behavior and morphological changes in the leech, *Placobdella parasitica*, during hypodermic insemination. *J. Morphol.* **57**, 617–48.

Nachtigall, W. (1974). Biological mechanisms of attachment. In *The comparative morphology and bioengineering of organs for linkage, suction, and adhesion*. Springer, Berlin. [Leeches: 87–8, 126–8.]

Nachtrieb, H.F. (1912). The leeches of Minnesota. Part I: General account of the habits and structure of leeches. *Geol. nat. Hist. Surv. Minnesota, Zool.* No. 5, 1–28.

Nadkarni, S.G., and Lobo, J.F. (1968). Alkaline phosphatase and mucopolysaccharides in the epididymis of the leech *Hirudinaria granulosa* (Savigny). *Indian J. exp. Biol.* **6**(3), 184–5.

Nagabhushanam, R., and Kulkarni, G.K. (1977a). Effect of some extrinsic factors on the respiratory metabolism of the freshwater leech, *Poecilobdella viridis:* Oxygen tension, pH and salt concentrations. *Hydrobiologia* **56**(2), 181–6. [1978BA:59072.]

—— —— (1977b). Thermal relations of the Indian leech *Poecilobdella viridis. Proc. Indian Acad. Sci. Sect. B* **86**(4), 229–34. [1978BA:52510.]

—— —— (1978a). Histomorphological studies on the brain neurosecretory profile of Indian leech, *Poecilobdella viridis* (Blanchard). *Proc. Indian Acad. Sci. Sect. B* **87**(9), 203–10. [1979BA:3110.]

—— —— (1978b). Hormonal control of oxygen consumption in the gnathobdellid leech *Poecilobdella viridis* (Blanchard). *Ind. J. Physiol. Allied Sci.* **32**(3), 116–19.

—— —— (1978c). Effects of some extrinsic factors on the respiratory metabolism of the freshwater leech, *Poecilobdella viridis* (Blanchard): desiccation and temperature. *Hydrobiologia* **61**(1), 3–7.

—— —— (1979). Effect of dehydration through desiccation on brain neurosecretory cells of the fresh-water Indian leech, *Poecilobdella viridis. J. anim. Morphol. Physiol.* **26**(1–2), 121–5.

—— —— (1980). Annual changes in the brain neurosecretory profile of the Indian freshwater leech, *Poecilobdella viridis* in relation to reproduction. *Hydrobiologia* **69**(1/2), 163–8. [1980BA:23787.]

—— —— and Hanumante, M.M. (1976). Hormonal control of spermatogenesis in the leech *Poecilobdella viridis. Marathwada Univ. J. Sci.* (Aurangabad) **158**, 197–203. [1977BA:93791.]

Nagao, Z. (1957). Observations on the breeding habits in a freshwater leech *Herpobdella lineata* O.F. Müller. *J. Fac. Sci. Hokkaido Un., Ser. VI Zool.* **13**, 192–6.

—— (1958). Some observations on the breeding habits of a freshwater leech *Glossiphonia lata* Oka. *Jap. J. Zool.* **12**(2), 219–23.

Nagata, M., and Kadota, K. (1976). Specificity of muscle response of the Japanese medical leech to acetylcholine. *Jap. J. Pharmacol.* **26**(5), 631–4. [1977BA:63429.]

Nagayoshi, Y. (1955). On leeches parasitizing the nasal cavity in man. *Chiryo-Yaku-Ho* No. 529, 19.

Nageotte, J. (1905). La structure fine du système nerveux. II. Les neurofibrilles et le réseau interne de Golgi. *Revue Idées* **2**, 108–29.

Nagel, M. (1976). New distributional records for piscicolid leeches in Oklahoma. *J. Parasitol.* **62**(3), 494–5.

Nakas, M., Penn, R.D., and Loewenstein, W.R. (1966). Disconnection of electrically connected nerve and gland cell junctions by calcium removal (Hirudinea, Diptera, Abstract). *Fed. Proc. Fedn Am. Socs exp. Biol.* **25**(1 Pt. 1), 270.

Nambudiri, P.N., and Vijayakrishnan, K.P. (1958). Neurosecretory cells of the brain of the leech *Hirudinaria granulosa* (Sav.). *Curr. Sci., India* **27**, 350–1.

Nanda, S., and Nanda, D.K. (1976). Neurosecretory neurons in the supra-oesophageal ganglia of gnathobdellid leech, *Hirudinaria granulosa* (Sav.). *Indian J. Physiol. all. Sci.* **30**(4), 152–7. [1978BA:14990.]

—— —— (1978). Neurosecretory system of the suboesophageal ganglia of *Hirudinaria granulosa* (Gnathobdellae, Hirudidae). *Proc. zool. Soc.* (Calcutta) **31**(1–2), 11–18.

Nass, M.M.K., Nass, S., and Afzelius. B.A. (1965). The general occurrence of mitochondrial DNA. *Expl cell. Res.* **37**, 516–39. [Mitochondrial DNA in intestinal cells of *Piscicola geometra*.]

Needham, A.E. (1966*a*). The tissue-pigments of some fresh-water leeches. *Comp. Biochem. Physiol.* **18**, 427–61.

—— (1966*b*). Absorption spectrum of haemoglobin of leeches. *Nature, Lond.* **210**, 427–8.

—— (1969). Absorption spectrum of haemoglobin of leeches: a correction. *Nature, Lond.* **221**, 572.

Needham, E.A. (1969). Protozoa parasitic in fishes. Ph.D. thesis, University of London.

Neill, R.M. (1938). The food and feeding of the Brown Trout (*Salmo trutta* L.) in relation to the organic environment. *Trans. R. Soc. Edinb.* **59**, 481–520.

Nekhaev, V.M. (1957). [The problem of the spermatogenesis in the officinal leech.] *Zh. Obshch. Biol. SSSR* **18**, 208–16. [In Russian.]

—— (1959*a*). *Tr. Molodich Uchenich Instituta Hidrobiologia, Kiev.* (Incomplete reference)

—— (1959*b*). [Annual developmental cycle of testes of the leeches *Hirudo medicinalis* L. and *Haemopis sanguisuga* Bergm.] *Zool. Zh.* **38**(2), 280–2. [In Russian; English summary.]

—— (1960). *Spermatogenes i hodichnia tsikl raevitiya semennikov meditsiiskoi piyavki v estestvennich vodoemach i pri iskusstvennom sodersanii, Avtoreph.* Kand. Diss., Kiev.

Nekipelov, M.I. (1961). [On the toxicity of nitrates for water organisms.]

Zool. Zh. **46**, 932–6. [In Russian; English summary.] [*Hirudo medicinalis.*]

—— (Incomplete reference). [The toxic effect of aluminium compounds on aquatic organisms.] [In Russian.] [Minimum lethal conc. of DL_{50} (aluminium nitrate) 11.8–17.9 mg/l. for leeches.]

Nenninger, U. (1948). Die Peritrichen der Umgebung von Erlangen mit besonderer Berücksichtigung ihrer Wirtsspezifität. *Zool. Jahrb. Abt. Syst. Oekol. Geogr. Tiere* **77**(3/4), 169–266.

Neumann, M.P., and Vande Vusse, F.J. (1976). Two new species of *Alloglossidium* Simer 1929 (Trematoda: Macroderoididae) from Minnesota leeches. *J. Parasitol.* **62**(4), 556–9. [1977BA:9032.]

Neumann, R.O. (1909). See Autrum (1939c).

Newell, R., and Canaris, A.G. (1969). Parasites of the pigmy whitefish, *Prosopium coulteri* (Eigenmann and Eigenmann) and mountain whitefish *Prosopium williamsoni* (Girard) from western Montana. *Proc. helminthol. Soc., Wash.* **36**(2), 274–6.

Newman, M.W. (1978). Pathology associated with *Cryptobia* infection in a summer flounder (*Paralichthys dentatus*). *J. Wildl. Dis.* **14**, 299–304.

Newton, L.C. (1972). Pharmacological studies of invertebrate ACh receptors. Ph.D. thesis, University of Southampton.

—— Walker, R.J., and Woodruff, G.N. (1970). The effect of nicotinic and muscarinic antagonists on single neurones of the leech, *Hirudo medicinalis. J. Physiol.* **210**, 54P–55P.

Nicholls, D., Cooke, J., and Whiteley, D. (1971). *Oxford book of invertebrates.* Oxford University Press. [Leeches: 116–19.]

Nicholls, J.G. (1980). Regeneration of synaptic connections by individual identified leech neurons. *Neurosci. Lett.* (Suppl. 5): S13. 4th European Neuroscience Meeting, Brighton, England, Sept. 16–19, 1980. [1980BA:26676.]

—— and Baylor, D.A. (1968a). Long-lasting hyperpolarization after activity of neurons in leech central nervous system. *Science, NY* **162**, 279–81.

—— —— (1968b). Specific modalities and receptive fields of sensory neurons in CNS of the leech. *J. Neurophysiol.* **31**(5), 740–56.

—— —— (1969). The specificity and functional role of individual cells in a simple central nervous system. *Endeavor* **28**, 3–7.

—— and Kuffler, S.W. (1964). Extracellular space as a pathway for exchange between blood and neurons in the central nervous system of the leech: ionic composition of glial cells and neurons. *J. Neurophysiol.* **27**, 645–71.

—— —— (1965). Na and K content of glial cells and neurons determined by flame photometry in the central nervous system of the leech. *J. Neurophysiol.* **28**(3), 519–25.

—— and Purves, D. (1970). Monosynaptic chemical and electrical connections between sensory and motor cells in the central nervous system of the leech. *J. Physiol.* **209**(3), 647–67.

—— —— (1972). A comparison of chemical and electrical synaptic transmission between single sensory cells and a motoneurone in the central nervous system of the leech. *J. Physiol.* **225**(3), 637–56.

—— and Van Essen, D. (1974). The nervous system of the leech. *Scient. Am.* **230**(1), 38–48.

—— and Wallace, B.G. (1978*a*). Modulation of transmission at an inhibitory synapse in the central nervous system of the leech. *J. Physiol.* **281**, 157–70. [1979BA:15799.]

—— —— (1978*b*). Quantal analysis of the transmitter release at an inhibitory synapse in the central nervous system of the leech. *J. Physiol.* **281**, 171–86. [1979BA:15800.]

—— and Wolfe, D.E. (1967). Distribution of ^{14}C-labelled sucrose, inulin, and dextran in extracellular spaces and in cells of the leech central nervous system. *J. Neurophysiol.* **30**(6), 1574–92.

—— *et al.* (1977). Regeneration of individual neurones in the nervous system of the leech. In *Synapses* (ed. C.A. Cottrell and P.N. Usherwood) pp. 249–63. Academic Press, New York.

Nigrelli, R.F. (1929). On the cytology and lifehistory of *Trypanosoma diemyctyli* and the polynuclear count of infected newts (*Triturus viridescens*). *Trans. Am. microsc. Soc.* **48**, 366–87.

—— (1941). Parasites of the Green Turtle, *Chelonia mydas* (L.), with special reference to the rediscovery of trematodes described by Looss from this host species. *J. Parasitol.* **27**(6), 15–16.

—— (1942). Leeches (*Ozobranchus branchiatus*) on fibroepithelial tumors of marine turtles (*Chelonia mydas.*). *Anat. Rec.* **84**(4), 539–40.

—— (1944). Trypanosomes from North American amphibians with a description of *Trypanosoma grylli* Nigrelli (1944) from *Acris gryllus* (Le Conte). *Zoologica* **30**, 47–56.

—— (1946). V. Parasites and diseases of the Ocean Pout, *Macrozoarces americanus*. *Bull. Bingham Oceanograph. Coll.* **9**, 187–221. [Hirudinea: 215. *Platybdella buccalis = Oceanobdella sexoculata.*]

—— and Smith, G.M. (1943). The occurrence of leeches, *Ozobranchus branchiatus*, on fibro-epithelial tumors of marine turtles, *Chelonia mydas* (Linnaeus). *Zoologica* **28**(2), 107–8.

—— Pokorny, K.S., and Ruggieri, S.J. (1975). Studies on parasitic kinetoplastids. II. Occurrence of a biflagellate kinetoplastid in the blood of *Opsanus tau* (toadfish), transmitted by the leech (*Piscicola funduli*). *J. Protozool.* **22**(3), 43A. [*Calliobdella vivida*]

Nikolaev, S.G. (1980). [Ecology and reproduction of *Herpobdella octoculata* and *Helobdella stagnalis* (Hirudinea) in the Sevan Lake, Armenian S.S.R., U.S.S.R.] *Zool. Zh.* **59**(9), 1421–5. [In Russian.]

Nikolsky, G.V. (1963). *The ecology of fishes.* Academic Press, London. [*Pisicicola geometra*, a serious pest in fish hatcheries.]

Nilsenn, J.P. (1980). Acidification of a small watershed in southern Norway and some characteristics of acidic aquatic environments. *Int. Rev. Ges. Hydrobiol.* **65**, 177–207. [*Hirudo* in Norway]

Nishioka, R.S., and Hagadorn, I.R. (1961). Neurosecretion and granules in neurones of the brain of the leech. *Nature, Lond.* **191**, 1013–14.

Nistri, A. (1974). Proceedings: effects of central nervous system depressants and stimulants on the acetylcholine concentration of leech ganglia *in vivo. Br. J. Pharmacol.* **52**(3), 438P–439P.

—— Cammelli, E., and de Bellis, A.M. (1978). Pharmacological observations on the cholinesterase activity of the leech central nervous system. *Comp. Biochem. Physiol. C Comp. Pharmacol.* **61**(1), 203–6. [1979BA:41819.]

—— de Bellis, A.M., and Cammelli, E. (1975). Drug-induced changes in behaviour and ganglionic acetylcholine concentration of the leech. *Neuropharmacology* **14**(8), 565–70. [1976BA:8207.]

Nobre, A. (1903*a,b*). See Autrum (1939*c*).

Nohynkova, E., and Kulda, J. (1974). The occurrence of *Trichomonitus batrachorum* in the horse leech *Haemopis sanguisuga. J. Protozool.* **21**(3), 458. (Abstr.)

Nöller, W. (1912). *Entamoeba aulastomi* n. sp., eine neue parasitische Amöbe aus dem Pferdeegel (*Aulastomum gulo* Moqu.-Tand.). *Arch. Protistenk.* **24**, 195–200.

—— (1913). See Nöller, W. (1913*a*) in Autrum (1939*c*).

Nonato, E.F. (1946). Sobre sanguesugas do genero *Liostoma. Bol. Fac. Fil. Cienc. Letr. Sao Paulo, Zool.* **11**, 288–356.

Norse, E.A., and Estevez, M. (1977). Studies on Portunid crabs from the eastern Pacific. I. Zonation along environmental stress gradients from the coast of Colombia. *Marine Bio.* **40**, 365–73. [Page 371, *Myzobdella*-like leeches with eggs on *Callinectes arcuatus* and *C. toxotes*, in brackish water of river mouths, Pacific side of Cauca State, Colombia.]

Nowak, G., and Markwardt, F. (1980). Influence of hirudin on endotoxin-induced disseminated intravascular coagulation in weaned pigs. *Exp. Pathol.* (Jena) **18**(7–8), 438–43.

Nowicki, E. (1940). Zur Pathogenität der *Trypanoplasma cyprini. Z. Parasitkd.* **11**, 486–73.

Nurminen, M. (1965). [The setiferous leech, *Acanthobdella peledina* Grube (Hirudinea).] *Luonnon Tutkija* **69**(3), 107–10. [In Finnish.]

—— (1966). Notes about *Acanthobdella peledina* Grube and *Branchiobdella pentodonta* Whitman (Annelida). *Ann. Zool. fenn.* **3**(1), 70–2.

Nusbaum, J. (1905). See Autrum (1939*c*).

Nusbaum, M.P., and Kristan Jr, W.B. (1981). Motor control in the leech: a newly identified interneuron subserving ventral flexion. *Neurosci. Abstr.* **7**, 137.

—— —— (1982). The swim-initiating ability of intersegmental serotonin-containing leech interneurons. *Neurosci. Abstr.* **8**, 161.

Nybelin, O. (1943). *Nesophilaemon* n. g. fur *Philaemon skottsbergi* L. Johansson aus den Juan Fernandez Inseln. *Zool. Anz., Leipzig* **142**(11/12), 249–50.

—— (1963). Har fiskigeln *Acanthobdella peledina* Grube en diskontinuerlig utbredning i Skandinavien? *Fauna flora* **58**, 239–46.

Obaid, A.L., Shimizu, H., and Salzberg, B.M. (1982). Intracellular staining with potentiometric dyes: optical signals from identified leech neurons and their processes. *Biol. Bull.* **163**(2), 388. (Abstr.)

Obermeier, P. (1974). Modern drugs for fish. Masoten ᴿ: ectoparasite control. *Vet. Med. Rev.* No. 2, 172–8. [Masoten [Metrifonate] at 0.2–0.4 p.p.m. effective against leeches.]

Obr, S. (date undetermined). Contribution to the study of the fauna of springs, lakes and torrents in the mountains Liptovska hole (Tatra-Czechoslovakia). *Mem. Soc. Zool. tcheosl.* **19**, 10–26.

Ocheretenko, E.E. (1948). [Leeches of south-eastern Ukraine.] *Rastitel'nii i Zhivotnii Mip Ugo-vostoka USSR Anim.* **2**(4), 55–8.

Ochmann, A. (1965). [How can leeches be induced to bloodsucking?] *Hippokrates* **36**, 678.

Oka, A. (1894). Beiträge zur Anatomie der *Clepsine. Z. wiss. Zool.* **58**, 79–151.

—— (1895*a*). Description d'une espèce d'*Ozobranchus. Zool. Mag. Tokyo* 7. (Incomplete reference)

—— (1895*b*). On some new Japanese land leeches (*Orobdella* n. g.) *J. Coll. Sci., Tokyo* 7 (1894), 275–306.

—— (1902). Über das Blutgefässsystem der Hirudineen. Vorl. Mitt. *Annot. Zool. Jap.* **4**, 49–60.

—— (1904). Über den Bau von *Ozobranchus. Annot. Zool. Jap.* **5**, 133–45.

—— (1910*a*). Synopsis der japanischen Hirudineen, mit diagnosen der neuen Species. *Annot. Zool. Jap.* **7**, 165–83.

—— (1912). Eine neue *Ozobranchus*-Art aus China (*Ozobranchus jantseanus* n. sp.). *Annot. Zool. Jap.* **8**, 1–4.

—— (1917*a*). *Ancyrobdella biwae* n.g.n.sp., ein merkwürdiger Rüsselegel aus Biwa-See. *Annot. Zool. Jap.* **9**, 185–93.

—— (1917*b*). Zoological results of a tour in the Far East. Hirudinea. *Mem. Asiat. Soc. Bengal* **6**, 157–76.

—— (1922). *Hirudinea* from the Inlé Lake, S. Shan States. *Rec. Ind. Mus.* **24**, 521–34.

—— (1923). Sur les deux genres *Mimobdella* Blanchard et *Odontobdella* nov. gen. *Annot. Zool. Jap.* **10**, 243–52.

—— (1925*a*). Notices sur les Hirudinées d'extrême Orient. I–IV. *Annot. Zool. Jap.* **10**, 311–26.

—— (1925*b*). Notices sur les Hirudinées d'extrême Orient. V–VII. *Annot. Zool. Jap.* **10**, 327–35.

—— (1926). Notices sur les Hirudinées d'extrême Orient. VIII. *Annot. Zool. Jap.* **11**, 59–62.

—— (1927*a*). Sur la morphologie externe de *Carcinobdella kanibir*. *Proc. Imp. Acad. Tokyo* **3**(3), 171–4.

—— (1927*b*). Sur la morphologie externe de *Trachelobdella okae*. *Proc. Imp. Acad. Tokyo* **3**(4), 239–41.

—— (1927*c*). Sur une nouvelle Ichthyobdellide parasite de l'huître. *Proc. Imp. Acad. Tokyo* **3**(6), 364–7.

—— (1927*d*). Sur la présence de l'*Ozobranchus margoi* au Japon, et description de cette Hirudinée. *Proc. Imp. Acad. Tokyo* **3**(7), 470–3.

—— (1928*a*). Description de l'*Hemiclepsis kasmiana*. *Proc. Imp. Acad. Tokyo* **4**(2), 64–6.

—— (1928*b*). Sur les Hirudinides de la Formose méridionale. *Proc. Imp. Acad. Tokyo* **4**(3), 122–4.

—— (1928*c*). Sur deux espèces nouvelles de *Whitmania*. *Proc. Imp. Acad. Tokyo* **4**(4), 169–71.

—— (1928*d*). Description de deux espèces Japonaises de *Glossiphonia*. (*Gl. smaragdina* et *Gl. lata*). *Proc. Imp. Acad. Tokyo* **4**(9), 543–6.

—— (1928*e*). Sur la morphologie et la variabilité de la *Callobdella livanovi*. *Proc. Imp. Acad. Tokyo* **4**(9), 547–9.

—— (1928*f*) Sur une nouvelle espécie d'*Hemiclepsis*, parasite d'un crabe fluviatile de Chine. *Proc. Imp. Acad. Tokyo* **4**(10), 607–8.

—— (1929*a*). Sur la présence de *Dinobdella ferox* en Formose. *Proc. Imp. Acad. Tokyo* **5**(5), 210–12.

—— (1929*b*). Sur une nouvelle *Placobdella*, parasite d'une crevette fluviatile de Formose. *Proc. Imp. Acad. Tokyo* **5**(6), 249–51.

—— (1929*c*). Révision des Herpobdelles d'extrême Orient. *Proc. Imp. Acad. Tokyo* **5**(7), 277–9.

—— (1930*a*). Sur une variété de l'*Haemadipsa zeylanica* s'attaquant aux oiseaux. *Proc. Imp. Acad. Tokyo* **6**(2), 82–4.

—— (1930*b*). Sur un nouveau genre d'Hirudinées provenant de l'Amérique du Sud. *Proc. Imp. Acad. Tokyo* **6**(6), 239–42.

—— (1930*c*). Sur l'anatomie de l'*Herpobdella formosana*. *Proc. Imp. Acad. Tokyo* **6**(7), 279–81.

—— (1931*a*). Sur l'anatomie de l'*Ichthyobdella uobir*. *Proc. Imp. Acad. Tokyo* **7**(2), 64–6.

—— (1931*b*). Sur une nouvelle espèce d'*Hemiclepsis* provenant de Chine. *Proc. Imp. Acad. Tokyo* **7**(3), 121–3.

—— (1931*c*). Révision du genre *Whitmania* d'extrême Orient. *Proc. Imp. Acad. Tokyo* **7**(10), 387–9.

—— (1932*a*). Sur une nouvelle espèce de *Placobdella*, *P. japonica* n. sp. *Proc. Imp. Acad. Tokyo* **8**(2), 51–3.

—— (1932*b*). Hirudinées extraeuropéenes du Musée Zoologique Polonais. *Ann. Mus. zool. Pol.* **9**, 313–28.

—— (1933*a*). Sur l'organisation intérieure de la *Carcinobdella kanibir. Proc. Imp. Acad. Tokyo* **9**, 188–90.

—— (1934*a*). Note sur les moers de la *Myxobdella sinanensis. Proc. Imp. Acad. Tokyo* **10**, 519–20.

—— (1934*b*). Comparaison des *Limnatis granulosa* provenant de la Formosa et de la Martinique. *Proc. Imp. Acad. Tokyo* **10**(5), 286–8.

—— (1935*a*). Description d'un nouveau genre d'Hirudinée de la famille des Glossiphonides, *Oligoclepsis tukubana* n. g. n. sp. *Proc. Imp. Acad. Tokyo* **11**, 66–8.

—— (1935*b*). Aperçu de la faune hirudinéenne japonaise. *Proc. Imp. Acad. Tokyo* **11**, 240–1.

—— (1938). Hirudinea of Jehol. *Report First Scient. Exped. Manchoukuo*, June–October, 1935. Sect. V, Div. i, **1**(3), 1–31 [In Japanese; English summary.]

[List of Oka's papers in *Annot. Zool. Jap.* **17**, 199–204 (1938).]

Okada, K. (1957) 1968. Annelida. In *Invertebrate embryology* (ed. M. Kumé and K. Dan) pp. 192–241. Bai Fu Kan Press, Tokyo. [Transl. in 1968 from Japanese by J.C. Dan; published for National Library of Medicine, Public Health Service and National Science Foundation, Washington, DC, by the NOLIT, Publishing House, Belgrade, Yugoslavia. Available from Clearinghouse, US Department of Commerce.]

Ökland, J. (1964). The eutrophic Lake Borrevann (Norway) — an ecological study on shore and bottom fauna with special reference to gastropods, including a hydrographic survey. *Folia limnol. scand.* **13**, 1–337.

Olive, J.H., and Dambach, C.A. (1973). Benthic macroinvertebrates as indexes of water quality in Whetstone Creek, Morrow County, Ohio (Scioto River Basin). *Ohio J. Sci* **73**(3), 129–49. [Page 145, pollution tolerant: *Helobdella stagnalis, H. 'fusca'*, and *Erpobdella punctata.*]

Oliver, D.R. (1958). The leeches (Hirudinea) of Saskatchewan. *Can. Field-Nat.* **72**(4), 161–5.

Oliver, L. (1950). Contribution à la connaissance de la faune aquatique de Puy-de-Dôme. *Rev. Sci. nat. Auvergne* (new series) **16**(1–4), 73–5.

Olson Jr, A.C., and Gadler, R.M. (In press). Larval trematodes from *Dina anoculata* (Hirudinea: Erpobdellidae) and its food habits in San Diego County, California. *J. Parasit.*

Olson, R.E. (1978). Parasitology of the English sole, *Parophrys vetulus* Girard in Oregon, U.S.A. *J. fish. Biol.* **13**(2), 237–48. [1979BA:17287.] [*Oceanobdella knightjonesi.*]

Olsson, P. (1896). Sur *Chimaera monstrosa* et ses parasites. *Mém. Soc. zool. Fr.* **9**, 499–512. [Two specimens of *Calliobdella nodulifera from* head of *Chimaera* from the Skagerrak.]

Onderíková, V. (1957). [*Batrachobdella paludosa* (Carena 1823). (Glossiphoniidae, Hirudinea) auf der Schutt-Insel.] *Biol. Bratislava* **12**, 776–9.]In Czechoslovak; German summary.]

—— (1958). [Faunistische Uebersicht über die auf der Schutt-Insel (Sudwestslowakei) vorkommenden Blutegel (Hirudinea).] *Biol. Bratislava* **13**(8), 628–31. [In Czechoslovak; German summary.]

Oostheer, R. (1970). Leeching. *S. Afr. med. J.* **44**, 678.

Oosthuizen, J.H. (1964). Collecting and preserving leeches (Hirudinea). *Newsl. Limnol. Soc. Sth. Afr.* **1**(2), 18–21.

—— (1978*a*). Two new *Batracobdella* species from Southern Africa and a redescription of *Batracobdella disjuncta* (Moore, 1939) comb. nov. (Hirudinea: Glossiphoniidae). *Madoqua* **II**(2), 89–106.

—— (1978*b*). A new *Glossiphonia* species from southern Africa (Hirudinea: Glossiphoniidae). *Ann. Transv. Mus.* **31**(13), 169–76.

—— (1979). Redescription of *Placobdella multistriata* (Johansson, 1909) (Hirudinea: Glossiphoniidae). *Koedoe* **22**, 61–79.

—— (1980). Problems concerning the evaluation of taxonomic characters in the Hirudinea with special reference to the leeches of Africa. *Indikator* **12** (1+2), 24. (Abstract). Zool. Soc. Southern Africa.

—— (1982). Redescriptions of *Placobdella stuhlmani* (Blanchard, 1897) and *Placobdella garoui* (Harding, 1932) (Hirudinea: Glossiphoniidae). *J. limnol. Soc. S. Afr.* **8**(1), 8–20.

—— Fourie, F. le R. (1985). Mortality amongst waterbirds caused by the African duck leech *Theromyzon cooperi*. *S. Afr. J. Wildl. Res.* **15**(3), 98–106.

Opalinski, K.W. (1971). Macrofauna communities of the littoral of Mikolajskie Lake. *Pol. Arch. Hydrobiol.* **18**(3), 275–85.

Orchard, I., and Webb, R.A. (1980). The projections of neurosecretory cells in the brain of the North American medicinal leech, *Macrobdella decora*, using intracellular injection of horseradish peroxidase. *J. Neurobiol.* **11**(3), 229–42. [1980BA:23790.]

Orr, R.T., and Eng, L.L. (1975). *Biological Studies of the Delta-Mendota Canal, Central Valley Project, California II.* Leeches, pp. 88–93. [*Mooreobdella microstoma* and *Helobdella stagnalis*.]

Orrhage, L. (1971). Light and electron microscope studies of some annelid setae. *Acta zool.* **52**, 157–69.

Országh, O., and Alföldy, J. (1940). A coagulative serum prepared by injecting hirudin into animals. *Lancet* **i**, 28–9.

Ort, C.A., Kristan Jr, W.B., and Stent, G.S. (1974). Neuronal control of swimming in the medicinal leech. II. Identification and connections of motor neurons. *J. comp. Physiol.* **94**, 121–54.

Osborne, J.A., Wanielista, M.P., and Yousef, Y.A. (1976). Benthic fauna species diversity in six central Florida lakes in summer. *Hydrobiologia* **48**(2), 125–9. [1976BA:13104.]

Osborne, L.L., Davies, R.W., and Rasmussen, J.B. (1980). The effect of total residual chlorine on the respiration rates of two species of freshwater leech (Hirudinoidea). *Comp. Biochem. Physiol.* **67C**, 203–5.

Osborne, N.N., Briel, G., and Neuhoff, V. (1972), The amine and amino acid composition in the Retzius cells of the leech *Hirudo medicinalis*. *Experientia* **28**, 1015–18.

—— Patel, S., and Dockray, G. (1982). Immunohistochemical demonstration of peptides, serotonin and dopamine-ß-hydroxylase-like material in the nervous system of the leech, *Hirudo medicinalis*. *Histochemistry* **75**(4), 573–84. [1983BA:65293.]

Osmanov, S.O. (1963). [Parasitic protozoa of fish in the Uzbek SSR.] *Vest. karakalpaksk. Fil. Akad. Nauk Uzbek SSR* **4**(14), 20–31. [In Russian.] [Appendix with leeches, hosts, and localities.]

—— (1964). [Parasites and their controls in fish in the Zeravschanck Reservoir). *Fil. Akad. Nauk Uzbek SSR* **4**(18), 28–45. [In Russian.]

—— (1971). [Parasites of fish of Uzbekistan.] *Tashkent: Izdatel'stvo 'FAN' Uzbekskoi SSR*. [In Russian.] [Review; three leech species.]

—— Arystanov, E., Ubaidullaev, K., and Yusupov, O. (1971). [Fisheries expedition to the Bugun'skii Bay and Uyaly Island.] *Vestnik Karakalpakskogo Filiala Akademii Nauk Uzbekskoi SSR* **3**(45), 92–4. [Leeches from Aral Sea.]

Osmanović, S.S. (1980). [Effect of ouabain on the electrical properties of the Retzius cell.] M.Sc. thesis. Medical Faculty, Beograd, Yugoslavia. [Serbian.]

—— and Beleslin, B.B. (1981). On the mechanism of ouabain induced depolarization on the Retzius nerve cells of the horse leech. *Periodicum biol.* **83**(1), 175–6.

—— —— (1982). Effect of the removal of external potassium on Retzius nerve cells of leech. *Periodicum biol.* **84**(2), 170–2.

Ostrovskii, I.S. (1980). [Changes in the bathymetric distribution of large zoo-benthic taxonomic groups relative to the drop in water level of Lake Sevan, Armenia.] *Biol. Zh. Arm.* **33**(3), 330–3. [In Russian.]

Overstreet, R.M. (1973). Parasites of some penaeid shrimps with emphasis on reared hosts. *Aquaculture* **2**, 105–40.

—— (1978). Marine maladies? Worms, germs, and other symbionts from the Northern Gulf of Mexico. Mississippi-Alabama. Sea Grant Consortium 78–021. Blossom Printing, Inc., Ocean Springs, Mississippi. 140 pp.

Owen, W.G., and Wagner, R.H. (1971). Preparation and properties of water-insoluble thrombin. *Am. J. Physiol.* **220**(6), 1941–3.

Özer, F., and Winterstein, H. (1949). Ueber die Beziehungen zwischen Sauerstoffverbraud und Kontraktion beim Blutegelmuskel. *Physiol. Comp. Oecol. Den Haag* **1**(3/4), 331–9. [English and French Summaries.]

Pai, G.G.A., and Prabhoo, N.R. (1980). Herpobdellid leech as a potential predator of larval *Culex fatigans* in Kerala, India. *Entomon.* **5**(3), 203–6.

Paja, S., and Varaporn, I. (1965). Field studies on the effectiveness of repellants against terrestrial leeches in southern Thailand. *Special Report Joint Thai–U.S. Military Research and Development Center*, Jan. 1–60.

Paldrock, A. (1896). Über die Benutzung ungerinnbaren Blutes zu Durchströmungen. *Arb. Pharm. Inst. Dorpat* **13**, 64. [Mentions Kruger's work on leeches.]

Palmieri, J.R., and James, H.A. (1976a). The effects of leech behavior on penetration and localization of *Apatemon gracilis* (Trematoda: Strigeidae) cercariae and metacercariae. *G. Basin Nat.* **36**(1), 97–100. [1977BA:21199.]

—— —— (1976b). Metacercarial composition and development in *Apatemon gracilis* (Trematoda: Strigeidae). *Iowa St. J. Res.* **50**(4), 409–17. [1976BA:55334.]

—— Kuzia, E.J., and James, H.A. (1913). Leeches for classroom study. *Carolina Tips* **36**(2), 5–6.

Paloumpis, A.A., and Starrett, W.C. (1960). An ecological study of benthic organisms in three Illinois River flood plain lakes. *Am. Midl. Natur.* **64**(2), 406–35.

Pan Len Shun (1959). [The leeches found in Nanking and vicinity.] *Zool. Mag.* **3**(5), 220–3. [In Chinese.]

Paperna, I., and Zwerner, D.E. (1974). Massive leech infestation on a white catfish (*Ictalurus catus*): a histopathological consideration. *Proc. helminthol. Soc. Wash.* **41**(1), 64–7. [*Myzobdella lugubris*, nec *Cystobranchus virginicus*.]

—— —— (1976). Parasites and diseases of striped bass, *Morone saxatilis* (Walbaum), from lower Chesapeake Bay. *J. fish. Biol.* **9**, 267–87.

Parat, M. (1928). Contribution à l'étude morphologique et physiologique de cytoplasme. Chondriome vacuome (appareil de Golgi), enclaves, etc. *Arch. Anat. microsc.* **24**, 73ff.

Parnas, I. (1981). Killing single cells. In: *The neurobiology of the leech.* (ed. K.J. Muller, J.G. Nicholls, and G.S. Stent) pp. 227–34. Cold Spring Harbor.

—— and Bowling, D. (1977). Killing of single neurones by intracellular injection of proteolytic enzymes. *Nature, Lond.* **370**, 626–8.

—— —— and Nicholls, J.G. (1978). Elimination of single neurons on the central nervous system of the leech. *Isr. J. med. Sci.* **14**(4), 497–8.

Parker, T.J., and Haswell, W.A. (1947). Hirudinea. In *A textbook of zoology*, 6th edn, Vol. I, pp. 355–70. Macmillan, London. [First edn: (1897) Hirudinea, Vol. I, pp. 465–81.]

Parukhin, A.M., and Epshtein, V.M. (1970). New data on the geographical distribution and the hosts of the leech *Trachelobdella lubrica*. All-Union symposium (1st) on parasites and diseases of marine animals, Sevastopol, 1970, Naukova Dumka, Kiev. pp. 104–5.

—— and Lyadov, V.N. (1981). [Parasite fauna of notothenioid fish from the Atlantic and Indian Oceans.] *Vestn. Zool.* **3**, 90–4. [In Russian; English summary.] [1983BA:28248.]

Paspaleff, G., Boschkow, D., Dokov, V.K., and Tehacarof, E. (Date undetermined). *Batracobdella algira* (Moquin-Tandon, 1846) — Überträger einer infektiösen Erkrankung an Fröschen in Bulgarien. *Biol. Bratislava* **18**(10), 781–6. [Infestation of *Batracobdella algira* causing sickness in frogs in Bulgaria.]

—— Dokov, V.K., Tehacarof, E., Boschkow, D., and Todorov, T. (1961). Durch rickettsiehähnliche Mikroorganismen hervorgerufene entzündlich — nekrotische Veränderungen in Seefrosch (*Rana ridibunda* Pall.). *Dokl. Bulg. AN* **14**(3), 317–20.

Pastisson, C. (1965). Recherches préliminaires sur l'ultrastructure du spermatozoïde de la sangsue, *Hirudo medicinalis* L. *C. r. hebd. Séanc. Acad. Sci. Paris* **261**, 2950–3.

—— (1966). Anatomie ultrastructurale du spermatozoïde de la sangsue, *Hirudo medicinalis* L. *Ann. Un. Ass. Reg. Etude Rech. Sci., Reims* **4**(2), 67–75.

—— (1968). Ultrastructure and histochemistry of the prostate of *Hirudo medicinalis*. *Ann. Sci. natur. Zool. Biol. anim.* **10**(12), 151–62.

—— (1977). L'Ultrastructure des celles séminales de la sangsue *Hirudo medicinalis* au cours de leur différenciation. *Ann. Sci. nat.*, Ser. 12, **19**, 315–47. [In French.]

Paterson, C.G., and Fernando, C.H. (1969). Macroinvertebrate colonization of the marginal zone of a small impoundment in eastern Canada. *Can. J. Zool.* **47**(6), 1229–38.

Paterson, W.B., and Desser, S.S. (1976). Observations on *Haemogregarina balli* sp. n. from the common snapping turtle, *Chelydra serpentina*. *J. Protozool.* **23**, 294–301.

Patrick, R., Cairns, Jr, J., and Roback. S.W. (1967). An exosystematic study of the fauna and flora of the Savannah River. *Proc. Acad. nat. Sci. Phil.* **118**(5), 109–407.

Patton, W.J., and Strickland, C. (1908). See Autrum (1939*c*).

Pavlichenko, V.I. (1977). [The role of *Hydropsyche angustipennis* Curt. (Trichoptera, Hydropsychidae) larvae in destroying black flies in flowing reservoirs of the Zaporozhye oblast.] *Ekoliogiya* **1**, 104–5. [In Russian.] [1978BA:1400.] [Black flies eaten by *Erpobdella octoculata*.] [See also: (1977). *Biol. Nauki* **20**(8), 44–6.]

Pawlowska, T. (1963). [Les sangsues du bassin de la Warta moyenne.] *Zesz. Nauk Uniw. Lodzkiego.* **2**(14), 123–32. [In Polish; French summary.] [See also Jazdzewska, T.]

Pawlowski, E.I. (1934). Parazitirovanie piyavok u cheloveka i u mlekopitauschikh v Turkmenii. *Tr. Karakalinskoi i Kz'l-Atrekskoi Eksped.*, Leningrad 1931, 149–54.

Pawlowski, L.K. (1934). *Drilophaga bucephalus* Vejdovsky, ein parasitisches Rädertier. *Mem. Acad. Pol. Cracovie, Ser. B Sci. Nat.* 95–104.

—— (1935). Beiträge zur Anatomie und Biologie von *Drilophaga delagei* de Beauchamp. *Arch. Hydrobiol. Suwalki* **9**, 1–30.

—— (1936a). Über den Verdauungskanals des Egels *Theromyzon tesselata* (O.F. Müller). *Fol. Morph.*, *Warszawa* **6**, 87–91.

—— (1936b). Pijawki (Hirudinea). *Fauna Slodkowodna Polski. Zeszyt* **26**, 3–176.

—— (1936c). Zur Ökologie der Hirudineenfauna der Wigryseen. *Arch. Hydrobiol. i Rybachtwa* **10**, 1–47.

—— (1937a). Bemerkungen über die ökologische Verbreitung der Hirudineen fauna in den Gebirgsbachen der polnischen Ostkarpathen. *Verh. int. Verein. Limnol.*, *Stuttgart* **8**, 187–97.

—— (1937b). Nowy gatunek pijawki na ziemiach polskich. *Przyr. ix Techn.*, *Lwów, Warsaw* **16**, 364.

—— (1938). *Theromyzon tesselatum* (O.F. Müller). Egel im menschlichen Auge. *Zool. Pol.* **2**, 181–3.

—— (1939). O rozmieszczeniu ekologicznym pijawek w potokach górskich Huculszczyzny. *Pam. XV. Zjazd. Lek. Przyr. Pol.*, *Lwów* 112–13.

—— (1947a). [Sur la biologie du *Cystobranchus fasciatus* (Kollar).] *Pr. Wydz. Mat.-Przyr, Lodz. TN, Lodz* 2.

—— (1947b). Przyczynek do znajomości biologii pijawki *Cystobranchus fasciatus* (Kollar). *Spraw. z czynn. i pos. Lodz. TN, Lodz* 1(1946), **1**, 61–2.

—— (1947c). Przyczynek do systematyki pijawek z rodzaju *Erpobdella* de Blainville. *Spraw. Lodz. TN, Lodz.* **1**(2), 30–4.

—— (1948a). Contribution à la systématique des sangsues du genre *Erpobdella* de Blainville. *Acta zool. oecol. Un. Lodz.* **1**, 1–56.

—— (1948b). Contribution à la connaissance des sangsues (Hirudinea) de la Nouvelle-Écosse, de Terre-Neuve et des îles françaises Saint-Pierre et Miquelon. *Frag. faun. Mus. zool. Pol.* **5**(20), 317–53.

—— (1949). Badania hydrobiologiczne i zoologiczne na Huculszczyźnie. *Spraw. z czynn. i pos. LTD, III* **1**, 72–4.

—— (1950a). Cas particulier d'anomalie dans la structure du corps de la sangsue *Erpobdella octoculata*. *Bull. Soc. Sci. Lett. Lodz. Cl. Sci. Math. Nat.* **3**(1), 1–3.

—— (1950b). Sur la biologie du *Cystobranchus fasciatus* (Kollar). *Bull. Soc. Sci. Lodz* **1** (1946–47), 79–80.

—— (1950c). Contribution à la systematique des sangsues du genre *Erpobdella* de Blainville. *Bull. Soc. Sci. Lodz* **1** (1946–47), 80–5.

—— (1951). [Leeches (Hirudinea) of the river pumps station and the waterworks of Warsaw.] *Frag. faun. Mus. zool. Pol.* **6**(10), 169–92. [In Polish.]

—— (1952a). Ciekawy pazypadek anomalii w budowie pijawki *Erpobdella octoculata* (L.). *Spraw. Lodz. TN, Lodz* 1949 **2**(8), 1–4. [See also: **2**(8), 78–80.]

—— (1952*b*). [Leeches (Hirudinea) of the river pumps station and the waterworks of Warsaw.] *Spraw. Lodz TN, Lodz* **4**(1949), **2**(8), 80–3. [In Polish; English summary.]

—— (1954*a*). [On the structure of somite and segmentation in leeches.] *Kosmos, Ser. Biol., Warsaw* 3, **4**(9), 455–7. [In Polish.]

—— (1954*b*). [Leeches from Wielki Staw at Snierka in the Karkonosz Mountains.] *Kosmos, Ser. Biol., Warsaw* 3,**5**(10), 622–3. [In Polish.]

—— (1955*a*). Revision des genres *Erpobdella* de Blainville et *Dina* R. Blanchard (Hirudinea). *Bull. Soc. Sci. Lett. Lodz* **6**(3), 1–14.

—— (1955*b*). Observations biologiques sur les sangsues. *Bull. Soc. Lett. Lodz* III **6**(5), 1–21.

—— (1958*a*). Hirudinées dans la collection du Docteur K. Lindberg. *Bull. Soc. Sci. Lett. Lodz* **9**(11), 1–13. [From Greece and France.]

—— (1958*b*). Wrotki (*Rotatoria*) rzeki Grabi Część I — faunisty czna. *Soc. Sci. Lódź., Section III*, No. 50.

—— (1959*a*). Aperçu sur la faune des hirudinées des Carpathes. *Pr. Wydz. III Nauk Mat. Przyr. Lodz. TN, Lodz* **55**, 1–45. *Soc. Sci. Lodz* 3(Section 3) (55), 1–49.

—— (1959*b*). Remarques sur la faune torrenticole des Carpathes. *Pr. Wydz. III Nauk Mat.-Przyr., Lodz TN, Lodz* **57**, 1–84.

—— (1960). Rotifères nouveaux et rares parmi la faune de la Pologne. *Bull. Soc. Sci. Lett. Lodz* **XI**, 6.

—— (1962). [Sur la présence de l'*Erpobdella octoculata* (L.). au Japon.] *Zeszyty Nauk Uniw. Lodz. Nauk. Matemat.-Przyr, Ser.* 2, **12**, 127–36. [In Polish; French transl.]

—— (1963). [Nomenclature of European leeches.] *Zesz. Nauk. Uniw. Lodz Nauk Matemat.-Przyr., Ser.* 2, **14**, 99–110. [In Polish; French summary.] [Page 103, taxonomic note on *Bakedebdella* Sciacchitano.]

—— (1968). Pijawki. Hirudinea *Katalog Fauny Polski, Warszawa* **11**(3), 3–94. [Comprehensive account of freshwater leeches of Poland; many references to Polish leeches not in this bibliography.]

—— (1972). *Salifa perspicax* Blanchard (Hirudinea) et quelques remarques sur les Pharyngobdelles. *Lodz. tow. Nauk, place wydzialu,* III **113**, 1–25.

—— (1973). Répartition géographique des sangsue du genre *Trocheta* Dutrochet et les glaciations du Pléistocene. *Soc. Sci. Lodz Sect.* III **114**, 1–35.

—— and Hoffmann, J. (1959). Note comparative sur la configuration des cocoons Piscicolidées: *Cystobranchus fasciatus* (Kollar) et *Cystobranchus respirans* (Troschel). *Arch. Inst. Grand-Ducal Sect. Sci. nat., Phys., Math. Luxembourg* (new series) **26**, 187–93.

—— and Jazdzewska, T. (1970). [The occurrence of the leech *Cystobranchus fasciatus* (Kollar) in Poland.] *Zesz. nauk. Uniw. Lodz. Ser.* 2 **40**, 19–29. [In Polish; English summary.]

Payne, M.J. (1963). *Trocheta subviridis* — a further Yorkshire record. *Naturalist*, Hull, 1963, 120.

Payton, B.W. (1981a). History of medicinal leeching and early medical references. I: *The neurobiology of the leech* (eds. K.J. Muller, J.G. Nicholls and G.S. Stent) Chap. 3, pp. 27–34. Cold Spring Harbor.

—— (1981b). Structure of the leech nervous system. In: *The neurobiology of the leech* (eds. K.J. Muller, J.G. Nicholls and G.S. Stent) Chap. 4, pp. 35–50. Cold Spring Harbor.

—— (1984). An historical survey of illustrations of the medicinal leech. *J. audiovisual Media in Medicine* 7, 105–12.

—— and Lowenstein, W.R. (1968). Stability of electrical coupling in leech giant nerve cells: divalent cations, propionate ions, tonicity and pH. *Biochim. biophys. Acta* 150, 156–8.

Pearce, R.A., and Friesen, W.O. (1984). Intersegmental coordination of leech swimming: comparison of *in situ* and isolated nerve cord activity with body wall movement. *Brain Res.* 299, 363–6.

Pearl, R., and Cole, L.J. (1902). [Experiments on light reactions in *Clepsine* sp.] *Rep. Mich. Acad. Sci.* 3. (Incomplete reference)

Pearse, A.S. (1924). The parasites of lake fishes. *Trans. Wisc. Acad. Sci.* 26, 437–40.

—— (1932). Parasites of Japanese salamanders. *Ecology* 13(2), 135–52. [Salamanders eating leeches and leech cocoons.]

—— (1936). Estuarine animals at Beaufort, North Carolina. *J. Elisha Mitchell Scient. Soc.* 52, 174–222.

Peaucellier, G. (1977). The life history, growth and age structure of *Nephelopsis obscura* (Hirudinoidea) in Alberta, Canada. *Expl Cell Res.* 106(1), 1–14. [1977BA:9101.]

Pellegrino, M., and Simonneau, M. (1982). Acetylcholine receptors on identified central leech neurons in culture. *Neurosci. Lett.* 380, 10.

Penn, R.D., and Lowenstein, W.R. (1966). Uncoupling of a nerve cell membrane junction by calcium-ion removal. *Science, NY* 151, 88–9.

Pennak, R.W. (1969). Colorado semidrainage mountain lakes. *Limnol. Oceanogr.* 14, 720–5. [*Glossiphonia complanata, Helobdella stagnalis*, and *Nephelopsis obscura* as indicator species.]

—— (1978). Hirudinea (leeches). In *Freshwater invertebrates of the United States*, 2nd edn, pp. 297–317. Wiley, New York. [1st edn (1953): pp. 302–21.]

Penner, L.R., and Sanjeeva Raj, P.J. (1977). Concerning the marine leech, *Pontobdella macrothela* Schmarda, 1861 (Piscicolidae: Hirudinea). *Parasitol. Mem. Doctor Eduardo Caballeroy Caballero. Un. Nac. Autonoma Méx. Inst. Biol. Publ. esp.* 4, 519–30. [Spanish summary.]

Penners, A. (1935). Die Bedeutung der Materialmenge für die Segmentierung und der Zusammenhang zwischen Zellgrösse und Körpergrösse bei Tubificiden und Triton. *Z. wiss. Zool.* 146, 463–516.

Pentreath, V.W. (1977). 5-Hydroxytryptamine in identified neurons (proceedings). *Biochem. Soc. Trans.* **5**(4), 854–8.

—— Seal, L.H., and Kai-Kai, M.A. (1982). Incorporation of tritium-labelled 2-deoxyglucose into glycogen in nervous tissues. *Neuroscience* **7**(3), 759–67. [1984BA:11876.]

Pérez, Ch. (1907). See Autrum (1939c).

Perez, H.V.Z. (1942). On the chromaffin cells in the nerve ganglia of *Hirudo medicinalis* L. *J. comp. Neurol.* **76**, 367–401.

Perez-Reyes, R. (1968). *Trypanosoma galba* n. sp. parasito de ranas mexicanas. Morfologia y ciclo en el vertebrado. *Rev. Latinoam. Microbiol. Parasitol.* **10**(2), 79–84. [Infection of the leech *Haementeria officinalis* was not obtained.]

Perkins, B.A., and Cottrell, G.A. (1972). Choline acetyltransferase activity in nervous tissue of *Hirudo medicinalis* (leech) and *Nephrops norvegicus* (Norway lobster). *Comp. gen. Pharmacol.* **3**(9), 19–21.

Perret, J.L. (1952a). *Cystobranchus respirans* (Trochsel) dans le lac de Neuchâtel et observations sur le genre *Trocheta* Dutrochet. *Rev. Suisse Zool.* **59**, 579–83.

—— (1952b). Les Hirudinées de la région neuchâteloise. *Bull. Soc. neuchâteloise Sci. nat.* **75**, 89–138.

Perti, S.L., and Agarwal, P.N. (1970). Synergism in insecticides. *Labdev J. Sci. Technol. B Life Sci.* **8**(2), 67–71.

Pertsovskii *et al.* (1976). *Vrach Delo.* 1976(1), 73–6. (Incomplete reference)

Peruzzi, P., and Colombini, N. (1938). Sulla reattivitá del muscolo dorsale di sanguisuga in un particolare periodo dell'ano. *Boll. Soc. Ital. Biol. sper.* **XIII**(10), 968–70.

Pessôa, S.B. (1969a). Experiencias sobre a transmissao do *Trypanosoma cruzi* por sanguessugas e de tripanosomas de vertebrados de sangue frio por triatomineos. *Rev. Saude publ.* **3**(1), 17–20. [*Trypanosoma* not reproduced in the leech *Haementeria lutzi.*]

—— (1969b). Formas evolutivas do *Hepatozoon leptodactyli* (LeSage, 1908) na sanguessuga *Haementeria lutzi* Pinto, 1920. *Separata Rev. Goiana Med.* **15**, 155–9.

—— (1970). Formas evolutivas do *Hepatozoon leptodactyli* (LeSage, 1908) na sanguessuga *Haementeria lutzi* Pinto 1920. *Rev. Goiana Med.* **16**(1/2), 35–9. [Developmental forms of *Hepatozoon leptodactyli* in the leech *Haementeria lutzi.*]

—— and Cavalheiro, J. (1969a). Notas sobre hemogregarinas de serpentes brasileiras: VII. Sobre a evolucão da *Hepatozoon miliaris* na sanguessuga *Haementeria lutzi. Rev. Brasil Biol.* **29**(4), 451–8.

—— —— (1969b). Notas sobre hemogregarinas de serpentes brasileiras. VIII: Sobre a evolucão da '*Haemogregarina miliaris*' na sanguessuga '*Haementeria lutzi*'. *Rev. Brasil Biol.* **29**(4), 451–8.

—— —— (1969c). Notas sobre hemogregarinas de serpentes brasileiras:

IX. Sobre a hemogregarina da *Helicops carinicauda*. (Wied). *Rev. Goiana Med.* **15**(3/4), 161–8. [*Hepatozoon carinicauda*, parasite of the snake *Helicops carinicauda* (Wied)., and its development in the leech *Haementeria lutzi* Pinto are described.]

—— and Cunha Neto, A.G. (1967). Notas sobre hemoparasitas de ras de Goiania. *Rev. Goiana Med.* **13**, 101–16.

—— Biasi, P. de, and Souza, D.M. de (1972). Esporulação do *Hepatozoon caimani* (Carini, 1909) parasita de jacaré-de-papoamarelo: *Caiman latirostris* Daud, no *Culex dolosus* (L.Arribálzaga). *Mem. Inst. Oswaldo Cruz* **70**, 379–83 [Multiple sporocystic oocysts in leeches.]

—— Sacchetta, L., and Cavalheiro, J. (1970). Notas sobre hemogregarinas de serpentes brasileiras: X. Hemogregarinas da *Hydrodynastes gigas* (Dumeril et Bibron) e sua evolucão. *Rev. Latinoam. Microbiol.* **12**(4), 197–200.

Peterson, D.L. (1982). Management of ponds for bait leeches in Minnesota. Minnesota Department of Natural Resources, Section of Fisheries, Investigational Report No. 375. 43 pp.

Peterson, E.L. (1983*a*). Visual processing in the leech central nervous system. *Nature* **303**, 240–2.

—— (1983*b*). Generation and coordination of heartbeat timing oscillation in the medicinal leech (*Hirudo medicinalis*): 1. Oscillation in isolated ganglia. *J. Neurophysiol.* **49**(3), 611–26. (See also 1981. *Neurosci. Abstr.* **7**, 137). [1983BA:49785.]

—— (1983*c*). Generation and coordination of heartbeat timing oscillation in the medicinal leech (*Hirudo medicinalis*): 2. Intersegmental coordination. *J. Neurophysiol.* **49**(3), 627–38. [1983BA:49786.]

—— (1983*d*). Frequency-dependent coupling between rhythmically active neurons in the leech. *Biophys. J.* **43**(1), 53–61.

—— and Calabrese. R.L. (1982). Dynamic analysis of a rhythmic neural circuit in the leech *Hirudo medicinalis*. *J. Neurophysiol.* **47**(2), 256–71. [1982BA:24987.]

Petersen, T.E., Roberts, H.R., Sottrup-Jensen, L., and Magnusson, S. (1976). Primary structure of hirudin, a thrombin-specific inhibitor. In: *Protides of the biological fluids.* (ed. H. Peters) pp. 145–9. Proc. 23rd Colloquium, Brugge, 1975. Pergamon, Oxford.

Petrushevskii, G.K., and Bauer, O.N. (1948*a*). Parazitarn'e zabolevaniya r'b Sibiri i ikh p'bokhozyaistvennoe i meditsinskoe znachenie. *Izv. Vsesouzn. Nauch.-Issl. Inst. Ozern. I Rechn. R'bn. Khoz.* **27**, 195–216. [In Russian.] [*Acanthobdella peledina* and *Cystobranchus mammillatus*.]

—— —— (1948*b*). Zoogeographicheskaya Kharakteristika parazitov r'b Sibiri. *Izv. Vsesouzn. Nauch.-Issl. Inst. Ozern. I Rekhn. R'bn. Khoz.* **27**, 217–31. [In Russian.]

—— and B'Khovskaya, I. (Pawlowskaya) (1935). Material' po parazitologii r'b Karelii. I. Parazit' p'b ozer paiona Konchezera. *Tr. Borodinsk. Biol. St.* **8**(1), 15–17. [In Russian.]

—— Mosevich, M.V., and Schupakov, I.G. (1948). Phauna parazitov r'b Rek Obi i Irt'sha. *Izv. Vsesouzn. Nauch.-Issl. Inst. Ozern. I Rechn. R'bn. Khoz* **27**, 67–96. [In Russian.]

Pettigrew, D.J., and Fried, B. (1973). Behavior of various helminths in a thermal gradient. *Proc. helminthol. Soc. Wash.* **40**(2), 178–80.

Petty, L.L., and Magnuson, J.J. (1974). Lymphocystis in age 0 bluegills (*Lepomis macrochirus*) relative to heated effluent in Lake Monona, Wisconsin. *J. fish. Res. Bd Can.* **31**(7), 1189–93. [1975BA:3199.] ['*Piscicolaria reducta* also were more prevalent in the heated area and may have caused increased lymphocystis incidence by creating foci for viral penetration or by acting as vectors.']

Pfannkuche, von O., Jelinek, H., and Hartung, E. (1975). Zur Fauna eines Süsswasserwattes im Elbe-Aestuar. *Arch. Hydrobiol.* **76**(4), 475–98. [Leeches: 487–8.]

Phillips, C.E., and Friesen. W.O. (1982). Ultrastructure ·of the water-movement-sensitive sensilla in the medicinal leech (*Hirudo medicinalis*). *J. Neurobiol.* **13**(6), 473–86. (See also 1981. *Neurosci. Abstr.* **7**, 371). [1983BA:65148.]

Pike, A.W. (1968). Notes on some cysticeroids from pulmonate molluscs and leeches in British freshwaters. *J. Helminth.* **42**, 131–8.

Pillai, P.B.K. (1954). Leech (*Dinobdella ferox*) in nasal passages of dog. *Ceylon vet. J.* **2**, 62.

Pillers, A.W.N. (1931). Notes on parasites in 1930. *Vet. Rec.* **11**, 668–70. ['*Hirudo*' *hilderbrandti* in mouth of ox.]

Pilsbry, H.A., and Bequaert, J. (1927). The aquatic mollusks of the Belgian Congo. *Bull. Am. Mus. nat. Hist.* **53**, 69–602.

Ping, C. (1931). Preliminary notes on the fauna of Nanking. *Biol. lab. Sci. Soc. China* **7**(4), 178–9.

Pinto, C. (1920*a*). Contribuicão ao estudo dos. I. Hirudineos do Brazil (*Haementeria lutzi*, nov. sp.), Nota previa. *Brazil-Med.* **34**(35), 567–70.

—— (1920*b*). Contribuicão ao estudo dos hirudineos do Brazil. II. *Trachybdella bistriata* n. gen., n. sp. *Brazil-Med.* **34**(38), 624–6.

—— (1920*c*). Contribuicão ao estudo dos hirudineos do Brazil. III. *Limnobdella brasiliensis* n. sp. *Brazil-Med.* **34**(43), 707–9.

—— (1923). Ensaio monographico dos Hirudineos. *Revista Mus. Paul.* **13**, 857–1118.

—— (1945). Hirudineos. In *Zoo-parasitos de interesse medico e veterinario*, 2nd edn, pp. 448–54. Rio de Janeiro.

Pinus, M., and Müller, H.E. (1980). (Enterobacteria of bats (Chiroptera)). *Zentralbl. Bakteriol. 1. Abt. Orig. A. Med. Mikrobiol. Infektionskr. Parasitol.* **247**(3), 315–22. [In German.]

Plate, G. (1970). Masoten [R] für die Bekämpfung von Ektoparasiten bei Fischen. *Arch. Fischereiwiss.* **21**(3), 258–67. [French summary.]

Poe, T.P. (1972). Leeches (Hirudinea) parasitizing Illinois fishes. M.S. thesis, Northern Illinois University. Dekalb.

—— and Mathers, C.K. (1972). First occurrence of the leech *Actinobdella triannulata* in Illinois. *Trans. Ill. St. Acad. Sci.* **65**(1), 87–8.

Poll, H. (1908). See Autrum (1939c).

—— (1909). Über Nebennieren der Wirbellosen. *Sitz. K. Preuss. Akad. Wiss. Jrg.* 1909. 2 Halbbd, 889–96.

—— and Sommer, A. (1903). Über phaeochrome Zellen im Zentralnervensystem des Blutegels. *Arch. Anat. Physiol.* 549–50.

Polls, I., Lue-Hing, C., Zenz, D.R., and Sedita, S.J. (1980). Effects of urban runoff and treated municipal wastewater on a man-made channel in northeastern Illinois, USA. *Water Res.* **14**(3), 207–16.

Poloni, A. (1951). L'acetilcolina nel liquor dei malati di mente. Mancanza di effetto acetilcolinico e azione curaro-simile del liquor di schizofrenici sul muscolo dorsale di sanguisuga. *Cervello* **27**, 81–104. [Samples of cerebrospinal fluid from schizophrenics relax leech muscle.]

—— (1955a). Serotonina e schizophrenia. Rilieva sperimentale in favore dell'ipotesi di una tossicossi da 5-idrossitryptamina della schizofrenia. *Cervello* **31**, 231–42. [5-HT the active principle in the cerebrospinal fluid which relaxes leech muscle.]

—— (1955b). Il muscolo dorsale di sanguisuga quale test biologico per l'evidenziamento dell'attivitá serotoninica nei liquidi organici. *Cervello* **31**, 472–6.

Polyanski, Yu. I. (1955). [Materials concerning the parasitology of fish in the nothern seas of the U.S.S.R. Parasites of fish in the Barents Sea.] *Tr. Zool. Inst. Akad Nauk SSSR* **19**, 5–170. [In Russian.] [Israel Program for Scientific Translations (1966).]

—— (1957). [Some problems of fish parasitology in Barents Sea.] *Tr. Murman Biol. St.* **3**, 175–83. [In Russian.]

Ponomarenko, V.A. (1960). Opyt bor'by s pijavkami vodoplavajuscej pticy. *Ptitsevodstvo* **10**(6), 31. [*Theromyzon* on birds; treatment.]

Poon, M. (1976). A neuronal network generating the swimming rhythm of the leech. Ph.D. thesis, University of California, Berkeley.

—— Friesen, W.O., and Stent, G.S. (1978). Neuronal control of swimming in the medicinal leech. V. Connexions between the oscillatory interneurones and the motor neurones. *J. exp. Biol.* **75**, 45–63.

Pope, E.C. (1963). [Photos of *Richardsonianus australis* travelling by a series of looping movements.] *Aust. nat. Hist.* **14**, 179.

Popow, M. (1904). *Operculria clepsinis* nov. sp. *Zool. Anz.* **27**, 340–3.

Poppius, B.R. (1898). Sukkajuotikkaan (*Acanthobdella*) esiintymisesta Lapissa. *Luonnon Ystava* **2**, 30–2.

Potter, D.W.B., and Learner, M.A. (1974). A study of the benthic macroinvertebrates of a shallow eutrophic reservoir in South Wales with emphasis on the chironomidae (Diptera); their life-histories and production. *Arch. Hydrobiol.* **74**(2), 186–226. [1975BA:30510.] [*Helobdella stagnalis*,]

Pough, F. (1971). Leech-repellent property of eastern Red-Spotted Newts, *Notophthalmus viridescens. Science, NY* **174**, 1144–5. [see also: (1972). *NY Food Life Sci. Q.* **5**(2), 4–7.]

Poulding, R.H. (1954). Parasitism of a herring gull by the duck leech. *Br. Birds* **47**, 306–7. [*Theromyzon tessulatum* associated with death of bird.]

Poupart, F. (1697). Hirudinis tradit anatomen, animalis chirurgici. Histoire Anatomique de la sangsue. Histoire naturelle de la sangsue. *Journal des Savans* 22 July 1697 **XXV**, 537, 539.

Prasad, S.B., and Sinha, M.R. (1983). Vaginal bleeding due to leech. *Postgrad. med. J.* (India) **59**, 272.

Pratt, H.S. (1925). *A manual of the common invertebrate animals*, pp. 315–21. McClury, Chicago. [Poor discussion and key (out of date).]

Prenant,M. (1935). Annelides. *Act. Scient. Industr., Paris* No. 196.

Prendel, O.R., Korenchevs'ka, G.O., Stachors'ka, N.I. (1957). [Ecology of leeches of the lower Dniester River.] *Pr. Odes'k. Derzh. Un-tu Im. I. I. Mechnikova*, 'pik' 93, T.147, *Ser. Biol. Nauk* **8**, 123–5. [In Russian.]

Prensier, G., and Malecha, J. (1974). Particularités ultrastructurales de la méiose chez *Branchellion torpedinis* (Sav.) (Hirudinée, Rhynchobdelle). *J. Microsc.* **20**, 81a–82a.

Press, A. (1948). [A case of haemorrhage due to application of leeches.] *Harefuah, Tel Aviv.* **34**, 50. [In Hebrew.]

Preu, Th. (1935). Untersuchungen zur Frage der Zellkonstanz bei den Rüsselegeln. *Z. wiss. Zool.* **146**, 517–46.

Price, A. (Date undetermined). Some notes on British leeches (Hirudinea). *Reading Nat.* **17**, 19–21.

Price, R. (1822). *A treatise on the utility of sangui-suction . . . including the opinions of eminent practitioners, ancient and modern.* London.

Prichard, J.W. (1971a). Pentylenetetrazol-induced increase in chloride permeability of leech neurons. *Brain Res.* **27**, 414–17.

—— (1971b). Effect of pentylenetetrazol on the leech Retzius cell. *Expl Neurol.* **32**(2), 275–86.

—— (1971c). Effect of strychnine on the leech Retzius cell. *Neurophar-macology* **10**, 771–4.

—— (1972a). Bemegride-induced paroxysmal discharges in leech ganglion. *Brain Res.* **45**(2), 594–8. [1974BA:37777.]

—— (1972b). Effect of phenobarbital on a leech neuron. *Neurophar-macology* **11**, 585–90.

—— (1972c). Events in leech neurons resembling 'paroxysmal depolari-zation shift'. *Neurology* **22**, 439.

—— and Glaser, G.H. (1972). Penicillin-induced paroxysmal discharge in leech ganglion. *Trans. Am neurol. Ass.* **97**, 323–5.

—— and Kleinhaus, A.L. (1974). Dual action of phenobarbital on leech ganglia. *Comp. gen. Pharmac.* **5**, 239–49.

—— —— (1976). Ph-dependent actions of pentobarbital on resting membrane properties of leech Retzius cells. *Neurosci. Abstr.* **II**, 183.

Principato, G.B., Rosi, G., Biagioni, M., and Giovannini, E. (1983). Kinetics study on the reaction mechanism of the proprionylcholinesterase from *Hirudo medicinalis*. *Comp Biochem. Physiol. C. Comp. Pharmacol.* **75**(1), 185–92. [1984BA:11675.]

Prisadkij, — (1914). Priedwaritielnyj otcziet po izsliedowanui ozier na wostocznom sklonie Urala. *Izw. Russkawo Geogr. Obszczestwa* **50**, V–VI. [*Glossiphonia paludosa* in Lake Tygisz, Ural Mts.]

Pronin, N.M. (1971). [Distribution of *Acanthobdella peledina* Grube 1851 (Hirudinea), a parasite of freshwater fishes, in waters of the USSR.] *Parazitologiya* **5**(1), 92–7. [In Russian; English summary.]

—— (1979). [The finding of subarctic leeches, *Acanthobdella peledina* and *Cystobranchus mammillatus*, in Lake Baikal Basin USSR and the factors for their absence from Baikal.] *Parazitologiya, Leningrad* **13**(5), 555–8. [In Russian.] [1980BA:38409.]

Prosser, C.L. (1935). Impulses in the segmental nerves of the earthworm. *J. exp. Biol.* **12**, 95–104.

Prost, M., and Studnicka, M. (1966). [Investigations on the use of organic esters of phosphoric acid in the control of external parasites of farmed fish. I. Control of the invasion of *Piscicola geometra* L.] *Med. Wet.* **22**(6), 321–30. [In Polish; Russian, English, French and German summaries.]

—— —— (1968). [Investigations on the use of esters of phosphoric acid in the treatment of external parasites of farmed fish. IV. Efficacy of "Foschlor".] *Med. Wet.* **24**(2), 97–101. [In Polish; Russian, French and German summaries.]

—— —— and Niezgoda, J. (1974). Efficacy of some methods controlling leeches in water. *Aquaculture* **3**(3), 287–94. [1974BA:26987.]

Prowazek, — (1904). Die Entwicklung von Herpetomonas. *Arb. kaiserl. Gesund.* **20**, 440 ["herpetomonads" found in gut of *Haemadipsa zeylanica*.]

Prozorovskii, V.I. (1978). [Injuries to corpses from animals.] *Sud. Med. Ekspert* **21**(3), 52–3. [In Russian.]

Prud'homme van Reine, W.J. (1941). Wat vind ik in sloot en plas? *Zutphen* 55–6.

Prusch, R.D., and Otter, T. (1977). Annelid transepithelial ion transport. *Comp. Biochem. Physiol. A Comp Physiol.* **57**(1), 87–92. [1977BA:14849.] [*Haemopis.*]

Pryor Jr, W.H., Bergner Jr, J.F., and Raulston, G.L. (1970). Leech

(*Dinobdella ferox*) infection of a Taiwan monkey (*Macacca cyclopis*). *J. Am. vet. Med. Ass.* **157**(11), 1926–7.

Przylecki, J. St. (1926). See Autrum (1939*c*).

Pucci, I., and Afzelius, B.A. (1962). An electron microscope study of sarcotubules and related structures in the leech muscle. *J. ultrastr. Res.* **7**(3–4), 210–24.

Puccinelli, I., and Mancino, G. (1964). Osservazioni cariologiche su *Pontobdella muricata* (Hirudinea, Piscicolidae) del Mar Tirreno. *Atti Soc. Toscana Sci. nat. Pisa Ser. B* **71**, 14–17.

—— —— (1965). Il corredo cromosomico e la spermatogenesi di *Erpobdella testacea* (Hirudinea, Erpobdellidae). *Boll. Zool.* **32**(2), 579–87.

—— —— (1967). Il cariotipo e la linea germinate femminile di *Erpobdella testacea* (Hirudinea, Erpobdellidae). *Mem. Soc. Tosc. Sci. nat.* **73B**, 106–12. [English summary.]

—— —— (1968). Osservazioni cariologiche sul genere *Trocheta* (Hirudinea, Erpobdellidae). *Atti Accad. Naz. Lincei Rend. Cl. Sci. Fis Mat. natur. Sez III* **45**(6), 597–605.

Puddu, S., and Pirodda, G. (1975). Catalogo sistematico ragionato della fauna cavernicola della Sardegna. *Rend. Sem. Fac. Sci. Un. Cagliari* **43**(3–4), 151–205. [English summary.] [Cave leeches in Sardinia.]

Puidak, U. (1965). [Occurrence of parasites in some fishes of the Estonian coastal waters.] *Izv. Akad. Nauk Est. SSR Ser. Biol.* **4**, 552–7. [One leech species.]

Pujatti, D. (date undetermined). *Placobdella ceylanica* Harding, probabile vettore di *Trypanosoma rotatorium* Mayer. *Doriana* **1**(41), 1–4.

Pumplin, D.W. *et al.* (1983). Distinctions between gap junctions and sites of intermediate filament attachment in the leech CNS. *J. Neurocytol.* **12**(5), 805–15.

Purves, D., and McMahan, U.J. (1972). The distribution of synapses on a physiologically identified motor neuron in the central nervous system of the leech: an electron microscope study after injection of the fluorescent dye Procion Yellow. *J. cell. Biol.* **55**, 205–20.

—— —— (1973). Procion Yellow as a marker for electron microscopic examination of functionally identified nerve cells. In *Intracellular staining in neurobiology* (ed. S.B. Kater and C. Nicholson). pp. 72–81. Springer, New York.

Pütter, A. (1907). Der Stoffwechsel des Blutegels (*Hirudo medicinalis* L.) Teil 1. *Z. allg. Physiol.* **6**, 217–86.

—— (1908). Der Stoffwechsel des Blutegels (*Hirudo medicinalis* L.) Teil 2. *Z. allg. Physiol.* **6**, 16–61.

Putz, R.E. (1972*a*). *Cryptobia cataractae* sp. n. (Kinetoplastida: Cryptobiidae), a hemoflagellate of some cyprinid fishes of West Virginia. *Proc. helminthol. Soc. Wash.* **39**(1), 18–22. [Vector is *Cystobranchus virginicus*.]

—— (1972*b*). Biologica studies on the hemoflagellates *Cryptobia cataractae* and *Cryptobia salmositica. US Sport Fish. Wildl. Tech. Pap.* **63**, 3–25.

Qadri, S.S. (1962). An experimental study of the life cycle of *Trypanosoma danilewskyi* in the leech *Hemiclepsis marginata. J. Protozool.* **9**, 254–8.

Qasim, S.Z. (1957). The biology of *Blennius pholis* L. (Teleostei). *Proc. Zool. Soc. Lond.* **128**, 161–208.

Quebec Game and Fisheries Department (1948). Control of leeches. *Sixth Annual Report, Biological Bureau. Quebec Game and Fisheries Department, Montreal,* pp. 85–7.

Quick, H.E. (1938). The medicinal leech *Hirudo medicinalis* Linn. in Breconshire, with notes on the other species of Hirudinea found in South Wales. *Proc. Swansea Sci. Field nat. Soc.* **2**, 12–14.

Quortrup, E.R., and Shillinger, J.E. (1941). 3000 wild birds autopsies on western lake areas. *J. Am. vet. med. Ass.* **99**, 382–7. [Occasional cases of verminous pneumonia developed in ducks infested with leeches in the bronchi.]

Radkevich, G. (1897). Spisok vodyan'kh myagkotel'kh i piyavok, sobrann'kh v Khar'kovskoi i Poltavskoi gub. *Tr. Khar'kovsk. Obsch. Isp't. Prirod', Prilozheniya* **12**(1), 1–2.

Radkiewicz, J. (1972*a*). [Three new collecting localities of *Haementeria costata* (Fr. Mull.) in the Zielona Góra district.] *Prz. Zool.* **16**(1), 38–40. [In Polish.] [1976BA:59542.]

—— (1972*b*). [*Hirudo medicinalis* L. in the west borders of the Zielona Góra district.] *Prz. Zool.* **16**(1), 40–3. [In Polish.]

—— (1974). [*Batracobdella slovaca* Košel, a new leech species of Southwest Slovakia (Hirudinoidea, Glossiphoniidae.] *Prz. Zool.* **18**(3), 361– 2. [In Polish; English summary.] [1975BA:49835.] [In Danube river.]

Radlowski, J., and Czechowicz, K. (1980). [Preliminary studies on origination of central nervous system and neurosecretory system in the leech *Herpobdella octoculata.*] *Pr. Nauk Uniw. Slask. Katowicach.* **375**, 159–72. [In Polish.]

Radulescu, I.L., Nalbant, T.T., and Angelescu, N. (1972). [New contributions to the knowledge of parasitic fauna of fishes of the Atlantic Ocean.] *Bul. Cercet. Piscic.* **31**(3/4), 71–6. [In Romanian; English, French, and Russian summaries.] [1975BA:4019.]

Raeside, J.R. (1964). A proving of *Hirudo medicinalis. Br. homoeopath. J.* **53**, 22–30.

Rafinesque, C.S. (1819). Prodrome de 70 nouveaux genres d'animaux découverts dans l'intérieur des Etats-Unis d'Amérique durant l'année 1818. *J. Phys. Chim. Hist. nat. Arts* **88**, 417–29.

Rahemtulla, F., and Lovtrup, S. (1975). The comparative biochemistry of invertebrate mucopolysaccharides — III. Oligochaeta and Hirudinea. *Comp. Biochem. Physiol.* **50B**, 627–9.

Raishite, D.I. (1967). [Characteristics of infection of leeches with cercariae of the trematode *Apatemon gracilii minor* (Yamaguti 1933).] *Zool. Zh.* **46**(12), 1846–9. [In Russian; English summary.]

—— (1969). [The degree of metacercarial infection of freshwater leeches in the Volga and Neman deltas.] *Prob. Parazit.* Part 1, 196–9.

Raj, P.J. Sanjeeva (1951). On a new species of *Ozobranchus* from Porto Novo, S. India. *J. Zool. Soc. India* **3**(1), 1–5. [*O. polybranchus.*]

—— (1953). First record of an ichthyobdellan leech *Branchellion* Savigny from the Indian waters. *Curr. Sci.* **22**, 310.

—— (1954*a*). A synopsis of the species of the genus *Ozobranchus* (de Quatrefages, 1852). (Hirudinea-Annelida). *J. Bombay nat. Hist. Soc.* **52**(2/3), 472–80.

—— (1954*b*). On a new species of marine leech of the genus *Branchellion* (Family Ichthyobdellidae) from the Indian coast. *Rec. Indian Mus.* **52**, 249–56. [*B. plicobranchus.*]

—— (1959*a*). Studies on the marine leech *Branchellion plicobranchus* Raj (Family Piscicolidae) from India. *J. zool. Soc. India* **11**(2), 152–61.

—— (1959*b*). Additions to the gilled-leeches of India, their geographic and parasite–host distribution. *Diamond Jubilee Souvenir of nat. Hist. Soc.*, 2–6.

—— (1959*c*). Occurrence of *Ozobranchus margoi* Apathy (Hirudinea, Annelida) in the Indian Seas. *Curr. Sci.* **28**, 496.

—— (1962). *Morphological and biological studies of two species of piscicolid leeches.* University Microfilms, Ann Arbor, Michigan. Order No. 62–4393. [*Calliobdela vivida* and *Myzobdella lugubris.*]

—— (1966). *Ozobranchus branchiatus* (Menzies, 1791) from Pulicat Lake, south India. *J. Bombay nat. Hist. Soc.* (1965). **62**(3), 582–4.

—— (1974). A review of the fish leeches of the Indian Ocean. *J. mar. Biol. Ass. India* **16**(2), 381–97.

—— and Penner, L.R. (1962). Concerning *Ozobranchus branchiatus* (Menzies, 1791) (Piscicolidae: Hirudinea) from Florida and Sarawak. *Trans. Am. microsc. Soc.* **81**, 364–71.

Rajak, R.L., Perti, S.L., and Agarwal, P.N. (1968). Recent advances on the biology of land leeches and protection against their ravages. *Labdev. J. Sci. Tech.* **6**(B)(3), 125–37.

—— Srivstava, A.P., and Perti, S.L. (1968). Studies on life-history and behaviour of land leeches. *Labdev. J. Sci. Tech.* **6**(B)(3), 165–8. [*Haemadipsa sylvestris:* life history and culture methods.]

Rajulu, G.S. (1965). Leeches (Hirudinea) as endo-parasites in centipedes (Chilopoda). *Curr. Sci.* **34**(13), 408–9.

—— Ramilangan, S., and George, G.S. (1968). Tanning of earthworm cocoons. *Indian J. exp. Biol.* **6**, 187–8.

Ramachandrau, P.K., Koshy, T., Sastry, K.G.K., Singh, S.P., Srinivasan, M.N., and Ganguly, S.K. (1971). Studies on leech repellents. *J. econ. Ent.* **64**(5), 1293–4. [Against *Haemadipsa sylvestris* N-benzoyl piperidine and N-toluyl piperidine worked best.]

Ramamurthi, R. (1962). Studies on the respiration of freshwater poikilotherms in relation to osmotic stress. Unpublished Ph.D. thesis, Sri Venkateswara University, Tirupati, Inda.

—— (1965). Metabolic response to osmotic stress in some freshwater poikilotherms. *Curr. Sci.* **34**, 351–2.

—— (1968). Oxygen consumption of the common Indian cattle leech *Poecilobdella granulosa* in relation to osmotic stress. *Comp. Biochem. Physiol.* **4**(1), 283–7.

Ramos, B., and Urdaneta-Morales, S. (1977). Hematophagous insects as vectors for frog trypanosomes. *Rev. Biol. Trop. Med.* **25**(2), 209–18.

Rang, H.P., and Ritter, J.M. (1969). *Molec. Pharmacol.* **5**, 394. (Incomplete reference)

Rao, P., Bailie, F.B., and Bailey, B.N. (1985). Leechmania in microsurgery. *Practitioner* **229**, 901–5.

Rashid, Ali S. (1968). Bottom fauna of the Korang Stream, Rawalpindi. *Pakist. J. Sci.* **20**, 266–70. [Productivity, including seasonal changes, of leeches.]

Raspopov, I.M. *et al.* (1968). [Hydrobiological characteristics of the bays in the western part of Shkhern region in the Ladoga Lake, suitable for aquatic birds breeding.] *Biol. Resurs. Ladoga Oz.* (*Zool.*) 71–104. [In Russian.]

Ratcliffe, D.A., ed. (1977). *A nature conservation review* Vol. 2, p. 203. Cambridge University Press, Cambridge, 274 pp.

Raubenheimer, O. (1923). Leeches: how to dispense them. *J. Am. pharm. Ass.* **12**, 338–40.

Raven, C.P. (1961). *Oogenesis: the storage of developmental information.* Pergamon, London.

Rawson, D.S. (1953). The bottom fauna of Great Slave Lake. *J. fish. Res. Bd. Can.* **10**, 486–520.

Ready, D.F., and Nicholls, J.G. (1979). Identified neurones isolated from leech CNS make selective connections in culture. *Nature, Lond.* **281**, 67–9.

Redeke, H.C. (1948). [Leeches.] *Hydrobiol. Nederl. Amsterdam* 225–9, 477.

Redondo, B.T., Garcia-Más, I., Beltrán, I.C., and Moreno, A.L.V. (1980). Estudio bacteriológico del tracto digestivo de *Dina lineata* (O.F. Müller, 1774) (*Hirudinea, Erpobdellidae*). *Bol. R. Soc. Española Hist. Nat.* (*Biol.*) **78**, 97–104.

Reeves, B. (1941). Leech infestation in Middle East. *Jl R. Army med. Corps* **77**, 205.

—— (1945). Leeches as temporary endoparasites in upper respiratory tract. *J. Laryngol. Otol.* **60**, 369–72.

Reibstein, H. (1931). Über Bau und Tätigkeit der Kiefer von *Hirudo medicinalis* L. *Zool. Jb. Anat.* **54**, 55–104.

Reichardt, W. (1961). Autocorrelation, a principle for the evaluation of sensory information by the central nervous system. In *Sensory communication* (ed. W.A. Rosenblith) pp. 303–17. MIT Press, Cambridge, Mass.

Reichenow, E. (1910). *Haemogregarina stepanowi.* Die Entwicklungsgeschichte einer Haemogregarine. *Arch. Protistenk.* **20**, 251–350.

—— (1921). Über intrazellulare Symbionten bei Blutsaugern. *Arch. Schiffs-Trop.-Hyg.* **25**, 366.

—— (1922). See Autrum (1939*c*).

Reilly, B.O., and Woo, P.T.K. (1982). Susceptibility of the leech *Batracobdella picta* to *Trypanosoma andersoni* and *Trypanosoma grylli* (Kinetoplastida). *Can. J. Zool.* **60**(6), 1441–5. [1983BA:34295.] [French summary.]

Reim, G. (1968). Experimentelle Untersuchungen zum physiologischen Farbwechesel bei Hirudineen. *Zool. Jb., Zool.* **74**, 198–232.

Reisinger, E. (1951). Lebenweise und Verbreitung des europäischen Landblutegels (*Xerobdella lecomtei* Frauenfeld). *Carinthia. II. Mitt. Nat. wiss. Ver.* **141**, 110–24.

—— (1953). Faunistiche Notizen aus Karnten. *Carinthia. II. Mitt. Nat. wiss. Ver.* **143**, 117–21.

Relini, G. (1962). Contribute allo studio della fauna bentonica del Golgo di Genova (Riviera di Ponente). *Doriana* (3)**117**, 1–6. [English summary.] [*Pontobdella muricata* in bottom fauna, Bay of Genoa, off Savona.]

Remane, A. (1952). Zur Verbreitung der am Menschen blutsaugenden Egel in Schleswig-Holstein. *Faun. Mit. Nordd.* **1**, 1–2. [*Hirudo* in Germany]

Remy, P. (1929). La faune d'ile Jan Mayen. *C.r. Soc. biogéogr., VI-ème année* No. 48. [Complete absence of freshwater leeches in the Arctic island, Jan Mayen, 71°N 9°W; frozen over from October to July.]

—— (1933*a*). Sur quelques Hirudinées des Balkans. *Ann. Soc. Linn. Lyon* **77**, 1–8.

—— (1933*b*). Présentation d'une sangsue nouvelle pour la faune française, *Cystobranchus respirans* Troschel. *Cahiers Lorrains, Strasburg* **12**, 111.

—— (1937). Sangsues de la Yougoslavie. *Bull. Soc. zool. Fr.* **62**, 140–8.

—— (1943). Notes faunistiques. *Bull. Soc. Linn. Lyon* **12**, 139–42. [*Cystobranchus respirans* in the Vair.]

—— (1953). Description des grottes yougoslaves. *Glasn. Prir. Muz. Srpske Zem. B* 5–6.

Resh, V.H., and Unzicker, J.D. (1975). Water quality monitoring and aquatic organisms: the importance of species identification. *J. Water Pollut. Control. Fed.* **47**(1), 9–19.

Rest, R.P.C. Du. (1963). Distribution of the zooplankton in the salt marshes of southeastern Louisiana. *Publ. Inst. mar. Sci. Un. Texas* **9**, 132–55. [Leeches: 139.]

Rettich, F. (1980). Field evaluation of permethrin and decamethrin against mosquito larvae and pupae (Diptera, Culicidae). *Acta entomol. Bohemoslov* **77**(2), 89–96.

Retzius, G. (1891). Zur Kenntnis des centralen Nervensystems der Würmer. *Biol. Unters.* (NF) **2**, 1–28.

—— (1898). Zur Kenntnis des sensiblen Nervensystems der Hirudineen. *Biol. Unters.* (NF) **8**, 94–7.

Reynoldson, Thomas B. (1952). A record of the leech *Hirudo medicinalis* from Islay, with brief mention of other species. *Scot. Nat.* **64**, 164–6.

—— (1956). An Anglesey record of the medicinal leech *Hirudo medicinalis* L. *Trans. Anglesey Antiquarian Soc. Field Club* 54.

—— (1966). The distribution and abundance of lake-dwelling triclads — towards a hypothesis. *Adv. ecol. Res.* **3**, 1–71.

Reynoldson, Trefor B. (1974). An investigation into some aspects of the biology of the Hirudinoidea. M.Sc. thesis, University of Calgary, Calgary.

—— and Davies, R.W. (1976). A comparative study of the osmoregulatory ability of three species of leech (Hirudinoidea) and its relationship to their distribution in Alberta. *Can. J. Zool.* **54**(11), 1908–11.

—— —— (1980). A comparative study of weight regulation in *Nephelopsis obscura* and *Erpobdella punctata*. *Comp. Biochem. Physiol.* **66A**, 711–14.

Reznik, P.A. (1939). [Leeches in the vicinity of Voroshilovska.] *Tr. Voroshilov. Gos. Ped. Inst.* **1**, 163–4.

Ribbands, C.R. (1946). Experiments with leech repellents. *Ann. trop. Med. Parasitol.* **40**, 314–19.

Richards, R. (1977). Diseases of aquarium fish — 1: The clinical approach. *Vet. Rec.* **101**(6), 111–13. [Skin lesions due to leeches.]

Richardson, L.R. (1942). Observations on the migratory behaviour of leeches. *Can. Field-Nat.* **46**(5), 67–70.

—— (1943). The freshwater leeches of Prince Edward Island and the problem of the distribution of leeches. *Can. Field-Nat.* **57**, 89–91.

—— (1947). A review of New Zealand leeches. *N. Zealand Sci. Congr.* 201–2.

—— (1948). *Piscicola punctata* (Verrill) feeding on the eggs of *Leucosomus corporalis* (Mitchell). *Can. Field-Nat.* **62**, 121–2.

—— (1949a). Studies on New Zealand Hirudinea: Part II. *Branchellion parkeri*, a new ichthyobdellid leech. *Zool. Publ. Victoria Un. Coll.* No. 1, 3–11.

—— (1949b). The occurrence of the leech *Batracobdella picta* (Verrill) in the dorsal sub-cutaneous lymph spaces of *Rana catesbiana*. *Can. Field-Nat.* **63**(2), 85–6.

—— (1950). Studies on New Zealand Hirudinea (Part I). *Pontobdella benhami* n. sp. *Trans. Proc. R. Soc. N. Zealand* **78**(1), 97–100.

—— (1953). Studies on New Zealand Hirudinea. Part III. *Bdellamaris eptatreti* n. g., n. sp., and notes on other Piscicolidae. *Trans. R. Soc. N. Zealand* **81**(2), 283–94.

—— (1959). New Zealand Hirudinea. IV. *Makarabdella manteri* n. g., n. sp., a new marine piscicolid leech. *Trans. R. Soc. N. Zealand* **87**(3/4), 283–90.

—— (1967). The suitability of Australian land-leeches as a source of experimental material in biological researches. *Aust. J. Sci.* **30**(3), 107.

—— (1968*a*). An annotated list of Australian leeches. *Proc. Linn. Soc. New S. Wales* **92**(3), 227–45.

—— (1968*b*). Observations on the Australian land-leech, *Chtonobdella limbata* (Grube, 1866). (Hirudinea: Haemadipsidae). *Aus. Zool.* **14**(3), 294–305. [*Quaesitobdella bilineata*, not *C. limbata.*]

—— (1969*a*). The rediscovery of *Hirudo elegans* Grube, 1867. *Mem. Queensl. Mus.* **15**(3), 191–203.

—— (1969*b*). On a distinctive new subequatorial Australian quadrannulate land-leech, and related matters. *Aust. Zool.* **15**(2), 201–13.

—— (1969*c*). A contribution to the systematics of the hirudinid leeches, with description of new families, genera and species. *Acta zool. acad. scient. hung.* **15**(1/2), 97–149. (Erratum: "The species I handled was not *plumbeus*", as published. Richardson, per. comm. 1983).

—— (1969*d*). The family Ozobranchidae redefined, and a novel ozobranchiform leech from Murray River turtles (class Hirudinoidea; order Rhynchobdelliformes). *Proc. Linn. Soc. New S. Wales* **94**(1), 61–80.

—— (1970*a*). Towards the new hirudinology. *J. Parasitol.* **56**(2), 237.

—— (1970*b*). *Bassianobdella victoriae* gen. et sp. nov. (Hirudinoidea: Richardsonianidae). *Mem. Natl. Mus. Victoria* **31**, 41–50.

—— (1970*c*). A new Australian '*Dineta/Barbronia*-like' leech, and related matters (Hirudinoidea: ?Erpobdellidae). *Proc. Linn. Soc. New S. Wales* **95**(3), 221–31.

—— (1970*d*). A note on marine piscicolid leeches from Port Phillip Bay, Victoria. *Aust. Zool.* **15**(3), 391–4.

—— (1970*e*). A contribution to the history of the Australian medicinal leech. *Aust. Zool.* **15**(3), 395–9.

—— (1970*f*). A new marine piscicolid leech from Newfoundland placed provisionally in the genus *Malmiana*. *Can. J. Zool.* **48**(4), 841–5. [*Malmiana nuda* = *Malmiana brunnea.*]

—— (1971*a*). A new species from Mexico of the Nearctic genus *Percymoorensis*, and remarks on the family Haemopidae (Hirudinoidea). *Can. J. Zool.* **49**(8), 1095–103.

—— (1971*b*). *Bassianobdella ingrami* sp. nov. from Tasmania (Hirudinoidea: Richardsonianidae). *Pap. Proc. R. Soc. Tasmania* **105**, 113–18.

—— (1971*c*). The relationship of the terrestrial jawed sanguivorous g. *Mesobdella* to the neotropical hirudiniform leeches (Hirudinoidea). *Proc. Linn. Soc. New S. Wales* **95**(3), 215–20.

—— (1971*d*). Gastrostomobdellidae f. nov. and a new genus for the gastroporous *Orobdella octonaria* Oka 1895, of Japan (Hirudinoidea: Arhynchobdellae). *Bull. nat. Sci. Mus. Tokyo* **14**(4), 585–602.

—— (1971*e*). *Habeobdella stagni*, a new genus and species from South-Western Australia (Hirudinoidea: Richardsonianidae). *Jl R. Soc. West. Aust.* **54**(2), 47–52.

—— (1972*a*). On the morphology and nature of a leech of the genus *Philobdella* (Hirudinoidea: Macrobdellidae). *Am. midl. Nat.* **87**(2), 423–33.

—— (1972*b*). *Bassianobdella fusca* sp. nov. (Hirudinoidea: Richardsonianidae), with an initial demonstration of systematic values in the lengths of annuli in the mid-nephric somites. *Rec. Aust. Mus. Syd.* **28**(8), 129–39.

—— (1972*c*). A new genus based on the seven-banded *Richardsonianus dawbini* Richardson 1969 (Hirudinoidea: Richardsonianidae). *Proc. Linn. Soc. New S. Wales* **97**(2), 130–40.

—— (1972*d*). *Quantenobdella howensis* Richardson 1969 of Lord Howe Island, with comment on dispersal by passive transport (Hirudinoidea: Richardsonianidae). *Mem nat. Mus. Victoria* **33**, 65–72.

—— (1972*e*). A genus and species of Sudan leech formerly confused with *Limnatis nilotica* (Hirudinidae s. 1.: Hirudinea). *Bull. Br. Mus. nat. Hist. (Zool.)* **21**(8), 349–57.

—— (1973). A new genus, *Priscabdella*, for aquatic jawed sanguivorous leeches from Flinders Island, north-east Tasmania, and southern South Australia. (Hirudinoidea: Richardsonianidae). *Rec. Queen Victoria Mus.* **47**, 1–18.

—— (1974*a*). *Domanibdella* gen. nov., a duognathous 5-annulate land-leech genus in New Guinea (Hirudinoidea: Haemadipsidae s.l.) with a discussion on general somital annulation. *Mem nat. Mus. Victoria* **35**, 97–109.

—— (1974*b*). A new leech from Papua representative of a third family of aquatic jawed sanguivores in the Australian Region (Hirudinoidea: Illebdellidae f. nov.) *Rec. Aust. Mus.* **29**(6), 187–96.

—— (1974*c*). *Amicibdella* and *Micobdella* gen. nov. of eastern Australia (Hirudinoidea: Haemadipsidae s. 1.). *Mem. Queensl. Mus.* **17**(1), 125–49.

—— (1974*d*). A new troglobitic quadrannulate land-leech from Papua (Hirudinoidea: Haemadipsidae s. 1.) *Proc. Linn. Soc. New S. Wales* **99**(1), 57–68. [1975BA:55566.] [*Leiobdella jawarerensis* feeding on bats in cave.]

—— (1975*a*). A contribution to the general zoology of the land-leeches (Hirudinoidea: Haemadipsoidea Superfam. nov.). *Acta zool. acad. sci. hung.* **XXI**(1–2), 119–52.

—— (1975*b*). A new species of terricolous leeches in Japan (Gastrostomob-dellidae, *Orobdella*) *Bull. Nat. Sci. Mus. Ser. A (Zool.)* **1**(1), 39–56.

—— (1975*c*). A convenient technique for killing and preserving leeches for general study. *J. Parasitol.* **61**(1), 78.

—— (1976*a*). Giantism in *Goddardobdella elegans* infesting the teat-cisters of cattle in northeastern Queensland (Hirudinea: Richardsonianidae). *J. Parasitol.* **62**(5), 847–8.

—— (1976*b*). The description of a Leiobdelline land-leech on Rennell Island (Haemadipsoidea: Domanibdellidae). *Nat. Hist. Rennell Island, British Solomon Islands* **7**, 96–106.

—— (1976*c*). On the nature of the genital primordia and their role in the development of the reproductive systems in Hirudinea. *Acta zool. acad. sci. hung.* **22**(1–2), 155–63.

—— (1977*a*). A system of intersomital dorsoventral muscles in the posterior body somites of land leeches (Hirudinea: Haemadipsoidea). *Aust. Zool.* **19**(2), 233–8.

—— (1977*b*). *Sibdella solomoni* Richardson, a 5-annulate domanibdelline land-leech from New Britain (Hirudinea, Haemadipsoidea, Domanib-dellidae). *Steenstrupia* **4**(15), 171–8. [1978BA:21875.]

—— (1978). On the zoological nature of land-leeches in the Séchelles Islands, and a consequential revision of the status of land-leeches in Madagascar (Hirudinea: Haemadipsoidea). *Rev. zool. Afr.* **92**(4), 837–66.

—— (1979*a*). On two land-leeches labelled as from New Zealand. *Tuatara* **24**(1), 41–8.

—— (1979*b*). *Anphilaemon kingi* gen. et sp. nov., a land leech on King Island, Bass Strait (Haemadipsoidea: Domanibdellidae). *Pap. Proc. R. Soc. Tasmania* **113**, 177–84.

—— (1981). The Papuan *Elocobdella novabritanniae*, the Oceanian *Abessebdella palmyrae* (Haemadipsoidea: Domanibdellidae) and an Oceanian barbronid (Hirudinea). *Rec. Aust. Mus.* **33**(14), 673–94. [1982BA:10617.] [Erratum: The barbronid was not from Palmyra Island. Locality unknown: L.R. Richardson, personal communication 1983.]

—— and Hunt, P.J. (1968). Trypanosomes in the crop of an haemadipsid leech. *Aust. J. Sci.* **30**(9), 374–5. [*Quaesitobdella bilineata*, not *Chtonobdella limbata.*]

—— and Meyer, M.C. (1973). Deep-sea fish leeches (Rhynchobdellae: Piscicolidae). *Galathea Rep.* **12**, 113–25.

Richmond, S.L. (1972). Leeches (Hirudinea) of northern Illinois, exclusive of the family Piscicolidae. M.S. thesis, Northern Illinois University, Dekalb.

Rieske, E., Schubert, P., and Kreutzberg, G.W. (1975). Transfer of radioactive material between electrically coupled neurones of the leech central nervous system. *Brain Res.* **84**(3), 365–82. [1975BA:61127.]

Riggs, M.R. (1980). Helminth parasites of leeches of the genus *Haemopis*. Master's thesis, Iowa State University.

Rindi, G. and Ferrari G. (1952*a*). Sul contenuto e sul significato di riboflavina e niacinamide nella cute di alcune specie di irudinei. *Boll. Soc. Ital. Biol. sper.* **28**(6), 1103–6.

—— —— (1952*b*). Contenuto in rame di alcune specie di irudinei. *Boll. Soc. Ital. Biol. sper.* **28**, 1512–13.

Ringuelet, R.A. (1942). Descripciones preliminares de neuvos Hirudíneos argentinos. *Nota Mus. La Plata, Zool.* **7**(59), 203–14.

—— (1943*a*). Refundicón de los géneros *Oxyptychus* Grube, *Diplobdella* Moore y *Argyrobdella* Cordero con una pequena monografía de las especies argentinas. *Notas Mus. La Plata, Zool.* **8**(65), 101–26.

—— (1943*b*). Nota sobre dos Hirudineos de sur de Chile: *Mesobdella gemmata* (Em. Bl.) y *Helobdella similis* Ring. *Physis* **19**(53), 262–378.

—— (1943*c*). Sobre la morfologia y variabilidad de *Helobdella triserialis* (Em. Bl.) (Hirudinea, Glossiphoniidae). *Notas Mus. La Plata, Zool.* **8**(69), 215–40.

—— (1944*a*). Sinopsis sistemática y zoogeográfica de los Hirudíneos de la Argentina, Brasil, Chile, Paraguay y Uruguay. *Rev. Mus. La Plata* (new series), *Zool.* **3**(22), 163–232.

—— (1944*b*). Notas sobre Hirudineos neotropicales. I. Caracteres y posicion sistematica del genero *Potamobdella* Caballero. *Not. Mus. La Plata, Zool.* **9**(73), 39–52.

—— (1944*c*). Notas sobre Hirudineos neotropicales. II. *Hygrobdella pelaezi* Cab., curiosa sanguijuela terrestre mexicana. *Not. Mus. La Plata, Zool.* **9**(74), 167–77.

—— (1944*d*). Revision de los Hirudíneos argentinos de los géneros *Helobdella* R. Bl., *Batracobdella* Vig., *Cylicobdella* Gr. y *Semiscolex* Kinb., *Rev. Mus. La Plata* (new series), *Zool.* **4**(25), 5–94.

—— (1945). Hirudíneos del Museo de la Plata. *Rev. Mus. La Plata* (new series), *Zool.* **4**(26), 95–137.

—— (1946). La sanguijuela medicinal argentina. *Prensa Méd. Argentina* **33**(9), 467–71.

—— (1947). Notas sobre Hirudíneos neotropicales. III. *Theromyzon propinquus* nov. sp., de la Argentina. *Not. Mus. La Plata, Zool.* **12**(100), 217–22.

—— (1948*a*). Notas sobre Hirudíneos neotropicales. IV. Una cuestion de nomenclatura: *Liostoma* versus *Cylicobdella*. *Not. Mus. La Plata, Zool.* **13**(109), 185–90.

—— (1948*b*). Notas sobre Hirudíneos neotropicales. V. Especies de la República del Paraguay. *Not. Mus. La Plata, Zool.* **13**(113), 213–44.

—— (1949). Notas sobre Hirudíneos neotropicales. VI. Presencia del género *Glossiphonia* en la Argentina y otras adiciones al conocimiento de la hirudofauna de los Países del Plata. *Not. Mus. La Plata, Zool.* **14**(122), 141–59.

—— (1953*a*). Notas sobre Hirudíneos neotropicales. VII. Un nuevo Haemadipsido del género *Mesobdella* Blanch. *Not. Mus. La Plata, Zool.* **16**(139), 187–93.

—— (1953*b*). Notas sobre Hirudíneos neotropicales. VIII. Algunas especies de Bolivia y Perú. *Not. Mus. La Plata, Zool.* **16**(142), 215–24.

—— (1953*c*). Notas sobre Hirudíneos neotropicales. IX. Rehabilitacion del género *Cyclobdella* Weyenbergh. *Not. Mus. La Plata, Zool.* **16**(143), 257–72.

—— (1954). La clasificación de los Hirudíneos. *Not. Mus. La Plata, Zool.* **17**(146), 1–15.

—— (1955). Sobre la sanguijuela de Juan Fernández (*Philaemon skottsbergii* Joh., Hirudinea). *Invest. Zool. Chilenas* **II** (9–10), 137–42.

—— (1958). Hirudíneos de lago Argentina (Santa Cruz, Argentina) coleccionados por le Dr. A. Willink. *Acta zool. Lilloana* **15**, 121–41.

—— (1959). Una colección de Hirudíneos del Perú. I. Sanguijuelas del lago Titicaca. *Physis* **21**(61), 187–99.

—— (1960). Clave sinoptica de los Hirudineos argentinos. *Physis* **21**(62), 337.

—— (1961). Hirudineos terrestres del Peru. *Actas y Trabajos del Primer congreso Sudamericano de Zoologia, La Plata*, 12–24 October, Vol. 2, pp. 251–6.

—— (1968). Llave o clave para el reconocimiento de las sanguijuelas conocidas de la República Argentina (Hirudinea) y apuntamientos sobre la Hirudofauna neotrópical y transicional mexicana. *Physis* **27**(75), 367–90.

—— (1972*a*). Sobre la identidad de *Blennobdella depressa* Ém. Blanchard, 1849, y la existencia de *Haementeria officinalis* de Filippi, 1849, en Estados Unidos (Hirudinea, Glossiphoniiformes). *Physis* **31**(82), 97–8.

—— (1972*b*). Algunos hirudíneos del Muséum d'Histoire Naturelle de Paris. *Physis* **31**(82), 99–103.

—— (1972*c*). Nuevos taxia de hirudineos neotropicos con la redefinicion de Semiscolecidae y la descripcion de Cyclobdellidae fam. nov. y Mesobdellidae fam. nov. *Physis* **31**(82), 193–201.

—— (1972*d*). Cylicobdellidae, nueva familia de Hirudíneos Erpobdelloideos. *Physis* **31**(83), 337–44.

—— (1972*e*). Hirudineos neotropicos de Colombia, Cuba y Chile con la descripcion de una nueva especie de *Oligobdella* (Glossiphoniidae). *Physis* **31**(83), 345–52.

—— (1974). Los hirudineos terrestres del genero *Blanchardiella* Weber del paramo nor-andino de Colombia. *Physis Secc. B Aguas Cont. Org.* **33**(86), 63–9. [English summary.] [1975BA:49831.] [Placed into family Cylicobdellidae; synonym of *B. paramoensis*.]

—— (1975*a*). Un nuevo hirudineo de Colombia parasito de la trucha arco iris. *Neotropica* **21**(64), 1–4. [English summary.] [*Batracobdella xenoica* on fish; Zapaquira, State of Cundinamarca, Colombia.]

—— (1975*b*). Sobre la macrotaxinomia de la familia Macrobdellidae Richardson y su division en subfamilias (Hirudiniformes, Hirudinoidea). *Neotropica* **21**(66), 113–18.

—— (1976*a*). Clave para las familias y géneros de sanguijuelas (Hirudinea) de aguas dulces y terrestres de Mesoamérica y Sudamérica. *Limnobios* **1**(1), 9–19.

—— (1976*b*). Dos hirudineos nuevos del genero *Patagoniobdella* (Hirudinoidea, Semiscolecidae) de los lagos andino patagonicos de la Republica Argentina. *Limnobios* **1**(3), 61–6.

—— (1976*c*). Un curioso Hirudineo alto andino del Perú (*Orchibdella peruviensis* nov. sp., Hirudiniformes, Cyclobdellidae). *Limnobios* **1**(4), 101–4.

—— (1976*d*). Los caracteres endosomáticos de *Haementeria officinalis* de Filippi, diagnosis del genero, y un estudio de antiquos ejemplares de *Nephelis mexicana* Duges, 1876 (Hirudinea). *Limnobios* **1**(4), 129–36.

—— (1976*e*). Una nueva *Diestecostoma* de la zona costera del Peru (*D. trujillensis* n. sp., Hirudinea, Diestecostomatidae) y diagnosis de esta familia. *Neotropica* **22**(68), 67–76.

—— (1977*a*). Sobre la presencia supuesta de un hirudineo del genero *Semiscolex* en Africa (*S. congolensis* Sciacchitano, 1939). *Limnobios* **1**(5), 165–6.

—— (1977*b*). Hirudinea. In: Hurlbert, S.H., *Biota acuát. sudam. austral.* pp. 121–9. San Diego, California.

—— (1978*a*). Hirudineos nuevos o ya descriptos de la Argentina y del Uruguay. *Limnobios* **1**(7), 258–68.

—— (1978*b*). Nuevos generos y especies de Glossiphoniidae sudamericanos basados en caracteres ecto y endosomaticos (Hirudinea, Glossiphoniiformes). *Limnobios* **1**(7), 269–76.

—— (1978*c*). *Tribothrynobdella andicola* nov. gen. y otros Glossiphoniidae nuevos del Perú (Hirudinea, Glossiphoniidae) recogidos por el Dr. Fortunato Blancas S. *An. Cient.* (*La Molina*) **16**(1–4), 7–10 [In Portugese; Spanish summary.] [1983BA:80835.]

—— (1980*a*). Aportes al conocimiento de las sanguijuelas de genero *Haementeria* de Filippi, 1849 (Hirudinea, Glossiphoniidae). *Limnobios* **2**(1), 50–3. [English summary.]

—— (1980*b*). Un hirudineo con marsupio de la region Andina de Jujuy, Argentina (*Maiabdella batracophila* n.g., n.sp., Glossiphoniidae). *Limnobios* **2**(1), 68–71. [English summary.]

—— (1980*c*). Hirudineos terrestres nor Andinos y alto Andinos de America del Sur. *Neotropica* **26**(75), 3–11. [English summary.]

—— (1980*d*). Biogeografia de los hirudineos de America del Sur y de Mesoamerica. *Obra Centenario Mus. Plata* **6**, 1–27. [English summary.]

—— (1981*a*). Some advances in the knowledge of neotropical leeches. *Limnobios* **2**(4), 226.

—— (1981*b*). [The Hirudinea of the Museum of Natural History of Montevideo, Uruguay.] *Comun. Zool. Mus. Hist. nat. Montev.* **11**(146), 1–37. [In Spanish; English summary.] [1983BA:88707.]

—— (In press.). [Synopsis of the leeches or Hirudinea of Chile.] [In Spanish; English summary.]

—— (In press). [Key to the leeches (Hirudinea) of Mexico.] *Anales Inst. Biol.* [In Spanish; English summary.]

—— (In press). [Analysis of the genus *Helobdella* Blanchard, 1893 (Hirudinea, Glossiphoniidae) with a world key to all species.] *Limnobios* [In Spanish; English summary.]

—— (In press). [Leeches (Hirudinea) from Costa Rica.] *Physis* [In Spanish; English summary.]

—— (In press). [The leeches (Hirudinea) of Mexico, Central America, West Indies and South America, a critical list with generic diagnosis and keys for all taxa.] *Limnobios* [In Spanish; English summary.]

(Ringuelet, R.A. 1982). Nota necrologica. Dr. Raúl A. Ringuelet (1914–1982). *Neotropica* **28**(79): 1–2.

Rio-Hortega, P. del (1917). Contribucion al conocimiento de las epiteliofibrillas. *Trab. Lab. invest. Biol. Madrid* **15**, 201–99.

Ritchie, C.I.A. (1979). *Insects, the creeping conquerors.* Elsevier, Amsterdam. [Condensed in *Science Digest*, August 1979, pp. 42–5.] [Includes leeching.]

Rizvi, S.K.A., Ali, S., Khan, M.e-A., Ahmad, M.M., and Kamal, A. (1975). A note on the abnormal male genital system in the leech *Hirudinaria granulosa* (Savigny). *Geobios, Jodhpur* **2**(6), 198.

Roberts, C.J., Nielsen, E., Krogsgaard-Larsen, P., and Walker, R.J. (1982). The actions of ibotenate, homoibotenate analogs and *a*-amino-3-hydroxy-5-methy-4-isoxazole propionic acid on central neurons of *Hirudo medicinalis, Limulus polyphemus* and *Helix aspersa. Comp. Biochem. Physiol. C Comp. Pharmacol.* **73**(2), 439–44. [1983BA:88637.]

—— and Walker, R.J. (1982). γ-D-Glutamylglycine as an antagonist of kainic acid on leech, *Hirudo medicinalis*, Retzius neurons. *Neuropharmacology* **21**(12), 1245–50. [1983BA:65147.]

Roberts, H.E. (1955). Leech infestation of the eye in geese. *Vet. Rec.* **67**, 203–4. [Cornea of geese became opaque after *Theromyzon tessulatum* fed at the conjunctiva.]

Roberts, R.J., and Shepherd, C.J. (1974). *Handbook of trout and salmon diseases.* Fishing News (books), West Byfleet. 168 pp.

Robertson, J.D. (1939). The natural history of Canna and Sanday, Inner Hebrides. Oligochaeta and Hirudinea. Turbellaria and Nematomorpha. *Proc. phys. Soc. Edinb.* **23**, 21–2.

Robertson, M. (1907). Studies on a trypanosome found in the alimentary canal of *Pontobdella muricata. Proc. R. Soc. Edinb.* **17**, 83–108.

—— (1908). A preliminary note on *Haematozoa* from some Ceylon reptiles. *Spolia zeylan.* **5**, 178–85.

—— (1909). Studies on Ceylon Haematozoa. I. The life cycle of *Trypanosoma vittatae. Q. Jl. microsc. Sci.* (new series) **53**(4), 665–95.

—— (1910*a*). Further notes on a trypanosome found in the alimentary tract of *Pontobdella muricata. Q. Jl. microsc. Sci.* (new series) **54**, 119–39.

—— (1910*b*). Studies on Ceylon *Haematozoa.* II. Notes on the life cycle of *Haemogregarina nicoriae* Cast. a. Willey *Q. Jl. microsc. Sci.* (new series) **55**, 741–62.

—— (1912). Transmission of flagellates living in the blood of certain freshwater fishes. *Phil. Trans. R. Soc. Lond.* **B202**, 29–50.

Robin, Y. (1954). Répartition et métabolisme des guanidines monosubstituées d'origine animales. Thèse de Doctorat de Sciences Naturelles, Paris.

—— Audit, C., and Landon, M. (1967). Biogénèse des dérivés diguanidiques chez la sangsue, *Hirudo medicinalis* L. II. Mécanisme de la double transamidination. *Comp. Biochem. Physiol.* **22**(3), 787–97.

—— —— and van Thoai, N. (1962). Biogénèse de l'arcaïne chez la sangsue, *Hirudo medicinalis* L. *C. r. Soc. Biol., Paris* **156**(7), 1232–5.

—— —— Zappacosta, S., and van Thoai, N. (1962). Biogénése de l'hirudonine — I. *Comp. Biochem. Physiol.* **7**(3), 221–5.

—— and Roche, J. (1965). Répartition biologique des guanidines substituées chez des vers terrestres et d'eau douce (Oligochaetes, Hirudinées, Turbellaries) recoltés en Hongrie. *Comp. Biochem. Physiol.* **14**(3), 453–61.

—— and van Thoai, N. (1961). Structure et synthèse de l'hirudonine [diamidinospermidine ou N-(3-guanidopropyl)-4-aminobutyl-guanidine]. *C r. hebd. Séanc. Acad. Sci., Paris* **252**(8), 1224–6.

—— —— van Thoai, N., and Pradel, L.A. (1957). Métabolisme des dérivés guanidylés. VII. Sur une nouvelle guanidine monosubstituée biologique: l'hirudonine. *Biochem. biophys. Acta* **24**(2), 381–4.

—— —— Roche, J. (1957). Sur la présence d'arcaïne chez la sangsue, *Hirudo medicinalis* L. *C. r. hebd. Séanc. Soc. Biol. Paris* **151**(12), 2015–17.

Robinson, G.L., and Jahn, L.A. (1980). Some observations of fish parasites in Pool 20, Mississippi River. *Trans. Am. microsc. Soc.* **99**(2), 206–12.

Roche, J., van Thoai, N., Robin, Y., and Pradel, L.A. (1956). Sur la présence d'un dérivé guanidique nouveau dans le muscle de la sangsue, *Hirudo medicinalis* L. *C. r. hebd. Séanc. Soc. Biol. Paris* **150**(10), 1684–6.

—— Bessis, M., and Thiery, J.P. (1960). Étude de l'hémoglobine plasmatique de quelques Annélides au microscope électronique. *Biochim. Biophys. Acta* **41**, 182–4. [EM of *Hirudo* haemoglobin.]

Rodhain, J. (1942). A propos de Trypanosomes de Poissons du Bassin du Fleuve Congo. *Rev. Zool. Bot. Afr.* **36**(4), 411–16.

Roever-Bonnet, H. de (1974). *Toxoplasma* infection in leeches (*Theromyzon tessulatum*). *Trop. geogr. Med.* **26**(3), 337. [*Toxoplasma* was isolated from *T. tessulatum* found in the ethmoid and on the skull under the skin of a mallard in the Netherlands; leech possible vector.]

Rohde, E. (1891). Histologische Untersuchungen über das Nervensystem der Hirudineen. *Zool. Beitr. Breslau* **3**, 1–68.

Röhlich, P. (1962). The fine structure of the muscle fiber of the leech, *Hirudo medicinalis*. *J. ultrastruct. Res.* **7**(5–6), 399–408.

—— and Török, L.J. (1964). Elektronenmikroskopische Beobachtungen an den Sehzellen des Blutegels, *Hirudo medicinalis* L. *Z. Zellforsch.* **63**, 618–35.

Roitrub, B.A., and Zlatin, R.S. (1983). [A method for a stable increase in sensitivity of a biological method of acetylcholine determination] [*Hirudo.*] *Fiziol. Zh.* (Kiev) **29**(2), 237–9. [In Russian.] [1984BA:13415.]

Rollinson, D.H.L., Soliman, K.N., and Mann, K.H. (1950). Deaths in young ducklings associated with infestations of the nasal cavity with leeches. *Vet. Rec.* **62**(15), 225–7. [*Theromyzon tessulatum* in nose and on body; Europe.]

Romanini, M.G. (1948*a*). Ricerche sul fattore diffusore negli irudinei. *Natura, Milano* **39**(3–4), 73–5.

—— (1948*b*). Azione del fattore diffusore de *Haemopis sanguisuga* su vari substrati. *Atti. Soc. Ital. Milano* **87**(3–4), 244–6.

Rosça, D.I. (1950). [Duration of survival and variations in body weight in *Hirudo medicinalis* placed in solutions of increasing salinity.] *Stud. Cercet. Acad. Romåne Fil. Cluj.* **1**, 211–22. [In Romanian; French summary.]

—— (1972). [Research on the hormonal regulation of osmotic exchanges in *H. medicinalis*.] *Stud. Un. Babes-Bolyai Ser. Biol. Cluj.* **17**(1), 115–32. [In Romanian; French and Russian summaries.]

—— and Oros, I. (1962). Cercetari asupra patrunderii fosfatului cu P^{32} in corpul lipitorilor (*Hirudo medicinalis*). (Résumé.) *Stud. Cercet. Biol. Cluj.* **13**(2), 347–53.

—— —— (1970). [Permeability of some leech tissues for $^{45}CaCl_2$.] *Stud. Un. Babes-Bolyai Ser. Biol. Cluj.* **15**(2), 123–7. [In Romanian; English and Russian summaries.] [*Hirudo medicinalis*.]

—— and Scheerer, I. (1963). [Modifications of the cholinesterase activity of the epitheliomuscular wall of *Hirudo medicinalis* by the influence of the osmotic factor.] *Stud. Un. Babes-Bolyai Ser. Biol. Cluj.* 117–22. [In Romanian; Russian and French summaries.]

—— Wittenberger, C., and Ruşdea, D. (1958). [Studies on the variations of salinity. XLV. Osmoregulation and the role of the nervous system in osmoregulation in *Hirudo medicinalis.*] *Stud. Cercet. Biol. Cluj.* **9**(1–2), 113–36. [In Romanian; French summary.]

Rosenbluth, J. (1967). *J. cell. Biol.* **34**, 15–33. (Incomplete reference)

—— (1972*a*). Myoneural junctions of two ultrastructurally distinct types in earthworm body muscle. *J. cell. Biol.* **54**, 566–79.

—— (1972*b*). Obliquely striated muscle. In: *The structure and function of muscle* (2nd edn) Vol. 1, pp. 389–420. Academic Press, London.

—— (1973). Postjunctional membrane specialization at cholinergic myoneural junctions in the leech. *J. comp. Neurol.* **151**(4), 399–406.

Rosenbaum, H. (1941). Beiträge zur Physiologie und Pharmakologie der Blutegelmuskulatur. *Arch. int. Pharmacol. Thér.* **66**, 475.

Rosenfeld, G., and Kelen, E.M.A. (1970). On the anticoagulant (fibrinogenolytic) activity of the salivary glands extract from a Brazilian leech (*Haementeria lutzi* Cesar Pinto 1920). (Abstr.) *13th Int. Congr. of Hematology*, Munich, p.145.

—— —— Rzeppa, H.W., and Vizotto, L.D. (1965). Observaçoes sobre a açao anticoagulante da glandula salivar de sanguessuga (*Haementeria brasiliensis*). *9th Congr. of the Brazilian Society of Hematology and Hemotherapy*, Belo Horizonte.

—— —— —— —— (1966). Observaçoes sobre a açao anticoagulante da glandula salivar de sanguessuga (*Haementeria brasiliensis* Weber 1915–Pinto 1920). *Ciência Cultura* **18**, 167; *Acta physiol. latinoam.* **16**, Suppl. 1, 113–14.

Ross, D.H., and Triggle, D.J. (1972). Further differentiation of cholinergic receptors in leech muscles. *Biochem. Pharmacol.* **21**(18), 2533–6.

Ross, M.S. (1983). The leech: of dermatologic interest? (Letter). *Arch. Dermatol.* **119**(4), 276–7.

Ross, W.N., and Reichardt, L.F. (1979). Species-specific effects on the optical signals of voltage sensitive dyes. *J. membr. Biol.* **48**(4), 343–56. [1980BA:39051.] [*Hirudo medicinalis.*]

Rothschein, J. (1973). Ueber den Einfluss der geplanten Donaukraftwerke auf die Hydrofauna des Tschechoslowakischen Donauabschnittes. *ZB Slov. Nar. Muz. Prir. Vedy.* **19**(1), 79–97. [1974BA:18909.] [Power plants; leeches increase as water flow decreases; substrates.]

Rousseau, E. (1912). Les Hirudinées d'eau douce d'Europe. *Ann. Biol. Lacustre* **5**, 259–95.

Rozhkova, E.K. (1974). [The effect of degeneration of motor nerves on cholinoreception of the dorsal muscle of the medicinal leech *Hirudo medicinalis.*] *Zh. Evol. Biokhim. Fiziol.* **10**(6), 628–31. [In Russian; English summary.] [English translation, pp. 573–5. Plenum, New York.] [1975BA:14434.]

Rubin, E. (1978). The caudal ganglion of the leech, with particular reference to homologues of segmental touch receptors. *J. Neurobiol.* **9**(5), 393–405. [1979BA:22418.]

Rucner, D. (1971). [Contributions to the knowledge of the fauna of some forest associations in Croatia.] *Larus* **23**, 129–203. [German summary.] [1972BA:36308.]

Rude, S. (1967). *Am. Zool.* **7**, 738. (Incomplete reference)

—— (1969). Monoamine-containing neurons in the central nervous system and peripheral nerves of the leech, *Hirudo medicinalis*. *J. comp. Neurol.* **136**, 349–72.

—— Coggeshall, R.E., and Van Orden, L.S. III (1969). Chemical and ultrastructural identification of 5-hydroxytryptamine in an identified neurone. *J. cell. Biol.* **41**, 832–54.

Ruffo, S. (1934). L'*Herpobdella atomaria* Carena var. *meyeri* Blanchard nella Grotta di Veja. *Atti Acc. Agr. Veronese* Ser. V, Vol. XII.

Rupp, R.S., and Meyer, M.C. (1954). Mortality among brook trout, *Salvelinus fontinalis*, resulting from attacks of freshwater leeches. *Copeia* **4**, 294–5.

Russell, D.A. (1973). The environments of Canadian dinosaurs. *Can. geogr. J.* **87**, 4–11.

Russev, B., and Janeva, I. (1976). [Review of the species composition, distribution, ecology and significance as biological indicators of leeches (Hirudinea) in Bulgaria.] *Khidrobiologiya, Sofia* No. 3, 40–56. [In Bulgarian.]

—— and Marinov, T. (1964). [Uber die Polychaten- und Hirudineen-fauna im bulgarischen Sektor der Donau.] *Izv. Zool. Inst. Sofiya* **15**, 191–7. [In Bulgarian; German summary.] [Leeches of Bulgarian Danube.]

Russo, R.A. (1975). Notes on the external parasites of California inshore sharks. *Calif. Fish Game* **61**(4), 228–32. [*Branchellion lobata* common on elasmobranchs of Tomales and San Francisco Bays.]

Rutschke, E. (1970). Zur Substruktur der Cuticula der Egel (Hirudinea). *Z. Morph. Ökol. Tiere* **67**(2), 97–105.

Rutty, J. (1772). *An essay towards a natural history of the County of Dublin.* Sleater, Dublin. [The medicinal leech was to be found in several localities in County Dublin; leeches imported from Wales to Dublin.]

Ryabov, I. (1854). *O pazvedenii vrachebnoi piyavki (Hirudo medicinalis) v Nizhnetagil'skom Zavode*, pp. 1–6. [In Russian.]

Rye Jr, R.P., and King Jr, E.L. (1976). Acute toxic effects of two lampricides to twenty-one freshwater invertebrates. *Trans. Am. fish. Soc.* **105**(2), 322–6. [1976BA:63775.]

Sabraze, J. (1933). I. Le renouveau des applications de Sangsues. Le Livre de Louis Vayson de Bordeaux. II. Sensibilisation et intolerance aux piques de Sangsues. III. L'hirudine ou l'hemophiline. IV. Incidents et accidents locauz. *Gaz. Sc. Méd.* **54**, 642.

Safarík, J. (1854). O Pijavici. *Ziva* **2**, 225–31, 268–77.

Safarov, R.A. (1972). [Species and quantitative content of the plankton and benthos of the Aggel' Lake.] *Izv. Akad. Nauk Az. SSR Ser. Biol. Nauk* **4**, 55–7. [In Russian; Azerb. summary.]

Safonova, T.A., and Zhuravlev, V.L. (1977). [Changes in electro-physiological properties of neurons of the horse leech at high intra-cellular Na$^+$ concentration.] *Biofizika* **22**(3), 456–60. [In Russian; English summary.] [1978BA:64761.] [English transl. *Biophysics* **22**(3), 469–74.]

Sage, B.L. (1958). On the avian hosts of the leech *Theromyzon* (*Proto-clepsis*) *tessellata* (O.F. Müller). *Br. ornithol. Club Bull*, **78**, 113–15.

Saha, S. *et al.* (1977). Unusual foreign body causing bleeding per urethra. *J. Indian med. Ass.* **69**(12), 286–7.

Sakharov, D.A. (1970). Cellular aspects of invertebrate neuropharma-cology. *A. Rev. Pharmacol.* **10**, 335–52.

Salzberg, B.M., Davila, H.V., and Cohen, L.B. (1973). Optical recording of impulses of individual neurones of an invertebrate central nervous system. *Nature, Lond.* **246**, 508–9.

Salzberger, (1928). *Laryngoscope* **38**, 27–32. (Incomplete reference)

Sampl, H. (1976). Tierwelt. In: *Die Natur Kärntens*, 2, (ed. by H. Sampl *et al.*) Verlag Johannes Heyn, Klagenfurt. [*Hirudo* in Austria]

Sanchez, D. (1909). El sistema nervioso de los hirudineos. *Trab. Labor. Invest. Biol. Un. Madrid* **7**, 31–187.

—— (1912). El sistema nervioso de los hirudineos. *Trab. Labor. Invest. Biol. Un. Madrid* **10**, 1–143.

Sandner, H. (1951). [Recherches sur la faune des sangsues.] *Acta zool. oecol. Un. Lodz., Lodz* **4**, 5–50. [In Polish; French summary.]

—— (1952). Badania nad fauna pijawek. *Spraw. Lodz. TN Lodz* **6**(1951), **1**(10), 52–4.

—— (1953a). [Studies on brackish water in Poland. Ecology of leeches (Hirudinea) found in the lakes Lebsko and Sarbsko.] *Ekol. Pol.* **1**(3), 55–72. [In Polish; English summary.]

—— (1953b). Z badan nad ekologia pijawek. *Kosmos, Ser. Biol. Warsaw 2* **2**(3), 88–9.

—— (1954). Recherches sur la faune de sangsues. *Bull. Soc. Sci. Lodz Cl. III Sci. Math.-Nat.* **5**, 1–16.

—— and Wilkialis, J. (1972). Leech communities (Hirudinea) in the Mazurian and Bialystok regions and the Pomeranian Lake District. *Ekol. Pol.* **20**(27), 345–65. [Polish summary.]

Sanfilippo, N. (1950). Le grotte della provincia di Genova e la loro fauna. *CAI Mem. Com. Scient. Cent.* No. 2. [*Haemopis sanguisuga.*]

Sanjeeva Raj, P.J. *see* Raj, P.J. Sanjeeva.

Sapkarev, J.A. (1963). Die Fauna Hirudinea Mazedoniens I. Systematik und Ökologie der Hirudinea des Prespa-Sees. *Bull. scient. Conseil Acad. RSFY* **8**(1/2), 7–8.

—— (1964). [Hirudineenfauna aus Mazedonians. Ein Beitrag zur Kenntnis der Systematik und Ökologie der Hirudinea des Dojran-Sees.] *Folia Balanica Inst. pisc. RSM* **2**(3), 1–8. [In Russian; German summary.]

—— (1968*a*). The taxonomy and ecology of leeches (Hirudinea) of Lake Mendota, Wisconsin. *Trans. Wis. Acad. Sci.* **56**, 225–53.

—— (1968*b*). [The fauna of Hirudinea of Macedonia. The taxonomy of leeches (*Hirudinea*) of Skopje's Valley.] *Fragm. Balcan. Mus. Macedonici Sci. Natur.* **6**(21), 189–93.

—— (1970*a*). The fauna of Hirudinea of Macedonia. The taxonomy and distribution of leeches of Aegean lakes. *Int. Revue ges. Hydrobiol.* **55**(3), 317–24.

—— (1970*b*). [Seasonal variations in the populations of *Erpobdella octoculata* in the large lakes of Macedonia (Dojran, Presplansko and Okhridsko).] *Godishen Zbornik Prirodno-matematicki Fakultet na Universitot Skopje* **22**, 19–31. [In Macedonian; English summary.] [From *Referatinvnyi Zhurnal, Biologiya* 7D85 (1971).]

—— (1975). Contribution to the knowledge of the earthworms (Lumbricidae) and leeches (Hirudinea) of Kosovo, Jugoslavia. *Ann. Fac. Sci. Un. Skopje* **27–8** (1974/75), 39–54.

Sapovalenko, L.I. (1938). Nabljudenija nad pigmentnymi Kletkemi *Haementeria costata* Müll. *Rab. Lab. Obsc. Biol. Zool. 3 Moscow Med. Inst., Moscow* **1**, 41–76.

Sarah, H.H. (1971). Leeches found on two species of *Helisoma* from Fleming's Creek, Michigan. *Ohio J. Sci.* **71**(1), 15–20.

Sargent, P.B. (1975). Transmitters in the leech central nervous system: analysis of sensory and motor cells. Ph.D. dissertation, Harvard University, Cambridge, Massachusetts.

—— (1977). Synthesis of acetylcholine by excitatory motoneurons in central nervous system of the leech. *J. Neurophysiol.* **40**(2), 453–60. [1977BA:27056.]

—— Yau King-Wai, and Nicholls, J.G. (1977). Extrasynaptic receptors on cell bodies of neurons in central nervous system of the leech. *J. Neurophysiol.* **40**(20), 446–52. [1977BA:27055.]

Sarma, D.N. (1972). Leech as pest on human host. *J. Indian med. Ass.* **58**, 262.

Sarojini, S.G., and Gowri, N. (1979). Presence of hyaluronic acid in the cuticle of leech *Hirudo medicinalis. Indian Zool.* **3**(1–2), 95–7.

Sauber, F., Reuland, M., Berchtold, J.-P., Hetru, C., Tsoupras, G., Luu, B., Moritz, M.E., and Hoffmann, J.A. (1983). Cycle de mue et ecdystéroïdes chez une sangsue, *Hirudo medicinalis. C.r. Seanc. Acad. Sci Ser III Sci. Vie.* **296**(8), 413–18. [English summary.] [1984BA:11674.] [Also note: Sauber, F, C. Kappler and J.A. Hoffmann. 1981. Communication au 5ᵉ Symposium Européen sur l'Ecdysone, Berne.]

Saubermann, A.J., and Riley, W.D. (1980). Determination of solute and water content of leech neurons and glia by X-ray micronanalysis of frozen hydrated sections. *J.cell. Biol.* **87**(2 Part 2), 83A. [1980BA: 59430.]

Savigny, J.C. (1822). See Autrum (1939c).

Sawada, M., and Coggeshall, R.E. (1975). The identification of a new leech inhibitory motor neuron and an analysis of its junctional potentials. *Neurosci. Abstr.* 565.

—— —— (1976a). Ionic mechanism of 5-hydroxytryptamine induced hyperpolarization and inhibitory junctional potential in body wall muscle cells of *Hirudo medicinalis*. *J. Neurobiol.* **7**(1), 63–76. [1976BA:8438.]

—— —— (1976b). A central inhibitory action of 5-hydroxytryptamine in the leech. *J. Neurobiol.* **7**(6), 477–82. [1977BA:44983.]

—— Wilkinson, J.M., McAdoo, D.J., and Coggeshall, R.E. (1976). The identification of two inhibitory cells in each segmental ganglion of the leech and studies on the ionic mechanism of the inhibitory junctional potentials produced by these cells. *J. Neurobiol.* **7**(5), 435–45. [1977BA:8613.] [Cell 119.]

Sawyer, Roy T. [see also: Budzynski, A.Z.; Daniels, B.A.; Elder, J.F.; Forrester, D.J.; Hetchel, F.; Jones, P.G.; Schaeffner, K.H.; Smiley, J.W.; Stent, G.S.; Wallace, R.K.; Weisblat, D.A.]

Sawyer, Roy T. (1967). The leeches of Louisiana, with notes on some North American species. *Proc. Louisiana Acad. of Sci.* **30**, 32–8.

—— (1968). Notes on the natural history of the leeches on the George Reserve, Michigan. *Ohio. J. Sci.* **68**(4), 226–8.

—— (1969). Studies on the Hirudinea. Ph.D. dissertation. University of Wales, Swansea.

—— (1970a). Observations on the natural history and behavior of *Erpobdella punctata* (Leidy) (Annelida: Hirudinea). *Am. Midl. Nat.* **83**(1), 65–80.

—— (1970b). The juvenile anatomy and post-hatching development of the marine leech, *Oceanobdella blennii* (Knight-Jones, 1940). *J. nat. Hist.* **4**, 175–88.

—— (1971a). The phylogenetic development of brooding behaviour in the Hirudinea. *Hydrobiologia* **37**(2), 197–204.

—— (1971b). Erpobdellid leeches as new hosts for the nematomorph, *Gordius* sp. *J. Parasitol.* **57**(2), 285.

—— (1971c). The rediscovery of the bi-annulate leech, *Oligobdella biannulata* (Moore 1900), in the mountain streams of South Carolina (Annelida: Hirudinea). *Ass. southeast. Biol. Bull.* **18**(2), 54.

—— (1972a). *North American freshwater leeches, exclusive of the Piscicolidae, with a key to all species.* pp. 155 University of Illinois Press, Urbana, Ill.

—— (1972b). A new species of 'tentacled' marine leech from the subantarctic Marion and Crozet Islands. *Hydrobiologia* **40**(3), 345–54.

—— (1972c). Observations on the marine leeches of South Carolina

(Annelida: Hirudinea). *Bull. S. Carolina Acad. Sci.* **34**, 103.

—— (1973*a*). Bloodsucking freshwater leeches: observations on control. *J. econ. Entomol.* **66**(2), 537.

—— (1973*b*). The rediscovery of *Glossiphonia swampina* (Bosc, 1802) in the coastal plain of South Carolina (Annelida: Hirudinea) *J. Elisha Mitchell scient. Soc.* **89**, 4–5.

—— (1974). Ecology of freshwater leeches. In: *Pollution ecology of freshwater organisms.* (eds. C.W. Hart and S.L.H. Fuller) Chapter 4, pp. 81–142. Academic Press, London.

—— (1978). Domestication of the world's largest leech for developmental neurobiology. *Yearb. Am. Phil. Soc.* pp. 212–13.

—— (1979*a*). Leeches of special concern from South Carolina. *Proc. of the First (1976) South Carolina Endangered Species Symposium.* (eds. B. Ezell and D. Forsythe) pp. 100–2. S.C. Wildlife and Marine Resources Department.

—— (1979*b*). The medicinal leech, *Hirudo medicinalis,* an endangered species. *Proc. of the First (1976) South Carolina Endangered Species Symposium.* (eds. B. Ezell and D. Forsythe) pp. 103–6. S.C Wildlife and Marine Resources Department.

—— (1981*a*). Leech biology and behavior. In: *The neurobiology of the leech* (eds. K.J. Muller, J.G. Nicholls and G.S. Stent) Chapter 2, pp. 7–26. Cold Spring Harbor Laboratory Publications, Cold Spring Harbor, New York.

—— (1981*b*). Why we need to save the medicinal leech. *Oryx* **16**(2), 165–8.

—— (1982). An expedition to Borneo to study the aggressive behavior of land leeches, with collateral analysis of their anticoagulants. *Yearb. Am. Phil. Soc.* pp. 43–44.

—— (1984*a*). Arthropodization in the Hirudinea: evidence for a phylogenetic link with insects and other Uniramia? *Zool. J. Linn. Soc. London* **80**, 303–22.

—— (1984*b*). Anticoagulant properties of leeches. VIII International Congress on Thrombosis, 4—7 June 1984. Istanbul, Turkey. p. 189 (Abstract)

—— (1985). Scientific rationale behind the medical use of leeches: a review of their biologically active secretions. *Thrombos. Haemostas.* **54**(1), 315. (Abstract)

—— (1986). Leeches. New roles for an old medicine. *Ward's Bulletin.* Spring 1986, pp. 1–4.

—— (In press). Scientific rationale behind the medical use of leeches. In *Animal venoms and hemostasis.* (ed. H. Pirkle and F.S. Markland Jr). Marcel Dekker, New York.

—— and Chamberlain, N.A. (1972). A new species of marine leech (Annelida: Hirudinea) from South Carolina, parasitic on the Atlantic menhaden, *Brevoortia tyrannus. Biol. Bull.* **142**(3), 470–79.

—— Damas, D., and Tomic, M. (1982). Anatomy and histochemistry of the salivary complex of the giant leech *Haementeria ghilianii* (Hirudinea: Rhynchobdellida). *Arch. Zool. exp. gén.* **122**(4), 411–25.

—— and Dierst-Davies, K. (1972*a*). Observations on the physiology and phylogeny of colour change in the Hirudinea (Annelida). *Hydrobiologia* **44**(2), 215–36.

—— —— (1972*b*). Some factors which influence physiological color change in leeches and observations on the evolution of the response (Annelida: Hirudinea). *Ass. southeast. Biol. Bull.* **19**(2), 97.

—— and Fitzgerald, S. (1981). Leech circulatory system. In: Vol. 1. *Invertebrate blood cells*. (eds. N.A. Ratcliffe and A.F. Rowley) Chapter 5, pp. 141–59. Academic Press, London.

—— and Hammond, D.L. (1973). Distribution, ecology and behavior of the marine leech *Calliobdella carolinensis* (Annelida: Hirudinea), parasitic on the Atlantic menhaden in epizootic proportions. *Biol. Bull.* **145**(2), 373–88.

—— and Kinard, W.F. (1980). A checklist and key to the marine and freshwater leeches (Annelida: Hirudinea) of Puerto Rico and other Caribbean islands. *Caribb. J. Sci.* **15**(3/4), 83–5.

—— Lawler, A.R., and Overstreet, R.H. (1975). The marine leeches of the eastern United States and the Gulf of Mexico with a key to the species. *J. nat. Hist.* **9**, 633–67.

—— and Pass, K.A. (1972). The occurrence of *Macrobdella decora* (Say, 1824), in the Appalachian Mountains of Georgia and South Carolina (Annelida: Hirudinea). *J. Elisha Mitchell Sci. Soc.* **88**(1), 34–5.

—— and R.H. Shelley, (1976). New records and species of leeches (Annelida: Hirudinea) from North and South Carolina. *J. nat. Hist.* **10**, 65–97.

—— and De Villeirs, A.F. (1976). Notes on two marine leeches (Annelida: Hirudinea) from subantarctic Marion Island, including a new record. *Hydrobiologia* **48**(3), 267–8.

—— and White, M.G. (1969). A new genus and species of marine leech parasitic on an Antarctic isopod. *Bull Br. Antarctic Surv.* No. 22, pp. 1–14.

—— Taylor, A., and Bin Sahat, M.J. (1982). The leeches of Brunei (Annelida: Hirudinea), with a checklist and key to the known and expected freshwater, terrestrial and marine leeches of Borneo. *J. Brunei Mus.* **5**(2), 168–201.

—— Le Pont, F., Stuart, D.K., and Kramer, A.P. (1981). Growth and reproduction of the giant glossiphoniid leech *Haementeria ghilianii*. *Biol. Bull.* **160**, 322–31.

Saxena, B.N., and Dubey, D.N. (1971). Studies on the life history and bionomics of land leech, *Haemadipsa sylvestris* Blanchard. *Armed Forces med. J., India* **27**(4), 542–50. [Forests of Assam.] [See also: (1969). *Pesticides Bombay* **3**(8), 27–9.]

—— Khalsa, H.G., and Pillai, K.R.M. (1969). Evaluation of repellents against land leeches. II. *Def. Sci. J.* **19**(2), 93–6. [See also: (1969). *Pesticides Bombay* **3**(5), 23–6.]

Scalet, C.G. (1971). Parasites of the Orangebelly Darter, *Etheostoma radiosum* (Pisces: Percidae). *J. Parasitol.* **57**(4), 900.

Schain, R.J. (1961). Effects of 5-hydroxytryptamine on the dorsal muscle of the leech (*Hirudo medicinalis*). *Br. J. Pharmacol. Chemother.* **16**(3), 257–61.

Schäperclaus, W. (1954). *Fischkrankheiten.* Berlin. [*Piscicola geometra* transmits *Cryptobia cyprini* and *C. borelli* to carp.]

Scharff, R.F. (1899). *The history of the European fauna.* W. Scott, London. [Irish leeches.]

Scharrer, B. (1954). Über sekretorischtätige Nervenzellen bei wirbellosen Tieren. *Naturwiss.* **25**, 131–8.

Schaeffner, K.H., Walz, B., and Sawyer, R.T. (1986). Organization and ultrastructure of the anterior salivary glands of the giant leech, *Haementeria ghilianii*. *Verh. Dtsch. zool. Ges.*

Schegolev, G.G. (1909). Zur Hirudinenfanua des Gebietes der Donkosaken. *Moskva Trd. Kruz izsl. russ prir.* **4**, 112–20.

—— (1912). Ein Beitrag zur Hirudineenfauna des Turkestan. *Arb. hydrob. Stat. Glubokoje* **4**, 163–92. [In Russian.]

—— (1914). Zur Kenntnis der Hirudineenfauna des Dnjepr. *Arb. biol. Dnjepr.-Stat. Kiew* **1**, 421–31 [In Russian.]

—— (1916). [About the fauna of the leeches in the Amur region.] *Rev. zool. Russe* **1**, 250–2. [In Russian; English summary.]

—— (1922a). Eine neue Egelart aus dem Baikalsee. *Russ. hydrobiol. Zhurn.* (Saratov) **1**(4), 136–42. [In Russian; German summary.] [*Torix baicalensis.*]

—— (1922b). [Hirudinea of the Oka River.] *Arb. biol. Oka-Stat.* **2**, 19–28. [In Russian.]

—— (1922c). [Bemerkungen uber die Hirudineen Russlands. I. *Glossiphonia octoserialis* nov. sp.] *Arb. biol. Wolgastat.* **6**, 233–40. [In Russian; German summary.]

—— (1922d). Verzeichnis der von der biologischen Wolga-Station eingesammelten Hirudineen. *Arb. biol. Wolgastat.* **6**, 241–6. [In Russian; German summary.]

—— (1925). Die Blutegel des Flusses Kljasma und benachbarter Wasserbehälter im Umkreise der Biologischen Station. *Bull. Stat. biol. Bolchewo* **1**, 7–14. [In Russian; German summary.]

—— (1927). Die Änderung der Färbung unter dem Einfluss des Lichtes bei *Protoclepsis tessellata* Braun 1805. *Rev. zool. Russe* **7**(3), 141–8. [In Russian; German summary: 149–67.]

—— (1928). Zur Hirudineenfauna des Wolgabassins. *Arb. biol. Wolgastat.* **10**, 28–9. [German summary: 30.]

—— (1938). [External morphology of *Trocheta subviridis* f. *danastrica* and comments on the subdivision of the annulus in leeches.] *Raboti Lavoratorii Obshchei i Zoologii 3-go Moskovskogo Meditzinskogo Inst.* **1**, 59–147.

—— (1945). [Of the size attained by the medicinal leech. On the question of the frequency of feeding.] *Zool. Zhurn.* **24**(5), 273–6. [In Russian.]

—— (1946). [On adaptive modification of the method of Lënera for the medicinal leech.] *Zool. Zhurn.* **25**(2), 111–14. [In Russian.]

—— (1948). [Observations on the frequency of cocoon deposition in the medicinal leech.] *Zool. Zhurn.* **27**(1), 13–16. [In Russian.]

—— (1949). Leeches, Hirudinea. In *Zhizn' Presnich Vod CCP* [*Life in fresh water of the SSSR*] Vol. 2, pp. 131–45. [In Russian.]

—— (1951). [Observations on mobility of *Hirudo medicinalis* in reservoirs.] *Zool. Zhurn.* **30**(5), 430–9. [In Russian.]

—— (1952). [When the medicinal leech is a lithophilic organism.] *Tr. Vsesouzn. Hidrobiol. Obsch.* **IV**, 151–86. [In Russian.]

—— (1953a). [Observation on feeding in the medicinal leech.] *Tez. Dokl. Na. Nauchn. Konfer. Pyazansk. Med. Inst. Im. Akad. I. P. Pavlova* 73–6. [In Russian.]

—— (1953b). [Observation on oxygen consumption of the medicinal leech and condition of maintenance.] *Tez. Dokl. Na Nauchn. Konfer. Pyazansk. Med. Inst. Im. Akad. I. P. Pavlova* 77–9. [In Russian.]

—— and Féodorova, S. (1955). [The medicinal leech and its application,] pp. 1–67. Moscow. [In Russian.]

—— and Perphil'eva, N.S. (1954). [On the maintenance of medicinal leech in turf.] *Tez. Dokl. Na XIV Nauchn. Konfer. Pyazansk. Med. Inst. Im. Akad. I. P. Pavlova* 18–20. [In Russian.]

—— and Schegoleva, Z.A. (1951). [Leeches of Turkmenistan.] *Tr. Murgabsk. Gidrobiol. St.* **1**, 77–102.

—— —— and Sidorenkova, N.V. (1964). [Effect of light, heat, and food on oögenesis and breeding in *Hirudo medicinalis*.] In [*Problems in modern embryology*] pp. 203–7. Moscow. [In Russian.]

[See also Utkina, O.T. (1958).]

Scherbina, G.V. (1973). [Live bodies in the respiratory organs.] *Zhurnal Ushnykh, Nosovykh i Gorlovykh Boleznei* No. 6, 78–9. [In Russian.] [*Limnatis nilotica*, Algeria.]

Schittenhelm, A., and Bodong, A. (1906). Beiträge zur Frage der Blutgerrinnung mit besonderer Berücksichtigung der Hirudinwirkung. *Arch. exp. Path.* **54**, 217–44.

Schleip, W. (1913). Die Furchung des Eies von *Clepsine* und ihre Beziehungen zur Furchung des Polychaeteneies. *Ber. naturf. Ges. Freiburg* **20**, 177–88.

—— (1914a). Die Entwicklung zentrifugierter Eier von *Clepsine sexoculata*. *Verh. dt. Zool. Ges.* **24**, 236–53.

—— (1914b). Die Furchung des Eies der Rüsselegel. *Zool. Jb. Anat.* **37**, 313–68.

—— (1939). Ontogenie der Hirudineen, In *Bronn's Klassen und Ordnungen des Tierreichs*, 4 Bd., 3 Abt., 4 Buch, Teil 2, pp. 123–319. Leipzig.

Schlenz, E. (1971). Thesis on leeches. [Two marine forms described: *Ozobranchus* and *Branchellion*.] Unpublished to date, so specific names unavailable. Universidade de Sao Paulo, Instituto de Biociências.

Schlue, W.R. (1975). Action of tetraethylammonium ion on accommodation in sensory cells in the leech. *Pflügers Arch. ges. Physiol. Suppl.* **355**, R82.

—— (1976*a*). Current excitation threshold in sensory neurons of leech central nervous system. *J. Neurophysiol.* **39**(6), 1176–83. [1977BA:57176.]

—— (1976*b*). Sensory neurons in leech central nervous system: changes in potassium conductance and excitation threshold. *J. Neurophysiol.* **39**(6), 1184–92. [1977BA:57175.] [Same title; (1976) *Verh. dt. Zool. Ges.* **69**, 279.]

—— (1977). Differential action of TEA on excitation threshold in sensory neurons of the leech central nervous system. *Comp. Biochem. Physiol. C Comp. Pharmacol.* **57**(1), 85–9. [1977BA:32885.]

—— and Appuhn, S. (1974*a*). Accommodation of sensory neurones in the central nervous system of the leech. *Arch. ges. Physiol. Suppl.* **347**, R52.

—— —— (1974*b*). Excitability of sensory neurons in the central nervous system of the leech. *Int. Congr. Physiol. Sci., 26th, New Delhi*, p. 159.

—— and Deitmer, J.W. (1980*a*). Extracellular potassium in neuropile and nerve cell body region of the leech central nervous system. *J. exp. Biol.* **87**, 23–43.

—— —— (1980*b*). Potassium concentration in extracellular spaces of the leech central nervous system. *Pflügers Arch. Eur. J. Physiol.* **384** (suppl.): R20, 77. [1980BA:15462.]

—— —— (1981). Intracellular potassium and conductance measurements of sensory neurones in the leech central nervous system. *Pflügers Arch. Suppl.* **391**, R33, 131.

—— —— (1984). Potassium distribution and membrane potential of sensory neurons in the leech nervous system. *J. Neurophysiol.* **51**, 689–704.

—— Thomas, R.G. (1985). A dual mechanism for intracellular pH regulation by leech neurones. *J. Physiol.* **364**, 327–38.

—— and Walz, W. (1979). Physiological properties of neuropil glial cells in the central nervous system of the leech. *Pflügers Arch. Eur. J. Physiol.* **379** (Suppl.), R39, 156.

—— —— (1984). Electrophysiology of neuropil glial cells in the central nervous system of the leech: a model system for potassium homeostasis in the brain. *Adv. Cellular Neurobiology.* **5**, 143–75.

—— and Wuttke, W. 1983. Potassium activity in leech neuropile glial cells changes with external potassium concentration. *Brain Res.* **270**, 368–72.

—— —— Deitmer, J.W. (1985). Ionic activity measurements in extracellular spaces, nerve and glial cells in the central nervous system of the leech. In *Ion measurements in physiology and medicine* (ed. Kessler, M. *et al.*). Springer-Verlag, Berlin, pp. 166–173.

—— Schliep, A., and Walz, W. (1980). Fluorescence marking of neuropil glial cells in the central nervous system of the leech *Hirudo medicinalis*. *Cell. Tiss. Res.* **209**(2), 257–70.

Schlüter, E. (1933). Die Bedeutung des Centralnervensystems von *Hirudo medicinalis* für Locomotion und Raumorientierung. *Z. wiss. Zool.* **143**, 538–93.

Schmidt, F. (1902, 1903, 1905, 1907). See separate bibliography on Branchiobdellida, p. 807.

Schmidt, G.A. (1917). Zur Entwicklung des Entoderms bei *Protoclepsis tesselata* (Müll., O.F.). *J. Sect. Zool. Soc. amateurs Sci. nat.* (new series) **4**, 1–22. [In Russian.]

—— (1921). Die Embryonalentwicklung von *Piscicola geometra* Blainv. *Zool. anz.* **53**, 123–7. (Gelehrte Schr. Astrachauer Univ. 1.).

—— (1923). Zur Embryonalentwicklung von *Archaeobdella esmontii* O. Grimm. *Rev. zool. Russe* **3**(3/4), 427–33. [In Russian; German summary: 434–5.]

—— (1924). Untersuchungen über die Embryologie der Anneliden. 2. Die Besonderheiten der Embryonalentwicklung der Ichthyobdellidae und ihre Entstehung. *Zool. Jb. Anat.* **46**, 199–244.

—— (1925*a*). Untersuchungen über die Embryologie der Anneliden. I. Die embryonalentwicklung von *Piscicola geometra* Blainv. *Zool. Jb. Anat.* **47**, 319–428.

—— (1925*b*). [Die polaren plasmatischen Massen im Ei von *Protoclepsis tesselata*.] *Rev. zool. Russe* **5**, 138–62. [In Russian; German summary: 163–4.]

—— (1927*a*). The polar plasmatic masses in the egg of *Protoclepsis tesselata*. *Proc. Congr. Zool. Anat. Histol. USSR*, Vol. 2, pp. 99–100.

—— (1927*b*). A new habitat of *Acanthobdella peledina* — the Lake Imandra. *Proc. Congr. Zool. Anat. Histol., Leningrad*, Vol. 3, 389–90. [In Russian.]

—— (1929*a*). [Observations on the biology of reproduction in *Protoclepsis tesselata*.] *Mem. Stat. Bolshevo, Zap. Bols. Biol. Stancii* **3**, 103–20. [In Russian; German summary, 120–1.]

—— (1929*b*). [On *Acanthobdella peledina* Grube from Lake Imandra.] *Trav. Stat. Biol. Murman*, **3**, 1–2. [In Russian.]

—— (1930). [Types des développements embryonnaires de *Crangonobdella murmanica* W.D. Selensk.] *Rev. Zool. Russe* **10**(4), 5–17. [In Russian; German summary.]

—— (1936). Gesetzmassigkeiten des Wechsels der Embryonalanpassungen. *Biol. Zhurn. Mosc.* **5**, 633–56.

—— (1937). Vergleichend embryologische Studien über die Typen der Embryonalanpassungen bei Hirudineen und Nemertinen. *Z. Morphol. Ökol* **32**, 650–71.

—— (1939). Dégénérescence phylogénétique des modes de developpement des organes. *Arch. Zool. exp. gén.* **81**(3), 317–70.

—— (1941*a*). [Research on the comparative embryology of leeches. 10. Early stages of development of the embryo in Ichthyobdellidae.] *Izd.*

Akad. Nauk SSR II 2 Moscow-Leningrad. *A la mémoire A.N. Sewert-zova, Moscow* **2**(2), 357–489. [In Russian.]

—— (1941*b*). [Division of the egg of *Protoclepsis tesselata* O.F. Müll.] *Tr. Inst. Tsitolog. Histolog. Embriolog. A.N. SSSR* **1**(1), 145–82. [In Russian.]

—— (1944). [Adaptive significance of the pecularities of the cleavage process in leeches.] *Sh. obshch. Biol.* **5**, 284–303. [In Russian; English summary.]

—— (1953). *Embriologija Zivotnych*, Vol. 2. Moscow.

—— (1963). [Conditions and modes of rudiments displacement in time.] *Izv. Akad. Nauk SSSR Biol.* 526–46. [In Russian; English summary.] [Osmotic function of the primary ectoderm of leech embryos.]

Schmidt, G.D., and Chaloupka, K. (1969). *Alloglossidium hirudicola* sp. n., a neotenic trematode (Plagiorchiidae) from leeches, *Haemopis* sp. *J. Parasitol.* **55**(6), 1185–6.

Schmidt-Nielsen, B., and Pagel, H.D. (Date undetermined). Micropuncture studies on the nephridia of the leech. *Bull. Mt. Desert Isl. Biol. Lab.* **9**, 58–9.

Schou, S.A. (1970). Igler og andre vilde dyr i sygdomsbehandling og farmaci. *Arch. Pharm. Chemi* **77**, 1081–110. [English abstr.][Illustrated history of leeching.]

Schoumaker, H., and Van Damme, N. (1971). Maillet's O_3O_4–$Z_N I_2$ fixative and Alcian blue staining the study of neurosecretion: Invertebrates. *Stain Technol.* **46**(5), 233–7.

Schramm, J.C., Hardman, C.H., and Tarter, D.C. (1981). The occurrence of *Myzobdella lugubris* and *Piscicolaria reducta* (Hirudinea: Piscicolidae) on fishes from West Virginia. *Trans. Am. microsc. Soc.* **100**(4), 427–8.

Schröder, O. (1914). Beiträge zur Kenntnis einiger Microsporidien. *Zool. Anz.* **43**, 320–7.

Schröder, P. (1976). [The phenology of *Hydropsyche instabilis* (Trichoptera: caddis flies) in Foehrenback (Black Forest, West Germany) with special emphasis on the larval stages.] *Beitr. Naturkd. Forsch. Südwestdtsch.* **35**, 137–48. [In German.]

Schuberg, A. (1899). Beiträge zur Histologie der männlichen Geschlechtsorgane von *Hirudo* und *Aulastomum. Z. wiss. Zool.* **66**, 1–15.

—— (1904). Über einen in den Muskelzellen von *Nephelis* schmarotzenden neuen Nematoden. *Myenchus bothryophorus* n. g. n. sp. (Vorläufige Mitteilung). *Verh. Naturh.-med. Ver. Heidelb.* **7**, 629–32.

—— and Kunze, W. (1906). Über eine Coccidienart aus dem Hoden von *Nephelis vulgaris* (*Herpobdella atomaria*), *Orcheobius herpobdellae* nov. gen. nov. sp. *Verh. Dt. Zool. Ges.* **16**, 233–49.

—— and Schröder, O. (1904). *Myenchus bothryophorus*, ein in den Muskelzellen von *Nephelis* schmarotzer der neuer Nematode. *Z. wiss. Zool.* **76**, 509–21.

Schubert, I. (1971). Beobachtungen über pathologische Erscheinungen bei Hirudineen. *Zool. Anz.* **187**(3/4), 219–22.

Schultze, E. (1892). See Autrum (1939c).

Schulze, P. See Herter, K. (1932).

Schürch, M., and Walter, J.E. (1978). Über die Diäten von Egeln (Hirudinea) und Plattwürmern (Turbellaria: Tricladida). *Arch. Hydrobiol.* **83**(2), 272–6. [English summary.] [1979BA:9526.] [*Erpobdella octoculata* and *Helobdella stagnalis*.]

Schuster, M. (1940). Die Entwicklungsgeschichtliche Bedeutung der Teloblasten bei *Glossiphonia complanata* L. *Z. wiss. Zool.*, *Leipzig* **153**, 393–461.

Schuurmans Stekhoven, J.H. Jr (1940). Wormen. *Wat leeft en groeit. Deel.*, *Utrecht.* **37**, 62–70.

Schwab, A. (1949). Über die Nerven- und Muskelphysiologie des Pferdeegels *Haemopis sanguisuga*. *Z. vergl. Physiol.* **31**(4), 506–26.

Schwartz, F.J. (1974). The marine leech *Ozobranchus margoi* (Hirudinea: Piscicolidae), epizootic on *Chelonia* and *Caretta* sea turtles from North Carolina. *J. Parasitol.* **60**(5), 889–90.

Schwarze, W. (1930). See Autrum (1939c).

Schweer, M. (1959). Untersuchungen zur vergleichenden Stoffwechsel-physiologie einheimischer Süsswasseregel (Hirudinea). *Z. vergl. Physiol.* **42**, 20–42.

Schweizer, G. (1936). See Autrum (1939c).

Schwert, D.P. (1979). Description and significance of a fossil earthworm (Oligochaeta: Lumbricidae) cocoon from post-glacial sediments in southern Ontario. *Can. J. Zool.* **57**, 1402–5.

Sciacchitano, I. (1931a). Sulla presenza della *Limnatis nilotica* in Cirenaica. *Atti Soc. nat. Mat.*, *Modena* (6)**10**, 16–19.

—— (1931b). Sugli Irudinei del Modenese e sulla distribuzione degli Irudinei in Italia. *Atti Soc. nat. Mat.*, *Modena* (6)**10**, 16–19.

—— (1932). Anellidi ed Irudinei d'Albania. *Atti Accad. scient. veneto-trent.-istriana* **22**, 87–91.

—— (1933a). Nuovo contributo alla conoscenza dell'elmintologia modense. *Atti Soc. nat. Mat.*, *Modena* **64**, 96–9.

—— (1933b). Missione Zoologica del Dott. E. Festa in Cirenaica. XVIII. Irudinea. *Boll. Musei Zool. Anat. comp. R. Un. Torino* (3)**43**, 43–5.

—— (1935). Sanguisughe del Congo Belga. *Revue Zool. Bot. afr.* **26**, 448–60.

—— (1936a). Altre sanguisughe del Congo Belga. *Revue Zool. Bot. afr.* **28**, 161–3.

—— (1936b). Anellidi cavernicoli d'Italia. *Boll. Zool.* **7**, 17–22.

—— (1936c). Una nuova sanguisuga del Congo Belga. *Boll. Mus. Lab. Zool. Anat. comp. R. Un. Genova* (2)**16**(89), 3–6.

—— (1937). Irudinei del Congo Belga. *Revue Zool. Bot. afr.* **29**, 426–9.

—— (1938*a*). Nuovo contributo alla conoscenza faunistica degli elminti cavernicoli d'Italia. *Boll. Zool.* **9**, 199–205.

—— (1938*b*). Elminti del Bresciano. *Boll. Zool.* **9**, 207–14.

—— (1938*c*). Su alcune sanguisughe del Museo di Verona e sulla distribuzione geografica degli Irudinei in Italia. *Boll. Zool.* **9**, 279–86.

—— (1939*a*). Irudinei dell'Africa orientale Italiana. *Boll. Pesca Piscic. Idrobiol.* **15**, 5–11.

—— (1939*b*). Nuovi Irudinei del Congo Belga. *Revue Zool. Bot. afr.* **32**, 348–67.

—— (1940). Quinto contributo alla conoscenza faunistica degli elminti cavernicoli d'Italia. *Boll. Zool.* **11**, 175–84.

—— (1941*a*). Le attuali conoscenze sugli Irudinei dell'Africa Italiana. *Riv. Biol. colon.* **4**, 161–70.

—— (1941*b*). Missione Biologica Sagan-Omo. *R. Accad. Ital. Irudinei, Roma.* (Incomplete reference)

—— (1943). Anellidi cavernicoli del Trentino. *Stud. trent. Sci. nat.* **24**, 33–7.

—— (1948*a*). Raccolte faunistiche compiute nel Gargano da A. Ghigi e F.P. Pomini. V. Anellidi. *Acta pontif. Acad. Sci.* **12**, 255–8.

—— (1948*b*). Fauna di Romagna (Collezione Zangheri). Oligocheti, Irudinei, Gordii. *Boll. Zool.* **15**, 99–104.

—— (1952*a*). Irudinei e Gordii cavernicoli in Italia. *Arch. zool. ital.* **37**, 439–43.

—— (1952*b*). Irudinei del Congo Belga. *Ann. Mus. r. Congo belge Sér. 8vo* (*Sci. zool.*) **16**, 1–87.

—— (1953). Fauna di Romagna (Collezione Zangheri). Oligocheti, Irudinei, Gordii (nota seconda). *Boll. Zool.* **20**, 3–11.

—— (1954). Contributo alla conoscenza degli Irudinei del Congo Belga. *Annl. Mus. r. Congo belge Sér. 4to* (*Zool.*) **1**, 278–83.

—— (1956). Irudinei e Gordii dei Monti Sibillini. *Mem. Mus. civ. Stor. nat., Verona* **5**, 189–90.

—— (1957). Contributo alla conoscenza degli Irudinei del Congo Belga (Nota seconda). *Revue Zool. Bot. afr.* **56**, 373–81.

—— (1958). Irudinei di Turchia. *Monit. zool. ital.* **66**, 10–17.

—— (1959). Hirudinea. In *South African animal life. Results of the Lund University Expedition in 1950–1951*, Vol. 6, pp. 7–11. Almqvist and Wiksells, Uppsala.

—— (1960*a*). Contributo alla conoscenza degli Irudinei del Congo Belga (Nota terza). *Revue Zool. Bot. afr.* **61**, 287–309.

—— (1960*b*). Sangsues de l'Afghanistan. Contribution a l'étude de la faune d'Afghanistan. In *Libro Homenaje al doctor Eduardo Caballero y Caballero, Jubileo 1930–1960*, No. 24, pp. 533–41. Editorial Politecnica, Mexico.

—— (1961*a*). Irudinei dell'Angola. *Publçoes cult. Co. Diam. Angola* **52**, 105–18.

—— (1961*b*). Irudinei dell'Afghanistan. *Monit. zool. ital.* **68**, 187–9.

—— (1961*c*). Hirudinées des Açores et de Madère. *Bolm Mus. munic. Funchal* **14**(42), 5–6.

—— (1962*a*). Su alcune sanguisughe del Parco Nazionale Kruger (Sud Africa). *Monit. zool. ital.* **69**, 145–8.

—— (1962*b*). Contributo alla conoscenza degli Irudinei dell'Africa Centrale. *Revue Zool. Bot. afr.* **65**, 376–81.

—— (1963*a*). Le attuali conoscenze sulla distribuzione geografica degli Irudinei nella Regione Etiopica. *Monit. zool. ital.* **70**–1, 175–84.

—— (1963*b*). A collection of Hirudinea submitted by F.M. Chutter (National Institute for Water Research, Pretoria), and the geographical distribution of Hirudinea in South Africa. *Ann. Transv. Mus.* **24**, 249–60.

—— (1964). Hirudinées des Açores. *Bolm. Mus. munic. Funchal* **18**(71), 121–2.

—— (1965). Nuovo contributo alla conoscenza degli Irudinei dell'Africa. *Revue Zool. Bot. afr.* **71**, 29–33.

—— (1967). Leeches (Hirudinea) from Ethiopa. *J. nat. Hist.* **1**, 189–94.

[See also Lanza, B. (1970).]

Scofield, A.M. (1981). A check list of the helminth parasites of domestic animals of the United Kingdom. Animal Health Division, Hoechst UK, Milton Keynes, UK [record of *Hirudo medicinalis*.]

Scordia, C. (1930). Le *Limnatis nilotica* Sav. (sin. *Haemopis vorax* M.T.) in Cirenaica. *Boll. Ist. Zool. R. Un. Messina* No. 4–5.

Scotchman, F.G.V. (Date undetermined). The leeches in the water meadow. *Rep. Marlboro. Coll. nat. Hist. Soc.* **101**, 18–23.

Scott, D.E. (1972). Host–parasite relationships: metazoan ectoparasites of the grey mullet, *Mugil cephalus. Ass. southeast. Biol. Bull.* **19**, 98.

Scott, S.A., and Muller, K.J. (1980). Synapse regeneration and signals for directed axonal growth in the central nervous system of the leech, *Hirudo medicinalis. Devl Biol.* **80**(2), 345–63.

Scriban, J.A. (1907). Notes histologiques sur les Hirudinées. *Arch. Zool. exp. gén.* (IV) **7**, 397–421.

—— (1910*a*). Contributions à l'anatomie et à l'histologie des Hirudinées. *Ann. sci. Un. Jassy* **6**, 147–286.

—— (1910*b*). Citologia celulei adipoase a Hirudineelor. *Publ. Acad. Roum. Bucuressti, Libr. Socec comp.* [In Romanian; French summary.] [*Piscicola geometra.*] (Incomplete reference)

—— (1911, 1923, 1924). See Autrum (1939*c*).

—— (1936). Über die Wandstruktur der ventralen Lakune bei *Haementeria costata* Fr. M. und das Ventralgefäss bei *Haemopis sanguisuga* L. *Zool. Anz.* **113**, 64–7.

—— and Autrum, H. (1934). Ordnung der Clitellata, Hirudinea, Egel. In *Kükenthal-Krumbach, Handb. Zool.* **2**(2), 119–352.

—— and Epure, E. (1934). Beobachtungen über das Gefässsystem der Herpobdelliden. *Bull. Soc. Sci. Cluj* **8**, 145–92.

Scudder, G.G.E., and Mann, K.H. (1968). The leeches of some lakes in the Southern Interior Plateau region of British Columbia. *Syesis* **1**(1/2), 203–9.

Scullion, J., and Edwards, R.W. (1980). Effects of coal industry pollutants on the macroinvertebrate fauna of a small river in the South Wales coalfield. *Freshwat. Biol.* **10**, 141–62.

Seddon, H.R. (1950). Diseases of domestic animals in Australia. *Comm. Dept. Hlth. Serv. Publ.* **5**, 1–223.

Seemüller, U. (1979). Inhibitoren aus dem Blutegel *Hirudo medicinalis* mit Hemmwirkung gegen chymotrypsin, subtilisin und die menschlichen Granulozytenproteinase elastase und kathepsin G. Ph. D. dissertation. University of Munich. 124 pp.

—— Eulitz, M., Fritz, H. and Strobl, A. (1980). Structure of the elastase — cathepsin G inhibitor of the leech *Hirudo medicinalis*. *Hoppe-Seyler's Z. Physiol. Chem.* **361**(12), 1841–6.

—— and Fritz, H. (1977). Inhibitors of human granulocytic proteases from leeches and other sources. *Hoppe-Seyler's Z. Physiol. Chem.* **358**(10), 1281–2. (Abstr.) [1977BA:51814.]

—— —— and Eulitz, M. (1981). Eglin: elastase-cathepsin G inhibitor from leeches. *Meth. Enzymol.* **80**, 804–16.

—— Meier, M., Ohlsson, K., Müller, H.P., and Fritz, H. (1977). Isolation and characterization of a low molecular weight inhibitor (of chymotrypsin and human granulocytic elastase and cathepsin G) from leeches. *Hoppe-Seyler's Z. Physiol. Chem.* **358**(9), 1105–18. [1978BA:46381.]

Seetaram, K. (1971). Use of anaesthesia for extraction of leech. *J. Indian med. Ass.* **57**, 395.

Segal, D.B., Humphrey, J.M., Edwards, S.J., and Kirby, M.D. (1968). Parasites of man and domestic animals in Vietnam, Thailand, Laos and Cambodia. *Exp. Parasitol.* **23**(3), 412–64.

Sehaiya, R.V. (1941). Tadpoles as hosts for the glochidia of the fresh-water mussel. *Curr. Sci.* **10**, 535–6. [Glochidia in leeches.]

Sekerak, A.D., and Arai, H.P. (1977). Some metazoan parasites of rockfishes of the genus *Sebastes* from the northeastern Pacific Ocean. *Syesis* **10**, 139–44. [British Columbia, Alaska.]

Selensky, W. (1906). Zur Kenntnis des Gefässsystems der *Piscicola*. *Zool. Anz.* **31**, 33–44.

—— (1907). Studien über die Anatomie von *Piscicola*. I. Die metamerie mit Berücksichtigung des Nervensystems. II. Das Gefässsystem. *Trav. Soc. Imp. nat. St. Pétersb.* **36**, Livr. 4, *Zool. Physiol.* 37–88. [In Russian; German summary: 89–111.]

—— (1914a). Über einige auf Arthropoden schmarotzende Ichthyobdelliden. *Zool. Anz.* **44**, 270–82.

—— (1914*b*). Notes sur la faune des Hirudinées de la côte de Mourman. *Trav. Soc. Imp. nat. St. Pétersb. C. R. Séanc.* I, **45**, 197–214. [In Russia; French summary.]

—— (1914*c*). Notes sur l'anatomie de l'*Oxytonostoma* Malm. *Trav. Soc. Imp. nat. St. Pétersb. C.R. Séanc.* **45**(1), 249–58. [In Russian; French summary: 261–4.]

—— (1915). [Études morphologiques et systématiques sur les Hirudinées. I. L'organisation des ichthyobdellides.] *Petrograd.* 1–256. [In Russian.] [See Moore, J.P. and Meyer, M.C. (1955). English transl.]

—— (1923). *Crangonobdella murmanica* n. g. n. sp., eine auf Sclerocrangon schmarotzende Ichthyobdellide. *Zool. Jb. Syst.* **46**, 397–488.

—— (1927). [On a new Ichyobdellid parasitizing on Mysidae (*Mysidobdella oculata* n. gen., n. sp.)] *Proc. Second Congr. Zool. Anat. Histol. USSR,* Vol. 2, pp. 32–3. [In Russian.] [= *M. borealis.*]

Serafinska, J. (1957). *Z zycia pijawek.* PWN, Warszawa.

—— (1931). Über die Gattung *Hemibdella* nebst einigen allgemeinen Bemerkungen über die Organisation der Ichthyobdelliden. *Pubbl. Staz. zool. Napoli* **11**, 1–21.

—— (1958). [Contributions to the Polish fauna of leeches (Hirudinea).] *Fragmenta Faun., Warszawa* **8**(3), 17–64. [In Polish.]

Seravin, L.N. (1964). [Factors promoting and preventing swelling in leech dorsal muscle in hypotonic media.] *Vestnik Leningrad. Un. Ser. Biol.* **19**(4), 34–8. [In Russian; English summary.] [English transl. *Fed Proc. Fedn Au. Socs exp. Biol.* **24**(6), Pt. II: T1105–7 (1965).]

—— (1965). Role of acetylcholinesterase in the water metabolism of the leech 'dorsal muscle.' *Dokl. Biol. Sci. Akad. Nauk SSSR* **160**, 116–18. [English transl. *Dokl. Akad. Nauk SSSR* **160**(2), 486–8. [In Russian.]

Sergeeva, S.S., and Bazanova, I.S. (1979). [Correlation between membrane potential level and oxygen tension over the Retzius cell in the leech (*Hirudo medicinalis*).] *Fitziol. Zh. SSSR IM.I.M. Sechenova* **65**(9), 1367–71. [In Russian.] [1980BA:64943.] [Close connection between energetics and electrogenesis.]

—— —— (1980). Change in the extracellularly recorded bioelectrical activity of the leech (*Hirudo medicinalis*) Retzius neuron after insertion of a microelectrode. *Dokl. biol. Sci.* **251**, 214–17. [English transl. of *Dokl. Akad. Nauk SSSR, Ser. Biol.*]

Shafer, M.R., and Calabrese, R.L. (1981). Similarities and differences in the structure of segmentally homologous neurons that control the hearts in the leech *Hirudo medicinalis. Cell Tiss. Res.* **214**(1), 137–54. (See also 1980. *Neurosci. Abstr.* **6**, 26).

Shankland, M. (1984). Positional determination of supernumerary blast cell death in the leech embryo. *Nature, Lond.* **307**, 541–3.

—— and Weisblat, D.A. (1984). Stepwise commitment of blast cell fates during the positional specification of the O and P cell lines in the leech embryo. *Developmental Biology* **106**, 326–42.

Shapovalenko, L.I. (1938). Nabludeniya Nad pigmentnimi Kletkami y *Haementeria costata*. *Raboti Lab. Obsch. Biol. Zool. 3-go Mosk. Med. In-ta* **1**, 41–57. [In Russian.]

—— (1940). Nabludeniya Nad aktivnost'u sekreta, prepyatstvooushchego svert'vaniu krovi, oo meditsiniskih piyavok paznogo vozrasta. *Bull. Mosk. Obshch. Isp. Prirod', Otd. Biol.* **49**(2), 53–7. [In Russian.]

—— (1947). Nabludeniya nad aktivnost'u protivosvert'raushchego ekstrakta iz golovok meditsinskih piyavok, snyat'h vo premya sosaniya. *Bull. Eksp. Biol. Med.* **24**(2), 143–6. [In Russian.]

Sharma, R.E., and Fernando, C.H. (1961). Leeches and their ways. *Malayan nat. J.* **15**, 152–9. [A hirudinid, possibly *Limnatis dissimulata*, attacking fish *Rasbora elegans* and *Tilapia mossambica*, at Batu Barendam, Malacca.]

Sharma, S., and Kaur, S. (1973). A case of abnormal genital organs in *Hirudinaria granulosa* (Savigny). *Zool. Pol.* **22**(4), 265–70.

Shelley, R.M. (1975). Distributional patterns of leeches in North and South Carolina (Annelida: Hirudinea). *Ass. Southeast. Biol. Bull.* **22**(2), 79. (Abstr.)

—— (1977). North Carolina's terrestrial leech. *Wildlife N. Carolina* **41**(9), 24.

—— and Braswell, A.L. (1981). Host record for the leech *Placobdella nuchalis* Sawyer and Shelley (Rhynchobdella: Glissiphoniidae). *J. Parasitol.* **67**(5), 748.

—— —— and Stephan, D.L. (1979). Notes on the natural history of the terrestrial leech, *Haemopis septagon* Sawyer and Shelley (Gnathobdella: Hirudinidae). *Brimleyana* **1**, 129–33.

Shen Chia Shui (1956). [The horse leech.] [*Gen. Biol. J.*] **11**, 5–9. [In Chinese.]

Sherstyuk, V., and Zimbalevskaya, L.N. (1973). [Calorific value of phytophagous invertebrates in Dnieper reservoirs. *Gidrobiol. Zh.* **9**(4), 83–7. [In Russian.]

Shevkunova, E.A. (1955). [Certain facts on the feeding biology of the leech *Limnatis nilotica* (Savigny, 1820).] *Med. parazitol. parazit. bolezni* **24**(4), 346–51.

—— (1958*a*). K boprosu o vozmojnosti povtorn'h invazii konskoi piyavki na slizist'e obolochki mlekopitauschih. *Med. parazitol. parazit. bolezni* **27**(5), 577–80. [*Limnatis nilotica.*]

—— (1958*b*). Eksperimental'naya model' dlya Izucheniya limnatioza. *Tr. Inst. Zool. A.N. Kazah SSR* **9**, 247–9. [*Limnatis nilotica.*]

—— and Kristman, V. (1962). Piyavki. *Bol'shaya Meditsinskaya Entsiklopediya* **24**, 794–800.

Shidlauskaite, L.A., and Danyulite, G.P. (1972). [Reactions of aquatic animals in electric fields. 1. Worms.] *Liet. TSR Mokslu Akad. Darb.*, *Ser. C* **1**(57), 83–90. [In Russian; Lithuanian and English summaries.] [*Hirudo medicinalis.*]

Shidlauskaite-Petrauskene, L.A. (1975). [Reactions of aquatic animals in electric fields: 3. The importance of neuronal structures in the manifestation of electrotaxis.] *Liet. Tsr Mokslu Akad. Darb.*, *Ser C* **4**(72), 95–102. [In Russian; Lithuanian and English summaries.] [1977BA:3483.] [Cathodal electrotaxis of *Hirudo medicinalis* observed after removal of suboesophageal and caudal ganglia.]

—— and Tereshkov, O.D. (1976*a*). [Reactions of aquatic animals in electric fields. 4. Responses of the lateral nerves in the leech segmental ganglion to extracellular stimulation with direct electric current.] *Liet. Tsr Mokslu Akad. Darb.*, *Ser C* **1**(73), 127–32. [In Russian; Lithuanian and English summaries.] [1977BA:44981.]

—— —— (1976*b*). [Reactions of aquatic animals in electric fields: 5. Responses of single neurons in the leech ganglion to the extracellular action of direct electric current.] *Liet. Tsr Mokslu Akad. Darb.*, *Ser C*. **3**(75), 159–67. [In Russian; Lithuanian and English summaries.] [1978BA:59065.]

Shigeru, K., Saito, K., Ohri, M., Oshima, T., Toshioka, S., Yamamoto, H., and Okuyama, Y. (1977). Scanning electromicrography of some species of leech (Hirudidae and Haemadipsidae): I. *Hirudo nipponica* from Japan. *Jap. J. Sanit. Zool.* **28**(4), 393–400. [1976BA:21706.]

Shipley, A.E. (1888). On the existence of communications between the bodycavity and the vascular system. *Proc. Cambr. phil. Soc.* **6**, 213–21.

—— (1915*a*). Leeches and the war. *Lancet* **188**, 225.

—— (1915*b*). The minor horrors of war. London. [Hirudinea: 123–61.]

—— (1927). Historical preface to Harding, W.A. and Moore, J.P., *The Fauna of British India: Hirudinea*, p. V–XXXII. London.

Shishido, I. (1897). [Anatomy of the medicinal leech, *Hirudo nipponica*.] *Zool. Mag.* **9**, 415ff. [In Japanese.]

Shlom, J.M., Amesse, L., and Vinogradov, S.N. (1975). Subunits of *Placobdella* haemoglobin. *Comp. Biochem. Physiol. B Comp. Biochem.* **51**(4), 389–92. [1975BA:54909.]

Shope, R.E. (1957). The leech as a potential virus reservoir. *J. exp. Med.* **105**, 373.

Shrivastav, H.O.P., and Shah, H.L. (1971). Occurrence of the leech *Hemiclepsis marginata asiatica* in the bile duct of a pig (*Sus scrofa domestica*). (Correspondence). *Indian vet. J.* **48**(2), 203–4. [In Jabalpur.]

Shubin, I. (date undetermined). [Neurosomite of *Pontobdella muricata*.] Quoted in Selensky (1915). [This study, under the guidance of N.A. Livanow of University of Kasan, was unpublished as of 1915, when a manuscript was seen and contents quoted by W.D. Selensky of the University of Leningrad. Possibly this important work was published in a Russian journal about that time and remains undiscovered.]

Shul'man, S.S., and Shul'man-al'bova, R.E. (1953). [*Parasites of White Sea fish.*] [In Russian.] (Incomplete reference)

Shumkina, O.B. (1951a). [Cleavage of the egg of the medicinal leech.] *Dokl. Akad. Nauk SSSR C. r. Acad. Sci. URSS* (new series) **77**(2), 353–6. [In Russian.]

—— (1951b). [Development and metamorphosis of the medicinal leech.] *Dokl. Akad. Nauk SSSR C. r. Acad. Sci. URSS* (new series) **77**(4), 761–4. [In Russian.]

—— (1951c). [The embryonic development of the head of the medicinal leech in relation to the rest of the embryo.] *Dokl. Nauk SSSR C. r. Acad. Sci. URSS* (new series) **78**(4), 821–4.]In Russian.]

—— (1951d). [Periods of intra-cocoon development of the medicinal leech.] *Dokl. Akad. Nauk SSSR C. r. Acad. Sci. URSS* **78**(6), 1259–62. [In Russian.]

—— (1953). [Embryonic development of the medicinal leech.] *Tr. Inst. Morf. Zhiv. Im. A.N. Servertsova* **8**, 216–79. [In Russian.]

Shuster, C.N. Jr, Smith, R.F., and McDermott, J.J. (1953). Leeches and their importance in the life cycle of the fishes. *New Jersey fish. Surv. (1951)* Report No. 2, 175–81.

Siegel, — (1903). See Autrum (1939c).

Sigalas, R. (1921). See Autrum (1939c).

Sikorowa, A. (1965). Fauna denna jeziora Kortowskiego w latach 1952–1954. *Zesz Nauk. WSR Olsztyn* **19**, 81–112.

Silva, P.H.D.H. de (1960). A key to the genera of the family Piscicolidae (Hirudinea). *Ceylon J. Sci. (Biol. Sci.)* **3**(12), 223–33.

—— (1963a). The occurrence of *Pontobdella* (*Pontobdellina*) *macrothela* Schmarda and *Pontobdella aculeata* Harding in the Wadge Bank. *Spol. Zeyl.* **30**(1), 35–8.

—— (1963b). A new leech, *Pontobdella taprobanensis* sp. nov., from the Wadge Bank. *Spol. Zeyl.* **30**(1), 39–46.

—— (1963c). *Zeylanicobdella arugamensis* gen. nov. and sp. nov. from Arugam Kalapu, Eastern Province, Ceylon. *Spol. Zeyl.* **30**(1), 47–53.

—— and Anderson, A.A. (1964). A record of *Dinobdella ferox* (Blanchard) (Hirudinea, Hirudinea) taken from the nasal cavity of man. *Ann. trop. Med. Parasitol.* **58**, 1–2.

—— and Burdon-Jones, C. (1961). A new genus and species of leech parasitic on the fish *Cottus bubalis*. *Proc. zool. Soc. Lond.* **136**(3), 343–57.

—— and Fernando, C.H. (1965). Three marine leeches (Piscicolidae, Hirudinea) from the Malay Peninsula. *Spol. Zeyl. Bull. nat. Mus. Ceylon* **30**(2), 227–32.

—— and Kabata, Z. (1961). A new genus and species of leech parasitic on *Drepanopsetta platessoides* (Malm), the long rough dab. *Proc. zool. Soc. Lond.* **136**(3), 331–41.

Silver, A. (1972). The localization of cholinesterases in the leech and some other invertebrates. *J. Physiol.* **226**, 27P.

Simmonds, P.L. (1870). The trade in leeches. *Pharm. J.* **3**, 521–2.

Simms, B. (1970). Junior aquarist: four easily recognized leeches. *Aquar. Pondkpr.* **34**, 324–5.

Sineva, M.V. (1938). [Biological observations on *Protoclepsis maculosa*.] *Raboty Lab. Obsjtsjej Biol. Zool 3. Moskovsk. Meditsinck Inst. Moscow.* **1**, 5–40. [In Russian.]

—— (1940). [Several observations on the influence of temperature on the growth and reproduction of *Protoclepsis maculosa*.] *Zool. Zhurn.* **19**, (pages undetermined) [In Russian.]

—— (1941). [Observations sur la réproduction de *Protoclepsis maculosa* Rathke 1862.] *Zool. Zh.* **20**(4–5), 528–73. [In Russian; French summary.]

—— (1944). [Observations on breeding the medicinal leech.] *Zool. Zh.* **23**(6), 293–303. [In Russian; French summary.]

—— (1949). [Biological observations on propagation of the medicinal leech.] *Zool. Zh.* **28**(3), 213–24. [In Russian.] [Effect of temperature and food on cocoon production.]

—— (1950). [The growth of the medicinal leech in relation to feeding.] *Moskovskoe Obshchestvo Ispytotelei Prirod' Bull. Otdil Biol.* **55**(6), 50–6. [In Russian.]

Singh, M. and Naim, A.F. (1979). Respiratory obstruction and haematemesis due to the leech. *Lancet* Dec. 22/29, 1979, p.1374.

Singhal, R.N., Davies, R.W., and Shah, K.L. (1985). The taxonomy and morphology of the leeches (Hirudinoidea: Glossiphoniidae) parasitic on turtles from the Beas River (India) including descriptions of two new species and one redescription. *Zool. Anz.* **215**, 147–55.

Sirasawa, M., and Makino, N. (1979). [Glucose-6-phosphatase activity in the intestinal epithelium of *Erpobdella lineata*.] *Zool. Mag. Tokyo* **88**(4), 646. (Abstr.) [In Japanese.]

Sivickis, P., and Likeviciene, N. (1961). [The land leech *Xerobdella lecomtei* Frauenfeld 1896 in the Lithuanian Soviet Republic.] *Acta parasitol. Lithuanica* **3**, 33–4. [In Lithuanian; Russian and English summaries.] [Found in the stomach of a mole; leech probably *Haemopis sanguisuga*.]

Sket, B. (1961). [Über die Höhlenfunde der Hirudineen.] *Drugi Jugoslavenski Speleoloski Kongres* (*Congrès Yougoslave de Spéléologie, Deuxième Session, Split* 1958) *Zagreb*, pp. 185–8. [In Serbo-Croatian; German summary.]

—— (1968). [Zur Kenntnis der Egel-Fauna (Hirudinea) Jugoslawiens.] *Acad. Sci. Art. Slovenica Cl. IV Pars Historiconaturales, Diss.* **9**(4), 127–78. [In Serbo-Croatian; German summary.]

—— (1981). Rhynchobdellid leeches (Hirudinea, Rhynchobdellae) in the

Relic Ohrid Lake region. *Biol. Vestn.* **29**(2), 67–90. [Slovak and Macedonian summaries.] [1983BA:18890.]

—— (1982). Ohridsko jezero se ni raziskanol. *Proteus* **44**, 200–2. [Colour photographs of Lake Ohrid rhynchobdellids.]

—— (Incomplete reference). Cave Hirudinea. In: *Stygofauna mundi.* (ed. L. Botosaneanu) Amsterdam.

Skinner, R. (1975). Parasites of the striped mullet, *Mugil cephalus,* from Biscayne Bay, Florida, with descriptions of a new genus and three new species of trematodes. *Bull. mar. Sci.* **25**(3), 318–45. [1975BA:67201.] [*'Piscicola'* sp.]

Skopichenko, N.F. (1966). [Effect of hirudotherapy on some blood coagulation indices.] *Vrachebnde Delo.* **5**, 10–14. [In Russian.]

Sleeper, B. (1984). Who likes leeches? *Ranger Rick Magazine.* March, 1984, 29–32.

Slinn, D.J. (1970). An infestation of adult *Lernaeocera* (Copepoda) on wild sole, *Solea solea,* kept under hatchery conditions. *J. mar. Biol. Ass. UK* **50**, 787–800. [*Hemibdella solea;* Celtic Sea.]

Sloka, Ya (1956). [Materials on the leech fauna (Hirudinea) of the Latvian SSR.] *Izv. Akad. Nauk Latv. SSR* **3**(104), 89–93. [In Russian.]

Slonov, M.N., and Makarov, V.A. (1970). [Parasitising of horse leech in man in the Algiers People's Democratic Republic.] *Medskaya Parazit.* **39**(6), 736. [In Russian; English summary.] [*Limnatis nilotica.*]

Sluiter, C. Ph., Swellengrebel, N.H., and Ihle, J.E.W. (1921). *De dierlijke parasieten van den mensch en van onze huisdieren,* 3rd edn, pp. 445–70. Amsterdam. [2nd edn (1912) pp. 415–32; 1st edn (1895) pp. 258–76.]

Smiley, J.W., and Chamberlain, S.A. (1976). The fatty acid composition of leech lipids. *Comp. Biochem. Physiol.* **55B**, 243–4. [1977BA:15812.]

—— and Sawyer, R.T. (1976). Osmoregulation in the leeches *Macrobdella ditetra* and *Haemopis marmorata. Ass. southeast. Biol. Bull.* **23**(2), 96.

Smirnova, K.V. (1954). [Diseases of fish caused by parasites in the River Don in the area of Tzimlyan water-reservoir before its completion.] *Trans. spec. probl. Confer.* (7th conf. on problems of parasitology), *Moscow,* Vol. 4, pp. 61–5. [In Russian.]

Smith, C. (1750). *The antient and present state of the county and city of Cork.* Reilly, Dublin. [Blarney Lake abounded with 'good leeches' and says 'from whence Cork and Dublin may be supplied with them, the latter city having them from Wales'.]

Smith, D.G. (1977). The rediscovery of *Macrobdella sestertia* Whitman (Hirudinea: Hirudinidae). *J. Parasitol.* **63**(4), 759–60.

—— and Taubert, B.D. (1980). New records of leeches (Annelida: Hirudinea) from the shortnose sturgeon (*Acipenser brevirostrum*) in the Connecticut River, USA. *Proc. helminth. Soc. Wash.* **47**(1), 147–8. [*Calliobdella vivida* in fresh water.] [1980BA:6831.]

Smith, E.N., Johnson, C.R., and Voight, B. (1976). Leech infestation of the American alligator in Texas. *Copeia* **4**, 842. [1977BA:32156.] [*Placobdella multilineata.*]

Smith, F. (1920). Leeches considered as Oligochaeta modified for a predatory life. *Trans. Am. microsc. Soc.* **39**, 86–8.

Smith, M.W. (1939). Copper sulfate and rotenone as fish poisons. *Trans. Am. fish. Soc.* **69**, 141–57.

Smith, P.A., and Walker, R.J. (1973). Studies on 5-hydroxytryptamine receptors of neurones from *Hirudo medicinalis. Br. J. Pharmacol.* **47**, 633P–634P.

—— —— (1974). The action of 5-hydroxytryptamine and related compounds on the activity of the Retzius cells of the leech, *Hirudo medicinalis. Br. J. Pharmacol.* **51**, 21–7.

—— —— (1975). Further studies on the action of various 5-hydroxytryptamine agonists and antagonists on the receptors of neurones from the leeches, *Hirudo medicinalis* and *Haemopis sanguisuga. Comp. Biochem. Physiol. C Comp. Pharmacol.* **51**(2), 195–204. [1975BA:61297.]

—— Fitzsimons, J.T.R., Loker, J.E., and Walker, R.J. (1975). 5-hydroxytryptamine as a possible inhibitory neurotransmitter in the central nervous system of the leech, *Haemopis sanguisuga. Comp. Biochem. Physiol. C Comp. Pharmacol.* **52**(1), 65–73. [1976BA:43240.]

—— Sunderland, A.J., Leake, L.D., and Walker, R.J. (1975). Cobalt staining and electrophysiological studies of Retzius cells in the leech, *Hirudo medicinalis. Comp. Biochem. Physiol.* **51A**, 655–61.

Smith, P.H., and Page, C.H. (1974). Nerve cord sheath receptors activate the large fiber system in the leech. *J. comp. Physiol.* **90**(3), 311–20. [1974BA:54824.]

Smith, R.I. (1942). Nervous control of chromatophores in the leech *Placobdella parasitica. Physiol. Zool.* **15**, 410–17.

Smolnikoff, V.P. (1972). 'Leech' and 'vrach' (nomenclature and comment). *Lancet* **i**, 687, 1010.

Smythies, B.E. (1953). Notes and queries on land-leeches. *J. Bombay nat. Hist. Soc.* **51**, 954–8.

—— (1959). Leeches of Borneo. *Sarawak Mus. J.* **9**(13–14), 279–94.

Snieszko, S.F. (1970). *A symposium on diseases of fishes and shellfishes.* Special Publication No. 5. American Fisheries Society. Washington, DC.

Sobari, S., Ladds, P.W., Flanagan, M., and Lee, C.G. (1976). Infestation of the bovine mammary gland with the leech *Goddardobdella elegans.* (Correspondence.) *Aust. vet. J.* **52**(4), 197–8. [Near Townsville.]

Solem, J.O. (1971). Om Lille-Jonsvannet og dets bunnfauna. *TAFOs Aarbok* 1970/71, 33–8. [*Theromyzon maculosum* from Norway.]

—— (1973). The bottom fauna of Lake Lille-Jonsvann, Tröndelag, Norway. *Norw. J. Zool.* **21**, 227–61.

—— (1975). [New records of the leeches *Piscicola geometra* and *Acanthobdella peledina.*] *Fauna, Oslo* **28**(4), 222–3. [In Norwegian; English summary.] [1976BA:55586.] [In Tröndelag area of Norway.]

Soliman, K.N. (1954). Observations on some helminth parasites from ducks in southern England. *J. Helminth.* **29**, 17–26. [*Erpobdella octoculata* as intermediate host for *Cotylurus cornutus* and *Hymenolepis parvula.*]

Sologub, M.I., and Kiss, T. (1971). Intracellular potentials during protracted recording from the neurones of *Hirudo medicinalis. Ann. Inst. Biol. (Tihany) Hung. Acad. Sci.* **38**, 97–105.

Soltesz, J., and Pusztay, F. (1965). Determination of hirudin hematologic in oinments. *Pharmazie* **20**(10), 656.

Soltys, M.A., and Woo, P.T.K. (1968). Leeches as possible vectors for mammalian trypanosomes. *Trans. R. Soc. trop. Med. Hyg.* **62**(1), 154–6.

Somadikarta, S. (1959). Morphologische, histologische, histochemische, physiologische und bakteriologische Untersuchungen an Landblute-geln der Gattung *Haemadipsa* Tennent 1861 und ein Vergleich mit *Hirudo medicinalis* Linne 1758. *Diss. Math. Nat. Fak. FUS* 1–67.

Somorowska, B. (1954). [Investigation into the origin of blood extracted from the digestive tract of the medicinal leech (*Hirudo medicinalis* L.).] *Kosmos Ser. Biol., Warsaw 3* **4**(9), 466. [In Polish.]

—— (1955). [Investigation into the origin of blood extracted from the digestive tract of the medicinal leech (*Hirudo medicinalis* L.).] *Zesz. Nauk. Uniw. Lodz, Ser. 2, Mat.-Przyr.,* Lodz **1**, 39–45. [In Polish.]

Somme, S. (1950). Hvirvellose dyr i ferskuann. In *Norges Dyreliv,* Bd. 4, pp. 229–61. J.W. Cappelens Forlag, Oslo.

Sonchai, C. (1965). Collection data on leeches in Thailand. Special Report. Joint Thai–US Research and Development Center, Bangkok.

Song Ta-Hsiang. (1974). (Preliminary notes on the behaviour of some paddy-field leeches from Ninghsien, Chekiang Province, China.) *Acta zool. sin.* **20**(1), 52–60. [In Chinese; English summary, pp. 52–60.] [1975BA:43721.]

—— and Yang Tong. (1978). [Leeches of China.] Science Press: Peking. 176 pp. [In Chinese.] [Excellent summary of the freshwater leeches of China.]

—— Zhang Jun, Tan Enguang, and Liu Liaren. (1977). [On some blood-sucking leeches from China.] *Acta zool. sin.* **23**(1), 102–8. [In Chinese; English summary.] [1978BA:27942.]

Soos, A. (1939). Hirudineen aus dem Komitat Bars. *Fragm. Faun. Hung., Budapest* **2**(3), 44–6.

—— (1958). [New data to the ecology and distribution in the Carpathian basin of *Trocheta bykowskii* Gedr. (Hirudinea).] *Ann. hist. nat. Mus. Hung.* **50**, 173–7.

—— (1963a). New leeches (Hirudinea) from the fauna of Hungary. *Ann. hist. nat. Mus. Hung.* **55**, 285–92.

—— (1963b). Identification key to the species of the genus *Dina* R. Blanchard, 1892 (emend. Mann, 1952) (Hirudinea: Erpobdellidae). *Acta Un. Szegediensis, Acta biol.* (new series) **9**, 253–61.

—— (1964*a*). A revision of the Hungarian fauna of Rhynchobdellid leeches (Hirudinea). *Opusc. zool. Inst. zoosyst., Budapest* **5**(1), 107–12.

—— (1964*b*). [Leech species to be shown in the fauna of Hungary?] *Allatt. Közlem.* **51**, 125–33. [In Hungarian; English summary.]

—— (1965). Identification key to the leech (Hirudinoidea) genera of the world, with a catalogue of the species. I. Family: Piscicolidae. *Acta zool. Hung.* **11**(3/4), 417–63.

—— (1966*a*). On the genus *Glossiphonia* Johnson, 1816, with a key and catalogue to the species (Hirudinoidea: Glossiphoniidae). *Ann. hist. nat. Mus. Nat. Hung.* **58**, 271–9.

—— (1966*b*). Identification key to the leech (Hirudinoidea) genera of the world, with a catalogue of the species. II. Families: Semiscolecidae, Trematobdellidae, Americobdellidae, Diestecostomatidae. *Acta zool. Acad. sci. Hung.* **12**(1/2), 145–60.

—— (1966*c*). Identification key to the leech (Hirudinoidea) genera of the world, with a catalogue of the species. III. Family: Erpobdellidae. *Acta zool. Sci. Hung.* **12**(3/4), 371–407.

—— (1967*a*). On the genus *Hemiclepsis* Vejdovsky, 1884, with a key and catalogue of the species (Hirudinoidea: Glossiphoniidae). *Opusc. Zool., Budapest* **7**(1), 233–40.

—— (1967*b*). On the leech fauna of the Hungarian reach of the Danube (Danubialia Hungarica XLII). *Opusc. Zool., Budapest* **7**(2), 241–57.

—— (1967*c*). On the genus *Batracobdella* Viguier, 1879, with a key and catalogue to the species (Hirudinoidea: Glossiphoniidae). *Ann. Hist-nat. Mus. nat. Hung.* **59**, 243–57.

—— (1967*d*). Identification key to the leech (Hirudinoidea) genera of the world, with a catalogue of the species. IV. Family: Haemadipsidae. *Acta zool. Acad. sci. Hung.* **13**(3/4), 417–32.

—— (1968*a*). A new leech genus: *Richardsonianus* gen. nov. (Hirudinoidea: Hirudinidae). *Acta zool. Acad. sci. Hung.* **14**(3/4), 455–9.

—— (1968*b*). Identification key to the species of the genus *Erpobdella* de Blainville, 1818. (Hirudinoidea: Erpobdellidae). *Ann. Hist.-nat. Mus. nat. Hung.* (*Zool.*) **60**, 141–5.

—— (1969*a*). Identification key to the leech (Hirudinoidea) genera of the world, with a catalogue of the species: V. Family: Hirudinidae. *Acta zool. Acad. sci. Hung.* **15**(1/2), 151–201.

—— (1969*b*). Identification key to the leech (Hirudinoidea) genera of the world, with a catalogue of the species. VI. Family: Glossiphoniidae. *Acta zool. Acad. sci. Hung.* **15**(3/4), 397–454.

—— (1970). A zoogeographical sketch of the freshwater and terrestrial leeches (Hirudinoidea). *Opusch. Zool., Budapest* **10**(12), 313–24.

—— (1973). [On the European land leeches (Hirudinoidea: Xerobdellidae).] *Különlenyomat az allattani Közlemenyek. Szamabol.* **60**, 103–9. [In Hungarian; English summary.]

Soota, T.D. (1955). Fauna of the Kashmir Valley: leeches. *Proc. Indian Sci. Congr. Ass.* 42nd **4**, 10. (Abstr.)

—— (1959). Fauna of the Kashmir Valley: leeches. *Rec. Ind. Mus.* **54**, 1–4.

—— and Ghosh, G.C. (1977). On some Indian leeches. *Newsletter, Zool. Surv. India,* Calcutta, **3**(6), 358–61.

—— and Saxena, M.M. (1982). Leeches of some waters of the Indian Desert and their ecology. Abstr. (4.26). First All India Conference on Limnology, 3–5 March 1982, Nagarjuna University.

—— —— (Incomplete reference). Leeches of some waters of Rajasthan and their ecology.

—— —— and Baskaran, S. (1981). Leeches of arid region of Rajasthan around Jodhpur, and their ecology. Nat. Symp. on Evaluation of Our Environment. University of Jodhpur, Jodhpur (27–29 April 1981), Abstr. No.83, p. 64.

—— —— —— (1982). Leeches of arid region around Jodhpur and their ecology. *Geobios New Rep.* **1**, 136–8.

Sotnikov. O.S., and Lagutenko, Yu. P. (1977). [Morphologic changes in axodendritic synapses of the leech during conductance block in media with low ionic concentration.] *Neurofiziologiya* **9**(6), 613–18. [In Russian; English summary.] [1978BA:21705.] [English transl. *Neurophysiology* **9**(6), 465–9.]

Southern, R. (1908). [Seven species of leech.] In *Handbook to the City of Dublin and the surrounding district.* Ponsonby Gibbs, Dublin.

Souza, A.M. de (1980). Aspectos biologicos de *Haementeria depressa* (É. Blanchard, 1849) (Hirudinea, Rhynchobdellae, Glossiphoniidae). *Bolm. Zool.,* Univ. S. Paulo **5**, 31–8. [English summary.]

—— (1982). A study about the coelom of 4 *Haementeria* species (Glossiphoniidae, Rhynchobdellae, Hirudinea). *Stud. Neotrop. Fauna Environ.* **17**(4), 219–29. [1983BA:41893.]

Spassov, A. (1965). Bekämpfung des Blutegelbefalls bei Büffeln. *Wien. tierärzt. Mschr.* **52**(8), 792. [Control of leeches attacking buffaloes.]

Specht, P.C., and Eckert, R. (1964). Electrical characteristics of three ganglion cells of the leech *Hirudo. Am. Zool.* **4**(4), 386.

—— —— (1967). Demonstrated differences in the electrogenic properties of two types of ganglion cells in leech. *J. gen. Physiol.* **50**, 2488.

Spelling, S.M., and Young, J.O. (1983). A redescription of *Nosema herpobdellae* (Microspora: Nosematidae), a parasite of the leech *Erpobdella octoculata* (Hirudinea: Erpobdellidae). *J. Invert. Pathol.* **41**(3), 350–68. [1983BA:88520.]

Spiess, C. (1903). Recherches morphologiques, histologiques et physiologiques sur l'appareil digestif de la sangsue (Hirudo medicinalis). *Rev. Suisse Zool.* **11**, 151–239.

—— (1904). Recherches anatomiques et histologiques sur l'appareil digestif de l'Aulastome (*Aulastoma gulo*). *Rev. Suisse Zool.* **12**, 585–647.

—— (1905). See Spiess, C. (1905*b*) in Autrum (1939*c*).

Srebrodol'skaya, N.I. (1972). [Parasite fauna (helminths and leeches) of Rallidae in the Volynskoe Poles'e, Ukranian SSR.] In *Problemy parazitologii. Trudy VII Nauchnoi Konferentsii Parazitologov USSR. Part II, Kiev, USSR; Izdatel'stvo 'Naukova Dumka'*, pp. 289–90. [In Russian.]

Srivastava, L.P. (1966). Three new leeches (Piscicolidae) from marine shore fishes (Cottidae) in British waters. *J. Zool.* **150**(3), 297–318.

—— (1978). A review of the piscicolid genus *Johanssonia* Selensky, 1914. *Proc. Indian Sci. Congr.* **65**(3)C, 204–5. (Abstr.).

Stammers, F.M.G. (1950). Observations on the behaviour of land leeches (*Haemadipsa*). *Parasitology* **40**, 237–46.

Stankovic, S. (1960). Chapter V. Biogeography, Hirudinea. In: *The Balkan Lake Ohrid and its living world*. Monographiae Biologicae, 9. Dr. W. Junk, The Hague.

Staubesand, J., Kuhlo, B., and Kersting, K.H. (1963). Licht- und electronmikroskopische Studien am Nervensystem des Regenwurms. I. Die Hüllen des Bauchmarkes. *Zeit. Zellforsch. Mikroskop. Anat.* **61**, 401–33.

Steffenhagen, K., and Andrejew, P. (1911). See Autrum (1939c).

Stelly, N., Stevens, B.J., and André, J. (1970). Étude cytochimique de la lamelle dense de l'enveloppe nucléaire. *J. microsc., Paris* **9**(8), 1015–28.

Stent, G.S. (1979). Second look: leeches and 'magic' in medicine. *Hum. Nat.* (Incomplete reference)

—— (1980). The leech embryo. In: *Genes, cells and behavior* (ed. N.H. Horowitz and E. Hutchings, Jr.) pp. 151–60. 50th Anniversary Symposium, Pasadena, Ca, 1–3 November, 1978. Freeman, San Francisco. [1980BA:37416.]

—— and Kristan Jr., W.B. (1981). Neural circuits generating rhythmic movements. In: *The neurobiology of the leech* (ed. K.J. Muller, J.G. Nicholls and G.S. Stent) Ch. 7, pp. 113–46. Cold Spring Harbor.

—— and Weisblat, D.A. (1981). Cell lineage in the development of the leech nervous system. *Trends neurosci.* (October, 1981). pp. 251–5.

—— —— (1982). The development of a simple nervous system. *Sci. Am.* **246**(1), 100–10.

—— —— (1985). Cell lineage in the development of invertebrate nervous systems. *Ann. Rev. Neurosci.* **8**, 45—70.

—— Thompson, W.J., and Calabrese, R.L. (1979). Neuronal control of heartbeat in the leech and in some other invertebrates: *Physiol. Rev.* **59**(1), 101–36.

—— Weisblat, D.A., Blair, S.S., and Zackson, S.L. (1982). Cell lineage in the development of the leech nervous system. In: *Neuronal development* (ed. N. Spitzer). Plenum, New York.

—— —— Kristan, W.B. Jr. (In press). Development of the leech nervous system. In: *Handbook of physiology* (Developmental neurobiology).

—— Kristan Jr, W.B., Friesen, W.O., Ort, C.A., Poon, M., and Calabrese, R.L. (1978). Neuronal generation of the leech swimming movement. *Science, NY* **200**, 1348–57.

Stenzel, K., and Neuhoff, V. (1976). Tryptophan metabolism and the occurrence of amino acids and serotonin in the leech (*Hirudo medicinalis*) nervous system. *J. neurosci. Res.* **2**(1), 1–9. [1978BA: 60958.]

Stephanides, T. (1948). A survey of the freshwater biology of Corfu and of certain other regions of Greece. *Prak. Hellenic hydrobiol. Inst.* **2**(2), 1–263.

Stephenson, J. (1930). *The Oligochaeta*. 978 pp. Oxford University Press. (Reprinted 1972 by Verlag Von J. Cramer, Lehre, Germany.)

Štěrba, O., and Holzer, M. (1977). Fauna der interstitiellen Gewässer der Sandkiessedimente unter der aktiven Strömung. *Věstnik Ceskoslovenske Spolecnosti Zoologické* **41**(2), 144–59. [*Helobdella stagnalis*, *Erpobdella octoculata*, and *Dina lineata*.]

Stern, H. (1915). Leeching. In *Theory and practice of bloodletting*. pp. 81–5. Rebmon, New York.

Steullet, R. (1931). Un cas de parasitisme de l'urèthre par une sangsue, *Limnatis nilotica* (Savigny 1820). *Arch. Inst. Pasteur Alger* **9**, 481–3.

Stewart, M. (1960). (posthumous) Isabel M. Stewart recalls the early years. *Am. J. Nursing* **60**, 1426–30. p. 1428, 'Florence Nightingale's first nursing school requirements; nurses must be proficient in applying leeches both externally and internally.'

Stewart, W.W. (1981). Lucifer dyes — highly fluorescent dyes for biological tracing. *Nature*, Lond. **296**(5818), 17–21. [Cover article: leech ganglion.]

for man. *Hyg. Lab. Bull.* **142**, 69–126.

Stirling, W., and Brito, P.S. (1882). See Autrum (1939c).

Stoia, I. (1932). Results obtained in therapy of septicemic conditions by injection of plasma obtained by application of leeches. *Spitalul.* **52**, 490.

Stolte, H.A. (1933–1955). *Oligochaeta*. Bronn's Klass. u. Ord. d. Tierreichs, 4. Band, 3. Abteilung, 3. Buch. (1933–1940, 4 sections, pp. 1–722; 1955, fifth section, pp. 723–890).

Stolyarov, V.P. (1954). [Parasitic fauna of fish in the Ribinsk Reservoir during the seven years of its existence.] *Trans. spec. probl. confer.* (7th *Conf. on problems of parasitology*), Vol. 4, pp. 54–6. [In Russian.]

Stone, M.D. (1976). Occurrence and implications of heavy parasitism on the turtle *Chelydra serpentina* by the leech *Placobdella multilineata*. *Southwest. Nat.* **20**(4), 575–6.

Storer, R.W. (1982). The hooded grebe (*Podiceps gallardoi*) on Laguna de los Escarchados, South America: Ecology and behaviour. *Living Bird* **19**, 51–68. (1980/1981). [pp. 53, 66: feeds on leeches and other invertebrates.][1983BA: 32693.]

Strand, E. (1942). Miscellanae nomenclatorica zoologica et paleontologica. X. *Folia zool. hydrobiol.* **11**, 386–402. [Proposes genus *Malmiana.*]

Straus, A. (1970). '*Lumbriculus*' sp. nov. (?), ein Wurm (Annelida) aus dem Pliozän von Willershausen. *Ber. naturhist. Gesell. Hannover* **114**, 75–6.

Street, M., and Titmus, G. (1982). A field experiment on the value of allochthonous straw as food and substratum for lake macroinvertebrates. *Freshwat. Biol.* **12**(5), 403–10. [1983BA:71092.]

Streit, B. (1978). Uptake, accumulation and release of organic pesticides by benthic freshwater invertebrates. 1. Reversible accumulation of atrazine from aqueous solution. *Arch. Hydrobiol.* Suppl. **55**(1), 1.

—— (1979a). Uptake, accumulation and release of organic pesticides by benthic invertebrates: 2. Reversible accumulation of lindane, paraquat and 2,4-D from aqueous solution by invertebrates and detritus. *Arch. Hydrobiol.* Suppl. **55**(3/4), 349–72. [1980BA:47703.] [*Glossiphonia complanata.*]

—— (1979b). Uptake, accumulation and release of organic pesticides by benthic invertebrates: 3. Distribution of carbon-14 atrazine and carbon-14 lindane in an experimental 3-step food chain microcosm. *Arch. Hydrobiol.* Suppl. **55**(3/4), 373–400. [1980BA:47704.] [*Glossiphonia complanata.*]

—— and Peter, H.M. (1978). Long-term effects of atrazine to selected freshwater invertebrates. *Arch. Hydrobiol.* Suppl. **55**(1), 62–77. [1979BA:77136.]

—— and Schroeder, P. (1978). [Dominant benthic invertebrates in the stony littoral zone of Lake Constance: phenology, feeding, ecology, and biomass.] *Arch. Hydrobiol.* Suppl. **55**(2), 211–34. [1979BA:72489.]

Strout, R.G. (1965). A new hemoflagellate (genus *Cryptobia*) from marine fishes of northern New England. *J. Parasitol.* **51**, 654–9. [Summary of Ph.D. dissertation, University of New Hampshire, Durham, 1961, University Microfilms, No. 61–3773; *Calliobdella vivida* illustrated on Plate 4.]

Struemper, H.J. (1963). Blutegeltherapie-gestern und heute. *Dt. Apoth.-Ztg.* **103**, 1457–66.

Strugach, M.B. (1958). [Benthos of the Gor'ki reservoir (based on research carried out in 1956).] *Nauchn.-Tekh. Byul. Vses. Inst. Ozernogo i Rechnogo Rybn. Khoz.* **6/7**, 23–5. [In Russian.] [Clay–sand bottom — leeches found; not found on sand bottom.]

Stuart, A.E. (1969). Excitatory and inhibitory motorneurons in the central nervous system of the leech. *Science, NY* **165**, 817–19.

—— (1970a). Physiological and morphological properties of motoneurones in the central nervous system of the leech. *J. Physiol.* **209**(3), 627–46.

—— (1970b). Excitatory and inhibitory motorneurons in the CNS of the leech. *Diss. Abstr.* **30B**, 3850–1.

—— Hudspeth, A.J., and Hall, Z.W. (1974). Vital staining of specific monoamine-containing cells in the leech nervous system. *Cell Tissue Res.* **153**(1), 55–61. [1976BA:25920.]

Stuart, D.K. (1981). Monoamine neurons in embryonic and adult leech. *Neurosci. Abstr.* **7**, 43.

—— Thompson, I., Weisblat, D.A., and Kramer, A.P. (1982). Antibody staining of embryonic leech muscle, blast cell migration and neuronal pathway formation. *Neurosci. Abstr.* 1982.

Stuerbecher, M. (1957). Zur Geschichte einer Berliner Blutegelzuchtanstalt. *Berl. Med.* **8**, 455–7.

Stundl, K. (1951). Zur Hydrographie und Biologie der österreichischen Donau. *Schweiz. Z. Hydrol.* **13**(1), 36–53. [Leeches in Austrian Danube.]

Suchenko, G.E., and Iskow, M. (1964). Effektivnoje sredstvo borby s Karpoedom w rybowodnych prudach. *Ryb. Choz.* **9**, 21. [Phosphoro-organic preparation to control leech infestation.]

Sudarikov, V.E. (1959). [The order Strigeidida (La Rue, 1926) Sudarikov, 1959. Part I. Trematody.] *Zhivotnykh Cheloveka* **16**, 219–629. [In Russian.]

—— Karmanova, E.M., and Bakhemeteva, T.L. (1962). [To the problem of the species composition of metacercariae from flukes of the order Strigeidida from leeches in the Volga-delta.] *Tr. Astrakh. za. povednika* **6**, 197–202. [In Russian.] [Tetracotyles in coelomic system of *Erpobdella octoculata*, *Glossiphonia complanata*, and *Helobdella stagnalis*.]

Sukatschoff, B. (1897). Contributions a l'étude du système nerveux de la *Nephelis vulgaris*. *Trav. Soc. nat. St. Pétersb., Zool. Physiol.* **28**, 1–8. [In Russian.] [French transl. 9–14.]

—— (1900). Beiträge zur Entwicklungsgeschichte der Hirudineen. I. Zur Kenntnis der Urnieren von *Nephelis vulgaris* Moqu.-Tand. und *Aulastomum gulo* Moqu.-Tand. *Z. wiss. Zool.* **67**, 618–39.

—— (1902). [Regeneration in young *Haemopis*.] In *Aus dem Gebiete der Regeneration*. II. *Über die Regeneration bei Turbellarien* (ed. E. Schultz). *Z. wiss. Zool.* **72**, 1–30.

—— (1903). Beiträge zur Entwicklungsgeschichte der Hirudineen. II. Über die Furchung und Bildung der embryonalen Anlagen bei *Nephelis vulgaris* Moqu.-Tand. (*Herpobdella atomaria*). *Z. wiss. Zool.* **73**, 321–67.

—— (1908). Beiträge zur Kenntnis der Fauna Turkestans auf Grund des von D.D. Pedaschenko gesammelten Materials (1904–1906). III. *Herpobdella atomaria* Car. in Turkestan. *Trav. Soc. nat. St. Pétersb. Zool. Physiol.* **37**, 193–201. [In Russian and German.]

—— (1911). Die Hirudineen des Sadjerwsees (Livland). *SB Naturf. Ges. Dorpat, Tartu* **20**, 1–11. [German: 12–23.]

—— (1912). Beiträge zur Anatomie der Hirudineen. I. Über den Bau von *Branchellion torpedinis* Sav. *Mitt. zool. Stat. Neapel* **20**, 395–528.

Suko, T. (1964). [Studies on the benthic animal fauna in the River Arakawa.] *Bull. Chichibu Mus. nat. Hist.* No. 12, 79–105. [In Japanese;

English summary.] [Japanese distribution of leeches: *Erpobdella lineata, Hirudo nipponia, Myxobdella sinensis.*]

Sulc, V. (1966). Preparation of some groups of invertebrates. *Ziva* **14**(1), 24–5. [Leeches.]

Sunderland, A.J. (1978). Morphological, pharmacological and electrophysiological properties of Retzius neurones in the leeches *Hirudo medicinalis* and *Haemopis sanguisuga*. PhD. thesis, Portsmouth Polytechnic.

—— (1980). A hitherto undocumented pair of neurons in the segmental ganglion of the leech *Haemopis sanguisuga* which receive synaptic input from mechanosensory cells. *Comp. Biochem. Physiol. A Comp. Physiol.* **67**(2), 299–302.

—— Leake, L.D., and Walker, P.J. (1979). The ionic mechanism of the dopamine response in Retzius cells of the leeches (*Hirudo medicinalis* and *Haemopis sanguisuga*). *Comp. Biochem. Physiol. C Comp. Pharmacol.* **63**(1), 129–34. [1980BA:3234.]

—— —— —— (1980). Evidence for an amine receptor on the Retzius cells of the leeches *Hirudo medicinalis* and *Haemopis sanguisuga*. *Comp. Biochem. Physiol. C Comp. Pharmacol.* **67**, 159–66.

—— —— —— (1982). Structure-activity studies of the amine receptor on the Retzius cells of the leeches, *Hirudo medicinalis* and *Haemopis sanguisuga*. *Comp. Biochem. Physiol. C Comp. Pharmacol.* **73**(2), 347–52. [1983BA:80727.]

—— Smith, P.A., Leake, L.D., and Walker, R.J. (1974). Neuronal geometry of Retzius cells in *Hirudo medicinalis*. *Experientia* **30**(8), 918–19. [French summary.]

Susruta Samhita. *The Susruta or system of medicine taught by Dhanwantari and composed by his Disciple Susruta* (ed. S.M. Guita). Calcutta. [*Hirudinea:* I, Vol. 13, pp. 98–105.] [see J.P. Moore (1927).]

Sutcliffe, D.W., and Carrick, T.R. (1973). Studies on mountain streams in the English Lake District. I. pH, calcium and the distribution of invertebrates in the River Duddon. *Freshwat. Biol.* **3**, 437–62.

Sutherland, D.R., and Holloway Jr, H.L. (1979). Parasites of fish from the Missouri, James, Sheyenne, and Wild Rice Rivers in North Dakota. *Proc. helminth. Soc. Wash.* **46**(1), 128–34. [1979BA:68193.]

Svendsen, I., Boisen, S., and Hejgaard, J. (1982). Amino acid sequence of serine protease inhibitor chymotrypsin inhibitor 1 from barley: Homology with barley inhibitor chymotrypsin inhibitor 2, potato inhibitor 1 and leech eglin. *Carlsberg Res. Commun.* **47**(1), 45–53. [1983BA:21230.]

Sviderskij, V.L. (1972). [Development of innervation relationships in the locomotory muscles of some invertebrates (Annelida, Arthropoda).] *Zh. evol. Biokhim. Fiziol.* **8**(3), 307–14. [In Russian; English summary.] [Transl. *J. evol. Biochem. Physiol.* **8**(3), 273–9.]

Szerb, J. (1962). The estimation of acetylcholine using a leech muscle in a microbath. *J. Physiol.* **158**, 8P–9P.

Szidat, L. (1928). See Autrum (1939*c*).

—— (1930). Beiträge zur Entwicklungsgeschichte der Holostomiden. III. Über zwei Tetracotylen aus Hirudineen und ihre Weiterentwicklung in Enten zu *Cotylurus cornutus* Rud. und *Apatemon gracilis* Rud. *Zool. Anz.* **86**, 133–49.

—— (1931). Beiträge zur Entwicklungsgeschichte der Holostomiden. IV. Die Cercarien des Entenparasiten *Apatemon* (*Strigea*) *gracilis* Rud. und ihre Entwicklung im Blutgefässsystem des Zwischenwirtes (*Herpobdella atomaria* Car.). *Z. Parasitenk.* **3**, 160–72.

—— (1965). Estudios sobre la fauna de parasitos de peces Antarticos. I. Los Parasitos de *Notothenia neglecta* Nybelin. *Republica Argentina, Secretaria de Marina, Servicio de Hidrografia Naval,* H. 910, pp. 1–84.

—— (1972). Über zwei Arten der Hirudineen-Gattung *Branchellion* Savigny, 1820, von der Haut des Elephanten-Fisches *Calorhynchus calorhynchus* L. der Chilenischen Pazifik-Küste. *Stud. neotrop. Fauna* **7**(2), 187–93. [In German; Spanish summary.] [Transl. from German by E.M. Burreson.]

Szokalski, W. (1857). *Chodowanie (tak!) pijawek lekarskich pod wzgledem utrzymania, sztucznego rozmnazania, polowu, przewozu, przechowania zapasow i najoszczednicjszogo ich uzycia, zwlaszcza w szpitalach i dobroczynnych zakladach.* Warszawa.

Taft, S.J. and Kordiyak, G.J. (1973). Incidence, distribution and morphology of the macroderoidid trematode *Alloglossidium hirudicola* Schmidt and Chaloupka, 1969, from leeches. *Proc. helminth. Soc. Wash.* **40**(2), 183–6.

Takahashi, S. (1931). [On the distribution and ecology of *Whitmania laevis*]. *Zool. Mag., Tokyo* **43**, 607–11. [In Japanese.]

—— (1933). [On the ecology of *Glossiphonia lata* Oka.] *Zool. Mag., Tokyo* **45**, 203–6. [In Japanese.]

—— (1934*a*). [On the ecology and distribution of Taiwan leeches.] *Rep. Jap. Sci. Ass.* **10**, 744–9. [In Japanese.]

—— (1934*b*). [On the distribution of freshwater leeches.] *Bot. Zool., Tokyo* **2**, 60–2. [In Japanese.]

—— (1935). [Leeches of Koshun Peninsula, Taiwan.] *Kagaku no Taiwan* **3**, 51–2. [In Japanese.]

Takayama, T. (1967). [Parasitic leeches in the liver of man.] *Nip. Vet. Zoo. Col.* **16**, 109. [In Japanese.] [1969BA:73610.] [This remarkable observation needs confirmation; cf. flukes.]

Talesara, C.L. (1966). A histochemical study of the body wall muscles of the leech (*Hirudinaria granulosa*). *J. anim. Morphol. Physiol.* **13**(1/2), 72–7.

—— and Kumar, P. (1975). Intracellular organization of leech body wall muscle fibres: histoenzymological profile (correspondence). *Curr. Sci.* **44**(16), 586–8.

Tam, S.W., Fenton, J.W., and Detwiler, T.C. (1979). Dissociation of thrombin from platelets by hirudin: evidence for receptor processing. *J. biol. Chem.* **254**(18), 8723–5. [1980BA:28029.]

Tanabe, M. (1924). Studies on haemoflagellates of the loach *Misgurnus anguillicaudatus*. *Kitasato Arch. exp. Med.* **6**, 121–38. [*Trypanosoma cobitis* in *Hirudo nipponia.*]

Tandon, R.S. (1977). [Eco-physiological studies on fish trypanosomes.] In *Abstracts of the 1st National Congress of Parasitology*, Baroda, 24–26 February 1977, pp. 53–4. Indian Society for Parasitology. [*Hemiclepsis marginata asiatica*, parasitic on *Wallago attu;* had trypanosomes from host fish in gut.]

Tan En-Guang. (1980). [Three new species of land leeches from Zhanjiang and Hainan, Guangdon Province.] *Acta zoota. sin.* **4**(4), 353–7. [In Chinese; English summary.]

Taticchi, M.I. (1968). Vicende stagionali delle communità littoranee del Lago Trasimeno (1963–1965). *Riv. Idrobiol.* **7**, 195–302. [*Hirudo* in Italy]

Taube, C.M. (1966). Leeches. *Mich. Dept. Conservation Res. Dev., Rept. No. 55, Inst. Fish. Res. Rept.* No. 1713, pp. 1–14.

Taxt, T. (1975). [The nerve system in the leech, *Hirudo medicinalis*, and an unusual nerve cell.] *Fauna, Oslo* **28**(3), 156–62. [In Norwegian; English summary.] [1976BA:37692.]

Taylor, E. (1943). A note on Dutrochet's leech, *Trocheta subviridis* Dutrochet, and its occurrence in Oxford. *Ann. Mag. nat. Hist., Ser. II* **10**, 431–2.

Taylor, G. (1952). Analysis of the swimming of long and narrow animals. *Proc. R. Soc. Lond.* **A214**, 158–83.

Taylor, T.P., and Erman, D.C. (1980). The littoral bottom fauna of high elevation lakes in Kings Canyon National Park. *Calif. Fish Game* **6**(2), 112–19.

Telo, W., and Panciroli, E. (1954). Osservazioni cliniche su due rari casi di ipersensibilita locale al sanguisugio. *Il Policlinico* **61**, 1674.

Tennent, J.E. (1860). Leeches. In *Ceylon. An account of the island, etc.*, 4th edn 2 Vols, i, pp. 301–7. London.

Terekhov, P.A. (1966*a*). [On the conditions and time of the sexual maturity onset in leeches *Piscicola geometra*.] *Zool. Zh. Acad. Nauk SSSR* **45**(11), 1721–3. [In Russian.]

—— (1966*b*). Parazitophauna molodi tarani Kubanskikh limanov. *Tr. Azovsk. Nauchno-Issledov. Inst. R'bn. Khoz.* **9**, 145–50. [In Russian.]

—— (1967*a*). [Study of the pathogenic role of *Piscicola geometra* to larvae of roach.] *Mater. Nauch. Konf. vses. Obshch. Gel'mint.* Year 1966, Part 5, 325–7. [In Russian.]

—— (1967*b*). [On the duration of life cycle of the leech, *Piscicola geometra.*] *Zool. Zh. Acad. Nauk SSSR* **46**(6), 846–9. [In Russian; English summary.]

—— (1967*c*). Pistsikolez molodi tarani Kubanskikh limanov. *Avtoreph. Diss.* (*Novocherkassk*) 1–19. [In Russian.]

—— (1968*a*). [On the effect of water salinity on *Piscicola geometra* (L.) (Hirudinea).] *Gidrobiol. Zh.* **4**(2), 62–3. [In Russian.]

—— (1968*b*). [Reproduction of the common fish leech *Piscicola geometra* (Hirudinea, Piscicolidae) in the Kuban estuaries.] *Zool. Zhurn.* **47**(7), 1091–5. [In Russian; English summary.]

—— (1968*c*). [Ecology of the common fish leech (*Piscicola geometra*) in the Kuban estuaries.] *Zool. Zh. Acad. Nauk SSSR* **47**(3), 463–6. [In Russian; English summary.]

Tereshkov. O.D. (1967). [The structural and functional aspects of the coordination of leech locomotion.] *Biofizika* **12**(6), 1071–4. [In Russian.]

—— (1968*a*). [On the methods of setting in motion the action potential in the giant secretory cells of the leech.] *Vest. Mosk. gos. Un.* **3**, 47–52. [In Russian.]

—— (1968*b*). [On the fluctuation of bioelectrical activity of the abdomen nerve-chain in the leech.] *Vest. Mosk. gos. Un. Ser. 6*, No. 1, 38–42. [In Russian.] [*Aulastoma gulo = Haemopis sanguisuga.*]

—— (1973). [Inhibiting effect of a sodium-free medium and choline on interneuronal electrical transmission in central nervous system of the leech.] *Dokl. Akad. Nauk SSSR* **213**(6), 1454–7. [In Russian.] [1976BA:25918.]

—— and Dolzhanov, A.I. (1974). [Electrical responses of the Retzius cells in leech to inhibition of the active ionic transport with ouabain.] *Byull. Eksp. Biol. Med.* **78**(9), 10–14. [In Russian; English summary.] [1975BA:61131.]

—— and Fomina, M.S. (1969). [Electrophysiological studies of the paired giant cells in the abdominal chain of the leech, *Aulastoma gulo.*] *Zh. evol. Biokhim. Fiziol.* **5**, 304–9. [In Russian; English summary.] [Transl. in: *J. evol. Biochem. Physiol.* **5**, 244–50.]

—— —— (1970). [On the widespread occurrence of electrotonic communication between neurons in the ganglion of the leech *Hirudo medicinalis.*] *Dokl. Acad. Nauk. SSSR* **194**(1), 235–7. [In Russian.] [Transl. in: *Dokl. Acad. Sci. USSR* (*Biol.*) **194**, 587–8.]

—— —— (1971) (1972). Electrical transmissiom between non-symmetrical neurons in the leech ganglion. *Neurosci. Transl.* **3**, 413–18. [Translated from *Neurofiziologiya* **3**, 550–6 (1971); in Russian.]

—— —— (1974). [Reaction of leech neurons to penetration and direct electric stimulation.] *Fiziol. Zh. SSSR IM. I. M. Sechenova* **60**(3), 362–9. [1975BA:25540.]

—— —— and Gurin, S.S. (1969). Electrophysiological properties of paired giant cells of the leech *Aulastoma gulo*. *Neurosci. Transl.* **9**, 77–80. [Transl. from *Biofizika* **14**, 86–90; in Russian.] [*A. gulo = Haemopis sanguisuga.*]

Ter-Grigoryan, M.A. (1950). Nekotor'e nabludeniya nad pasprostraneniem meditsinskoi piyavki v Armenii. *Zool. Sb., Krevan* **7**, 122–6. [In Russian.]

Terio, B. (1950). Sulla presenza di reti nervose diffuse nel tratto faringogastrico di *Hirudo medicinalis*. *Boll. Zool.* **17**, 25–7.

Tertrin, C., de la Llosa, P., and Jutisz, M. (1966). Étude de quelques constantes physico-chimiques de l'hirudine. *Biochim. biophys. Acta* **124**, 380–8. [In French; English summary.]

—— —— —— (1967). Effet des modifications chimiques de l'hirudine sur son action inhibitrice de l'activité enzymatique de la thrombine. *Bull. Soc. Chim. Biol.* **49**, 1837–43.

Tetry, A. (1939). Contribution à l'étude de la Faune de l'Est de la France (Lorraine). *Bull. Mens. Soc. Sc. Nancy Mém.* No. 3. (Incomplete reference)

Tewari, H.B., and Tyagi, H.R. (1978). On the distribution patterns of phosphatase and non-specific esterase and their functional aspects amongst the constituents of the photoreceptors of the Indian medicinal leech, *Poecilobdella granulosa* (Savigny, 1822) *Acta morphol. neerlscand.* **16**(4), 225–40. [1979BA:9556.]

Thai Tran Bai (1965). [Locomotory apparatus of leeches.] *Avtoreph. Diss.*, *Moscow* 1–12. [In Russian.]

—— (1966). [On the locomotory apparatus of leeches.] *Zool. Zh.* **45**(4), 500–8. [In Russian; English summary.]

Theobald, S. (1875). An improved method of applying the artificial leech. *Am. J. med. Sci., Phila.* (new series) **19**, 139–42.

Thienemann, A. (1950). Verbreitungsgeschichte der Susswassertierwelt Europas. *Binnengewässer* **18**, 156–9. [Dispersal of *Glossiphonia complanata* by birds.]

Thijsse, J.P. (1944*a*). De medicinale bloedzuiger. *Levende Nat.* **49**, 12.

—— (1944*b*). Onze enquêtes. De bloedzuigers. *Levende Nat.* **49**, 21.

Thomas, A.E. (1969). Mortality due to leech infestation in an incubation channel. *Progr. Fish Cult.* **31**(3), 164–5.

Thomas, L. (1979). On magic in medicine. *Hum. Nat.* [Leech jar.] (Incomplete reference)

Thomas, P. (1806). Mémoire pour servir à l'histoire naturelle des sangsues. Paris. 151 pp. [In WIMH library.]

Thompson, A.J. (1977). A study of the growth and life-cycles of some freshwater leeches. Unpublished PhD. thesis, University of Hull.

Thompson, D.H. (1927). An epidemic of leeches on fishes in Rock River. *Bull. Ill. nat. hist. Surv.* **17**, 195–201. [Probably *Myzobdella lugubris*.]

Thompson, D'Arcy W. (1961). See D'Arcy Thompson (1961).

Thompson, I., and Jones, D.S. (1980). A possible onychophoran from the middle Pennsylvanian Mazon Creek beds of North Illinois. *J. Paleontology* **54**(3), 588–96.

Thompson, R.W., and Porter, L. (1966). Shock escape learning in the leech *Haemopis grandis*. Paper read at 1966 meeting of Ohio State Psychological Assoc.

Thompson, W. (1844). [Irish leeches.] *Ann. Mag. nat. Hist.* **13**, 430–40.

Thompson, W.J., and Stent, G.S. (1976a). Neuronal control of heartbeat in the medicinal leech. I. Generation of the vascular constriction rhythm by heart motor neurons. *J. comp. Physiol. A* **111**, 261–79.

—— —— (1976b). Neuronal control of heartbeat in the medicinal leech. II. Intersegmental co-ordination of heart motor neuron activity by heart interneurons. *J. comp. Physiol. A* **111**, 281–307.

—— —— (1976c). Neuronal control of heartbeat in the medicinal leech. III. Synaptic relations of the heart interneurons. *J. comp. Physiol. A* **111**, 309–33.

—— —— (1976d). A leech motor neuron with a peripheral and a central spike initiation zone. *Acta physiol. scand.* Suppl. **440**, 89. [1977BA: 34784.]

Thorndike, T.R. (1927). A history of bleeding and leeching. *Boston med. surg. J.* **197**, 474–7.

Threlfall, W. (1969). Some parasites from elasmobranchs in Newfoundland. *J. fish. Res. Bd Can.* **26**(4), 805–11.

Thut, R.H. (1969). A study of the profundal bottom fauna of Lake Washington. *Ecol. Monogr.* **39**(1), 79–100.

Tiegs, O.W., and Manton, S.M. (1958). The evolution of the Arthropoda. *Biol. Rev.* **33**, 255–337.

Tillman, D.L., and Barnes, J.R. (1973). The reproductive biology of the leech *Helobdella stagnalis* (L.) in Utah Lake, Utah. *Freshwat. Biol.* **3**, 137–45. [See also (1972) *Am benth. Soc.*, Abstr.]

Tilloy, R. (1937). La fonction athrocytaire chez les Hirudinées (essai d'explication de la sélection des colorants). *Bull. Mém. Soc. Sci. Nancy* (new series) **7**, 199–225.

Timm, V. (1967). [Notes on the benthos in the Narva River.] *Loodusuur Seltsi Aastar.* **58**, 154–63. [In Estonian; Russian and English summaries.]

—— and Timm, T. (1968). [Die Bodenfauna des Paunküla-Stausees in den Jahren 1961–1963.] *Loodusuur Seltsi Aastar.* **59**, 105–21. [In Estonian; Russian and German summaries.]

Timmers, S.F. (1979). *Alloglossidium schmidti*, new species (Trematoda: Macroderoididae) from hirudinid leeches (*Haemopis grandis*) in Manitoba, Canada. *Proc. helminth. Soc. Wash.* **46**(2), 180–4. [1980BA:17431.]

Timon-David, J. (1943). Sur la présence en Camargue et le développement expérimental de *Cotylurus cornutus* (Rud.) (Trématode, Strigeidé). *Bull. Mus. Hist. nat., Marseille* **3**, 17–21. [Tetracotyles from leeches considered to be the larvae of *C. cornutus*.]

Ting Ju-Nan (1938). Morphology and anatomy of the giant freshwater leech, *Whitmania laevis* (Baird). *Peking nat. Hist. Bull.* **13**(1), 29–33.

Tinsley, R.C., and van der Lande, V. (1973). The life history of *Marsupiobdella africana* (Hirudinea: Glossiphoniidae) with reference to leech phylogeny. (Br. Soc. Parasitol., Southampton, 3–5 April 1973.) Abstr. *Parasitology* **67**(2), xxvii.

Tiras, Kh.P., and Aslanidi, K.B. (1981). [Device for graphic recording of planaria behaviour.] *Zh. Vyssh. Nervn. Deyat. IM. I. P. Pavlova* **31**(4), 874–7. [In Russian.] [1982BA:14795.]

Titova, S.D. (1954). [Parasites of fishes in Teletz Lake.] *Trans. spec. Probl. Confer. (7th Conf. on Problems of Parasitology)*, Moscow Vol. 4, pp. 79–84. [In Russian.] [Leeches.]

Tolp, O. (1958). [Über das Benthos des Flusses Emajogi.] *Loodusuur Seltsi Aastar.* **49**, 143–60. [German summary.] [Emajogi river, Estonia.]

—— (1969). [On the bottom fauna in the Narva reservoir.] *Loodusuur Seltsi Aastar.* **60**, 67–79. [In Estonian; English summary.] [Influence of warm-water canal from Baltic thermal power plant and of Narva river on leeches in benthos of Narva reservoir.]

Tomic, M.T. (1979). Development of the salivary glands of *Haementeria ghilianii*. Bachelor's degree dissertation, Department of Molecular Biology, University of California, Berkeley.

Torska, I.V., and Byelova, L.M. (1969). [Neural contacts in the central nervous system of horse leeches.] *Fiziol. Zh. (Akad. Nauk. UKR RSR)* **15**(4), 496–501.

Tóth, A. (1968). [The last bloodletters of Heviz.] *Orv. Hetil.* **109**, 2334–6. [In Hungarian.]

Townsend, C.R., and McCarthy, T.K. (1980). On the defence strategy of *Physa fontinalis*, a freshwater pulmonate snail. *Oecologia, Berl.* **46**(1), 75–9.

Traub, R., and Wisseman Jr, C.L. (1952). Preliminary observations on a repellent for terrestrial leeches. *Nature, Lond.* **196**, 667–8.

Trauger, D.L., and Bartonek, J.C. (1975). Leech parasitism of waterfowl in North America. *Wildfowl* **28**, 143–52. [Good review.]

Trenina, G.V. (1968). [A characterization of the benthic fauna of Lake Lacha.] *Tr. Karel'sk. Otdel. Gos. Nauch-Issled. Inst. Ozern. Rechn. Rybn. Khoz.* **5**(1), 243–56.

Triebel, H., and Walsmann, P. (1966). Die Bestimmung des Molekulargewichtes von Hirudin in der Ultrazentrifuge. *Biochim. biophys. Acta* **120**, 137–47.

Troschel, F.H. (1850). *Piscicola respirans* nov. sp. *Arch. Naturg.* **16**(1), 17–26.

Truszowski, R. (1928). See Autrum (1939c).

Tubangui, M. (1932). Observations on possible transmission of surra by the land leech, *Haemadipsa zeylanica*. *Philippine J. Sci.* **48**, 115–27.

Tucker, D.S. (1958). The distribution of some fresh-water invertebrates in ponds in relation to annual fluctuations in the chemical composition of the water. *J. anim. Ecol.* **27**(1), 105–23.

Tucker Jr, J.W. (1973). Ecto-commensals and ecto-parasites as indicators of silt pollution. *Bull. S. Carolina Acad. Sci.* **35**, 108–9.

Tulp, A.S. (1967). Hydrobiologische notities over de Grote Wielen. *Levende Nat.* **70**, 27–41. [Records 11 species of leeches.]

Tulsi, R.S, and Coggeshall, R.E. (1971). Neuromuscular junctions on the muscle cells in the central nervous system of the leech, *Hirudo medicinalis*. *J. comp. Neurol.* **141**(1), 1–16.

Tümpling, W. von (1965). Untersuchungen über Neurosekretion bei Hirudineen. *Z. wiss. Zool.* **171**(1/2), 1–43.

Tur'ani, A. (1974). Case report: vaginal bleeding in a child due to a leech. *Jordan med. J.* **9**(1), 52–3.

Turk, J.L. and Allen, E. (1983). Bleeding and cupping. *Ann. R. Coll. Surg. Eng.* **65**, 128–31.

Turner, F.M. (1969). Pharyngeal leeches. *Lancet* **ii**, 1400–1. [In Burundi — 'occasionally fatal anaemia in young children; get from gourds'.]

Tuten'kov, S.K. (1959). Bentos ozera Balkhash i ego znachenie v pitanii r'b. *Sb. Rabot Po Ikhtiol. I. Gidrobiol., Alma-Ata* **2**, 45–79. [In Russian.]

Tuzet, O. (1959). Neurosecretion In *Traité de zoologie* (ed. P.P. Grassé) Vol. 5, pp. 544–6. Masson, Paris. [See also (1950) *Rev. Suisse Zool.* **57**, 433.]

Tvermyr, S. (1965). Legeiglen (*Hirudo medicinalis* L.) finnes enna frittle vende: Aust-Agder. *Fauna, Oslo* **18**(3), 136–9. [English summary.]

Tweed, J.J. (1850). A description of the apparatus for employing the mechanical leeches. *Med. Times* **21**.

Tyler, M.J., Parker, F., and Bulmer, R.N.H. (1966). Observations on endoparasitic leeches infesting frogs in New Guinea. *Rec. S. Aust. Mus.* **15**(2), 356–9. [*Leiobdella singularis*, in *Hyla angiana* and *Nyctimystes papua*.]

Uchida, H., Yamada, M., and Takeuchi, I. (date undetermined). The benthic invertebrates in the fishing ground of King Crab (*Paralithodes camtschatica*) off the west coast of Kamchatka Peninsula, 1957–64. Part I. *Bull. Hokkaido reg. fish. Res. Lab.* **35**, 119–59. [In Japanese; English summary.][*Notostomum cyclostomum* off W. coast of Kamchatka.]

Uexküll, J. von (1905). Studien über den Tonus III. Die Blutegel. *Z. Biol.* **46** (NF 28), 372–402.

Uludag, O.Ş. (1939). Tip tarihimizde sülük. *Türk Tip Tarihi Arşivi* **4**;14 (see also 1938, **3**;10). [Leech use in Turkey]

Unchis, O.N. (1966). Parazitophauna p'b kovdozerskogo vodohranilischa i Imandr'. *V Kn.: P'b' Murmanskoi Oblasti., Murmansk* 130–4. [In Russian.]

Ushakov, B.P., and Sleptsova, L.A. (1968). [Changes in muscle tissue thermostability of leeches during thermal acclimation.] *Tsitologiya* **10**(2), 259–62. [In Russian; English summary.]

Usov, S. (1859). [Natural history of the leech.] *Zapiski Komiteta Akklimat. Zhivotnyx* No. 1. [Notes of the Comm. Acclimat. of Animals.]

Ussing, H.J. (1979). Om nogle sjaeldne og lidet kendte danske Hirudinea. *Vid. Meddel. Dansk Naturh. Foren.* **88**, 203–20.

Utkina, O.T. (1958). [Scientific, pedagogical and social activity of Prof. G.G. Schegolev (1882–1956).] *Byul. Nauchn. Tr. Ryazansk. OTd. Vses. Obsch. Anat., Gistol., Embriol.* **2**, 3–6. [Transl. *Ref. Zh., Biol. 1957*, No. 91165.]

Vaillant, L. (1867). Remarques sur trois espèces d'hirudinées du Mexique. *C.r. hebd. Séanc. Soc. Biol., Paris* (4) **3**, 89–91.

—— (1870). Contribution à l'étude anatomique du genre Pontobdelle. *Ann. Sci. nat.* (*Zool.*) (5) **13**, Art. 5, 1–71.

Vale, T.W. (1963). *Trocheta subviridis* in Yorkshire. *Naturalist*, Hull. 1963, 29.

Valentin, G. (1936). Ueber den Verlauf und die letzten Enden der Nerven. *Kais. Leop.-Carol. Akad. Naturforscher.* **XVIII**, 1. [Cited by Muller, J. (1837). *Arch. Anat. Physiol. wiss. Med. I–CXXXIII.*] [Nerve cytology of leeches.]

Van Bemmel, A.C.V., Peters, J.C., and Zwart, P. (1960). Reports on births and deaths occurring in the gardens of the Royal Rotterdam Zoo during the year 1958. *Veeteelt Zuivelber.* **3**, 1203–13. [*Theromyzon* causing deaths in birds.]

Van Damme, N. (1974). Organogénèse de l'appareil génital chez la sangsue *Erpobdella octoculata* L. (Hirudinée; Pharyngobdelle). *Arch. Biol.* **85**(3), 373–97. [In French; English summary.]

—— (1976). Maturation génitale précoce par destruction des ganglions buccaux chez la sangsue *Erpobdella octoculata* L. (Hirudinée–Pharyngobdelle). *C. r. hebd. Séanc. Acad. Sci. Paris* **283**, 967–9. [1977BA:30221.]

—— (1977). Étude histologique et ultrastructurale des ganglions buccaux chez la sangsue *Erpobdella octoculata* L. (Hirudinée–Pharyngobdelle). *Arch. Biol., Bruxelles* **88**, 31–52.

—— (1978). Rôle des ganglions buccaux dans la reproduction chez la sangsue *Erpobdella octoculata* L. (Hirudinée–Pharyngobdella). *Arch. Biol., Bruxelles* **89**, 451–82. [In French.]

Vande Vusse, F.J. (1980). Revision of *Alloglossidium* Simer 1929 (Trematoda: Macroderoididae) and description of *A. microspinatum* sp. n. from a leech. *J. Parasitol.* **66**(4), 667–70.

—— Fish, T.D., and Neumann, M.P. (1975). Adult digenetic trematodes from Minnesota hirudinid leeches. *Am. Soc. Parasitol.* 63. (Abstr.)

—— and Neumann, M.P. (1971). Two new species of *Alloglossidium. J. Parasitol.* **62**(4), 556–9.

Van Der Kloote, W.G. (1967). Goals and strategy of comparative pharmacology. *Fedn Proc. Fedn Am. Socs exp. Biol.* **26**, 975–80.

Van Der Lande, V.M. (1966). Observations on the nutrition of certain British leeches (Annelida: Hirudinea). A thesis presented for the degree of Doctor of Philosophy, Zoology Department, University of Leeds.

—— (1968). Esterase activity in certain glands of leeches (Annelida: Hirudinea). *Comp. Biochem. Physiol.* **25**(2), 447–56.

—— (1972). Observations on the histochemical 'aminopeptidase' reaction in the intestine of certain species of leech (Annelida: Hirudinea), with particular reference to *Erpobdella octoculata* (L.). *Comp. Biochem. Physiol.* **41A**(4), 813–24.

—— (1983). Observations on the growth and development of the duck leech, *Theromyzon tessulatum* (Hirudinea: Glossiphoniidae), as a function of feeding. *J. Zool.* **201**, 377–93.

—— and Tinsley, R.C. (1976). Studies on the anatomy, life history and behaviour of *Marsupiobdella africana* (Hirudinea; Glossiphoniidae). *J. Zool., Lond.* **180**, 537–63.

Van Der Veen, B. (1968). List of scientific publications by Professor Dr. H. Engel. *Beaufortia Ser. Misc. Publ. Zool. Mus. Un. Amsterdam* **15**(180), 7–14. [Includes leeches.]

Van Duijn, C. (1973). *Diseases of fishes.* (3rd edn) Thomas, Springfield. Ill, 372 pp.

Van Essen, D.C. (1973). The contribution of membrane hyperpolarization to adaptation and conduction block in sensory neurones of the leech. *J. Physiol.* **230**(3), 509–34.

—— and Jansen, J.K.S. (1976). Repair of specific neuronal pathways in the leech. In: *The synapse.* Cold Spring Harbor Symposium on Quantitative Biology. Vol. XL, pp. 495–502. Cold Spring Harbor Laboratories. [1977BA:16785.]

—— —— (1977). The specificity of re-innervation by identified sensory and motor neurons on the leech. *J. comp. Neurol.* **171**(4), 433–54.

Van Geertruyden, and Bernard, M. (1921). Contribution à l'étude de l'action d'hirudine sur les accidents anaphylactiques. Paris. [Incomplete]

Van Harreveld, A., Khattab, F.I., and Steiner, J. (1969). Extracellular spaces in the central nervous system of the leech *Mooreobdella fervida*. *J. Neurobiol.* **1**(1), 23–40.

Van Obberghen-Schilling, E., Pérez-Rodriquez, R., and Pouysségur, J. (1982). Hirudin, a probe to analyse the growth-promoting activity of thrombin in fibroblasts, evaluation of the temporal action of competence factors. *Biochem. Biophys. Res. Commun.* **106**(1), 79–86.

Vartiainen, A. (1933). See Autrum (1939*c*).

Vasiliev, E.A. (1935). [Supplement to Russian translation of *Leech types.*] [In Russian.]

—— (1939). The Ichthyobdellidae of the Far East. *Tr. Karel'ske Gos. Ped. Inst. I Biol. Ser., Petrozavudsk* **1**, 25–68. [An important systematic and anatomical work largely unknown outside Russia until V.M. Epshtein informed M.C. Meyer of its existence. Translated into English by Antoria Glasse, edited by Meyer, who obtained a photocopy in 1964 through the Helminthological Abstracts of the Commonwealth Agricultural Bureaux, St. Albans. The Abstracts obtained the photocopy from a Russian library because there was apparently no original available outside Russia.]

Vassal, J.J. (1906). See Autrum (1939c).

Vavrouškova, K. (1952). [Colour change in the leech *Protoclepsis tesselata* (O.F. Müller).] *Véstn. Českosl. Zool. Spol.* (*Mém. Soc. Zool. Tchéosl.*) **16**, 334–53. [In Czechoslovak.]

Vera, M., and Loeb, L. (1914). Immunization against the anticoagulating effect of hirudin. *J. biol. Chem.* **19**(3), 305–21.

Vereresaev, V. (ed.) (1933). Gogol' V. Zhizni. [Gogol in Life.] Academia, Moscow. [Article on Gogol's death by A.T. Tarasenkov.] [Graphic account of the application of leeches to Nicolai Gogol toward end of his life.] [In Russian.]

Vereschagin, G.U. (1940a). Teoeticheskie vopros', svyazann'e s pazpabotkoi prodlem' proiskhozhdeniya i istorii Baikala. *Tr. Baikal. Limnol. Sta.* **10**. 1–72. [In Russian.]

—— (1940b). [Origine et histoire du Baikal, de sa faune et de sa flora.] *Tr. Baikal. Limnol. Sta.* **10**, 73–239. [In Russian.]

Verin, P., Sekkat, A., and Morax, S. (1973). Oeil et sangsue de l'hirudinose oculaine et de l'hirudinotherapie. *Ann. Ocul.* **206**(1), 21–35.

—— Yacoubi, M., Skirrjed, F., Berbich, A., Morax, S., and Sekkat, A. (1971). [Rare parasitological dead-end. Ophthalmologic attack by a leech.] *Bull. Soc. Ophthalmol. Fr.* **71**(11), 1018–21. [In French.] ['*Hirudo troctina.*']

Vérma, P.L. *et al.* (1970). Leech: a rare cause of epistaxis. *J. Laryngol. Otol.* **84**, 91–2.

Verriest, G. (1950a). Contribution à l'étude des hirudinées des eaux douces de la Belgique. *Biol. Jaarb.* **17**, 200–43.

—— (1950b). Note sur la recolte des hirudinées d'eau douce et terrestres, leur fixation et leur conservation. *Biol. Jaarb.* **17**, 244–6.

Verrill, A.E. (1872). Descriptions of North American fresh water leeches. *Am. J. Sci.* (*3*) **3**, 126–39.

—— (1874). Synopsis of the North American fresh water leeches. *Rep. US Fish Comm.* 1872/73, 666–89.

Vialli, M. (1934). Le cellule cromaffini dei gangli nervosi negli Irudinei. *Atti Soc. Ital. Mus. Civico Storia nat. Milano* **73**, 57–73.

Vickerman, K. (1977). DNA throughout the single mitochondrion of a kinetoplastid flagellate: observations on the ultrastructure of *Cryptobia vaginalis* (Hesse, 1910). *J. Protozool.* **24**(2), 221–33. [1977BA:63058.]

Vignal, W. (1883). See Autrum (1939*c*).

Vik, R. (1962). Borsteiglen funnet i Norge. *Fauna, Blindern* **15**(1), 31–6. [In Norwegian; English summary.] [Norwegian leeches; *Acanthobdella peledina* parasitizing grayling and trout.]

Villiers, A.F. de (1976). Littoral ecology of Marion and Prince Edward Islands (Southern Ocean). *S. Afr. J. Antarctic Res.* Suppl. 1, pp. 1–40. Hirudinea: pp. 2, 19, and 40.]

Vincent, B. (1979). Étude du benthos d'eau douce dans le haut-estuaire du Saint-Laurent (Quebec). *Can. J. Zool.* **57**, 2171–82.

—— and Vaillancourt, G. (1977*a*). Addition à la faune des Hirudinée (Annelida: Hirudinoidea) du Quebec. *Nat. Can.* **104**(3), 269–72. [1978BA:21307.]

—— —— (1977*b*). Le peuple des invertebrés du fleuve Saint-Laurent près des installations nucléaire de Gentilly (Quebec). *Trav. Lab. Hydribiol.* **66–8**, 95–118. (See also 1978. *Can. J. Zool.* **56**, 1585–92).

—— —— (1978). Les groupements benthiques du fleuve Saint-Laurent près des centrales nucléaires de Gentilly (Quebec). *Can. J. Zool.* **56**(7), 1585–92.

—— —— (1980). Les sangsues (Annelida: Hirudinea) benthiques du Saint-Laurent (Québec). *Nat. Can.* **107**(1), 21–33. [English summary] [Physical factors more important than chemical factors in determining leech distribution.]

Vinogradov, S.N., Hall, B.C., and Shlom, J.M. (1976). Subunit homology in invertebrate hemoglobins (Hb): a primitive heme binding chain? *Comp. Biochem. Physiol. B Comp. Biochem.* **53**(1), 89–92.

Viosca, P. (1962). Observations on the biology of the leech *Philobdella gracile* Moore in southeastern Louisiana. *Tulane Stud. Zool.* **9**, 243–4.

Voeltzkow, A. (1891). See Autrum (1939*c*).

Vogel, K.G., and Kelley, R.O. (1977). Cell surface glycosaminoglycans: identification and organization in cultured human embryo. *J. cell. Physiol.* **92**(3), 469–80. [1978BA:7516.] ['Leech hyaluronidase (4 µg/ ml) removed only hyaluronic acid from the human embryo fibroblast cell surface.']

Vogt, R.C. (1979). Cleaning feeding symbiosis between Grackles (*Quiscalus:* Icteridae) and map turtles (*Graptemys:* Emydidae). *Auk* **96**, 608–9.

—— (1980). Natural history of the map turtles *Graptemys pseudogeographica* and *G. ouachitensis* in Wisconsin, USA. *Tulane Stud. Zool. Bot.* **22**(1), 17–48. [1980BA:42408.] [Grackles, *Quiscalus quisala*, were observed removing leeches, *Placobdella parasitica*, from basking turtles.]

Voinov, V. (1928). See Autrum (1939*c*).

Vojtek, J. (1971). Beitrag zur Kenntnis des Entwicklungszyklus von *Cyathocotyle opaca* (Wisniewski, 1934) n. comb. (Trematoda: Cyathocotylidae). *Z. Parasitkde.* **36**(1), 51–61. [Metacercaria in

Czechoslovakian *Haemopis sanguisuga, Erpobdella,* and *Hirudo medicinalis.*]

—— Opravilovà, V., and Vojitová, L. (1967). The importance of leeches in the life cycle of the order Strigeidida. *Folia Parasitol., Prague* **14**(2), 107–19.

Von Bary, S. (1979). In: Habilitationsschrift. Medizinische Fakultät der Universität München.

Von Lukowicz, M. (1973). Zur Anwendung von Masoten. Ein Beitrag zur Bekämpfung von Fischparasiten. *Allg. Fischerei Ztg.* **3**, 146–7. [Phosphoro-organic preparation to control leech infestations.]

Vorobiov, M.M., and Kotel'nikov, G.O. (1959). P'yavki — paraziti vodoplavnoi ptitsi. *Cotsial. Tvarinnitstvo* **4**, 53–4.

Vos, A.P.C. de (1941). Zoölogische resultaten van een tocht rond het Ijsselmeer van 5–8 Juli 1937. *Biol. Zuiderzee tijd. drooglegging. Mededeeling Zuiderzee Comm.* **5**, 37–41.

Vovsi, B.M. *et al.* (1967). [Eye injury caused by a leech.] *Oftal. Zh.* **22**, 462–3. [In Russian.]

Waffle, E.L. (1963). An ecological study of the Iowa Glossiphoniidae (Annelida: Hirudinea) with emphasis on feeding and reproductive habits. Master's thesis, State University of Iowa.

Wagler, E. (1927). Der Blaufelchen des Bodensees. (*Coregonus wartmanni* Block). *Int. Rev. Hydrobiol., Leipzig* **18**, 129–230.

Wagner, C.C. (1969). Incidence of leech infestations on blacknose dace. *Am. fish. Soc. Trans.* **98**(1), 115–16. [*Myzobdella lugubris.*]

Wainer, T., and Yagil, R. (1978). *Limnatis nilotica* infestation in a dog. *Refuah Vet.* **35**(1), 14.

Waldschmidt-Leitz, F., Stadler, P., and Steigerwaldt, F. (1929). Über Blutgerinnung, Hemmung und Beschleunigung. *Z. Physiol. Chem.* **183**, 39–66.

Wales, J.H., and Wolf, H. (1955). Three protozoan diseases of trout in California. *Calif. Fish Game* **41**, 183–7.

Walker, G. (1814). 'The Leech Gatherer.' In *The Costume of Yorkshire.* Robinson and Sons, Leeds. [See Einstein (1978).]

Walker, J.W. (1943). Leech infection, case (with hypochromic anemia). *E. Afr. Med. J.* **20**, 114.

Walker, M.J.A., and Peng Nam Yeoh (1974). The *in vitro* neuromuscular blocking properties of sea snake (*Enhydrina schistosa*) venom. *Eur. J. Pharmacol.* **28**(1), 199–208. [1975BA:11416.]

Walker, R.J. (1967). Certain aspects of the pharmacology of *Helix* and *Hirudo* neurons. In *Neurobiology of invertebrates* (ed. J. Salanki) pp. 227–53. Plenum, New York. [Symposium: Hungarian Academy of Sciences, Budapest.]

—— (1982). Current trends in invertebrate neuropharmacology. *Verh. Dtsch. Zool. Ges.* 1982, 31–59.

—— and Kerkut, G.A. (1978). The first family (adrenaline, noradrenaline, dopamine, octopamine, tyramine, phenylethanolamine, and phenylethylamine). *Comp. Biochem. Physiol. C Comp. Pharmacol.* **61**(2), 261–6. [1979BA:73827.]

—— and Roberts, C.J. (1981). Actions of kainic acid and related analogues on *Hirudo, Helix* and *Limulus* central neurones. Eighth International Congress of Pharmacology IUPHAR, 19–24 July 1981, Tokyo. (Abstr.)

—— and Smith, P.A. (1973). The ionic mechanism for 5-hydroxytryptamine inhibition on Retzius cells of the leech *Hirudo medicinalis*. *Comp. Biochem. Physiol. A Comp. Physiol.* **45**(4), 979–93.

—— James, V.A., and Roberts, C.J. (1981). The action of FMRF-amide and Proctolin on *Helix, Hirudo, Limulus,* and *Periplaneta* neurones. In: *Advances in physiological science*, Vol. 20, *Advances in animal and comparative physiology* (ed. G. Pethes and V.L. Frenyo) pp. 411–16. Akademiai Kiado, Budapest.

—— Roberts, C.J., and Krogsgaard-Larsen, P. (1981). Actions of homoibotenate analogues on L-glutamate receptors of *Hirudo* Retzius neurones. Eight International Congress of Pharmacology IUPHAR, 19–24 July 1981, Tokyo. (Abstr.: 599)

—— Woodruff, G.W., and Kerkut, G.A. (1968). The effect of acetylcholine and 5-hydroxytryptamine on electrophysiological recordings from muscle fibres of the leech, *Hirudo medicinalis. Comp. Biochem. Physiol.* **24**(3), 987–90.

—— —— —— (1970). The action of cholinergic antagonists on spontaneous excitatory potentials recorded from the body wall of the leech, *Hirudo medicinalis. Comp. Biochem. Physiol.* **32**(4), 691–701.

—— James, V.A., Roberts, C.J., and Kerkut, G.A. (1981). Studies on amino acid receptors of *Hirudo, Helix, Limulus* and *Periplaneta*. In: *Advances in physiological science*, Vol. 22, *Neurotransmitters in Invertebrates* (ed. K.S. Rozsa) pp. 161–90. Akademiai Kiado, Budapest.

Wallace, B.G. (1980). Selective neurite atrophy during development of cells in the leech CNS. *Neurosci. Abstr.* **6**, 679.

—— (1981*a*). Neurotransmitter chemistry. In: *The neurobiology of the leech* (ed. K.J. Muller, J.G. Nicholls, and G.S. Stent) Ch. 8, pp. 147–73. Cold Spring Harbor.

—— (1981*b*). Distribution of AChE in cholinergic and non-cholinergic neurons. *Brain Res.* **224**. (Incomplete reference)

—— and Nicholls, J.G. (1977). Modulation of transmitter release at an inhibitory synapse in the CNS of the leech. *Neurosci. Abstr.* **III**, 1165, 520.

—— Adal, M.N., and Nicholls, J.G. (1977). Sprouting and regeneration of synaptic connexions by sensory neurones in leech ganglia maintained in culture. *Proc. R. Soc. Lond.* **B199**, 567–85.

Wallace, R.K. Jr, and Sawyer, R.T. (1977). Occurrence of *Malmiana philotherma* (Annelida: Hirudinea) on *Hemiramphus brasiliensis* (Pisces: Hemiramphidae) in Puerto Rico. *Bull. mar. Sci.* **27**(2), 347–8.

Walsmann, P. (1964). Über die Standardisierung von Hirudin-Handelspräparaten. *Blut* **10**, 203–4.

—— and Vogel, G. (1970). Über die Anwendung von Hirudin im Klinischen Gerinnungslaboratorium. *Folia Haem., Leipzig* **94**, 143–52.

—— *et al.* (1981). [Biochemical and pharmacological aspects of the thrombin inhibitor hirudin.] *Pharmazie* **36**(10): 653–60. [In German.]

Walter, J.E., and Geiser-Barkhausen, A. (1980). Konkurrenz zwischen *Herpobdella octoculata* (Hirudinea) und *Dugesia polychroa* (Turbellaria) im Labor. *Archs Hydrobiol.* **88**, 458–62.

Walther, J.B. (1963). Intracellular potentials from single sense cells in the eyes of the leech, *Hirudo medicinalis*. *Proc. 16th Int. Cong. Zool.*, Washington, DC, Vol. 2, p. 71.

—— (1964). Untersuchungen zur Temperaturabhängigkeit des Generatorpotentials einzelner Sehzellen in Augen von *Hirudo medicinalis*. *Verhandlg. Dt. Zool. Ges. Kiel.* 353–8.

—— (1965). Untersuchungen über die Erregungsvorgänge einzelner Lichtsinneszellen in den Augen des Blutegels, *Hirudo medicinalis*. *Habilitationsschr. Math.-Nat. Fakültat Un. Göttingen*. (Incomplete reference)

—— (1966). Single cell responses from the primitive eye of an annelid (*Hirudo medicinalis*). *Proceedings of the International Symposium on the Functional Organization of the Compound Eye* (ed. C.G. Bernhard) 25–27 October 1965, Stockholm, Vol. 7, pp. 329–36. Pergamon Press, London.

—— (1970). Widerstandsmessungen an Sehzellen des Blutegels, *Hirudo medicinalis* L. *Verh. Dt. Zool. Ges.* **64**, 161–4. [English summary.]

Walton, A.C. (1945). Miscellaneous parasites of amphibia. *Contribution from the Biological Laboratories of Knox College (Galesburg, Ill.)* No. 100.

Walton, B.C. (1955). The 'nasal leech' *Dinobdella ferox* from Borneo and Malaya. *J. Parasitol.* **41**(6), 32. [In nasal passages of *Rattus mulleri*, Malaya; and of dogs, Sabah.]

—— Traub, R., and Newson, H.D. (1956). Efficacy of the clothing impregnants M-2065 and M-2066 against terrestrial leeches in North Borneo. *Am. J. trop. Med. Hyg.* **5**(1), 190–6.

Walz, B. (1979a). A comparison of receptive and non-receptive plasma membrane areas of photoreceptor cells in the leech, *Hirudo medicinalis*. *Cell. Tissue Res.* **198**(2), 335–48. [1980BA:78640.]

—— (1979b). Subcellular calcium localization and ATP-dependent calcium ion uptake by smooth endoplasmic reticulum in an invertebrate photoreceptor cell: an ultrastructural, cytochemical and X-ray microanalytical study. *Eur. J. cell Biol.* **20**(1), 83–91. [1980BA:64957.]

—— (1980). Sub-microvillar smooth endoplasmic reticulum sequesters calcium in leech photoreceptors. *Invest. Ophthalmol. vis. Sci.* 1980 (Suppl.), 243–4. [1980BA:37415.]

—— (1982*a*). Calcium-sequestering smooth endoplasmic reticulum in an invertebrate (*Hirudo medicinalis*) photoreceptor: 1. Intracellular topography as revealed by osmium tetroxide-ferricyanide staining and in situ calcium accumulation. *J. cell. Biol.* **93**(3), 839–48. [1982BA: 83664.]

—— (1982*b*). Calcium-sequestering smooth endoplasmic reticulum in an invertebrate (*Hirudo medicinalis*) photoreceptor: 2. Its properties as revealed by microphotometric measurements. *J. cell. Biol.* **93**(3), 849–59. [1982BA:83665.]

Walz, W. (1982). Do neuronal signals regulate potassium flow in glial cells? Evidence from an invertebrate central nervous system. *J. Neurosci. Res.* **7**(1), 71–80. [1982BA:26505.]

—— and Schlue, W.R. (1980). Neuropil glial in the central nervous system of the leech: dependence of membrane potential on external chloride concentration. *Pflügers Arch. Eur. J. Physiol.* **384** (Suppl.) R18, 70. [1980BA:59426.]

—— —— (1981*a*). Interaction of 5-hydroxtryptamine with a glia membrane in the neuropile of the leech. *Pflügers Arch. Suppl.* **389**, R24, 95.

—— —— (1981*b*). The Na$^+$/K$^+$ pump in an invertebrate glial cell. *Pflügers Arch. Suppl.* **389**, R25.

—— —— (1982*a*). External ions and membrane potential of leech neuropil glial cells. *Brain Res.* **239**, 119–38.

—— —— (1982*b*). Ionic mechanisms of a hyperpolarizing 5-hydroxytryptamine effect on leech (*Hirudo medicinalis*) neuropil glial cells. *Brain Res.* **250**(1), 111–22. [1983BA:10950.]

—— Wuttke, W., and Schlue, W.R. (1983). The sodium, potassium pump in neuropil glial cells of the medicinal leech. *Brain Res.* **267**(1), 97–100. [1984BA:11673.]

Wanke, H. (1951). Blutegelbehandlung im Rahmen der Konstitutions-therapie. *Hippokrates* **22**, 70–3.

Wareham, D.C. (1972). Leeches. *Aquarist Pondkpr.* **37**(4), 148–9. [Breeding habits of *Trocheta subviridis*.]

Warwick, T. (1961). The vice-county distribution of the Scottish freshwater leeches and notes on the ecology of *Trocheta bykowskii* (Gedroyc) and *Hirudo medicinalis* L. in Scotland. *Glasgow Nat.* **18**(3), 130–5.

—— and Mann, K.H. (1960). The fresh water leeches of Scotland. *Ann. Mag. Nat. Hist.* (13)**3**, 25–34.

Washbourn, R., and Jones, R.F. (1935–36). Percy Sladen Expedition to Lake Huleh, Palestine. *Nature, Lond.* 1935, p. 538; 1936, pp. 852–4.

Washizu, Y. (1967). Electrical properties of leech dorsal muscle. *Comp. Biochem. Physiol.* **20**(2), 641–6.

Wass, M.L. (1961). A revised preliminary check list of the invertebrate fauna of marine and brackish waters of Virginia. (Incomplete reference)

Waterston, A.R. (1935). A leech *Glossiphonia heteroclita* (Linn.), new to the Scottish fauna. *Scott. Nat.* **55**, 98.

—— (1936). Further records of the distribution of the leech *Glossiphonia heteroclita* (Linn.) in Scotland. *Scott. Nat.* **56**, 163.

Watson, D.G., Davis, J.J., and Hanson, W.C. (1966). Terrestrial invertebrates. (ed. J.J. Wilimovsky and J.N. Wolfe) in *Environment of the Cape Thompson Region, Alaska* pp. 565–84. US Atomic Energy Commission, Division of Technical Information. [*Theromyzon rude*, Ogoturuk Valley region, Alaska.]

Watson, R.A., and Dick, T.A. (1979). Metazoan parasites of whitefish, *Coregonus clupeaformis* (Mitchill) and *Cisco coregonus artedii* Lesueur, from southern Indian Lake, Manitoba, Canada. *J. fish Biol.* **15**(5), 579–88. [See also 1980, **17**, 255–261.]

Waxman, L. (1975). The structure of annelid and mollusc hemoglobins. *J. biol. Chem.* **250**(10): 3790–5.

Webb, G.J.W., and Manolis, S.C. (1983). *Crocodylus johnstoni* in the McKinlay River area, Northern Territory (Australia): 5. Abnormalities and injuries. *Aust. Wild. Res.* **10**(2): 407–20. [1983BA:86027.]

Webb, R.A. (1980*a*). Intralamellar neurohemal complexes in the cerebral commissure of the leech, *Macrobdella decora:* an electron microscope study. *J. Morphol.* **163**(2), 157–66. [1980BA:16850.]

—— (1980*b*). Spermatogenesis in leeches. I. Evidence for a gonadotropic peptide hormone produced by the supraoesophageal ganglion of *Erpobdella octoculata. Gen. comp. Endocr.* **42**(3), 401–12. [Probably *Erpobdella punctata* nec *E. octoculata.*]

—— and Omar, F.E. (1981). Spermatogenesis in leeches. II. The effect of the supraoesophageal ganglion and ventral nerve cord ganglia on spermatogenesis in the North American medicinal leech *Macrobdella decora. Gen. comp. Endocr.* **44**(1), 54–63.

—— and Orchard, I. (1979). The distribution of putative neurosecretory cells in the central nervous system of the North American medicinal leech, *Macrobdella decora. Can. J. Zool.* **57**(10), 1905–14. [1980BA:44567.]

—— —— (1980). Octopamine in leeches. I. Distribution of octopamine in *Macrobdella decora* and *Erpobdella octoculata. Comp. biochem. Physiol.* **67C**, 135–40.

—— —— (1981). Octopamine in leeches. II. Synthesis, release and reuptake of octopamine by the ventral nerve cord of *Erpobdella octoculata. Comp. Biochem. Physiol.* **70C**, 201–7.

Weber, E. (1967). Biologie der Donau: (C) Stauregion. *Limnol. Donau* **3**, 272–83. [Ecology of leeches in dams of Danube.]

Weber, E.H. (1846). Ueber die Entwicklung des medicinischen Blutegels und der Clepsine. *Arch. Anat., Phys. wiss. Med* 429–34. [Includes a letter from G. Kunze.]

Weber, Maurice (1913). Hirudinées colombiennes. In *Voyage d'exploration scientifique en Colombie. Mém. Soc. Sci. nat. Neuchâtel* **5**, 731–47.

—— (1915). Monographie des Hirudinées Sud-Américaines. Thèse Fac. Sci. Neuchâtal.

—— (1916). Hirudinées péruviennes. *Zool. Anz.* **48**, 93–6, 115–22.

Weber, Max (1906). Mededeeling. *Tijdschr. Ned. Dierkundige Ver.* (2) *X* **3**, pp. xx (moet zijn, p. xl).

Weeks, J.C. (1980*a*). The implications of recent experimental results for the validity of modelling studies of the leech swim central pattern generator. *Behav. Brain Sci.* **3**, 562–3.

—— (1980*b*). The roles of identified interneurons in initiating and generating the swimming motor pattern of leeches. Ph.D. dissertation. University of California, San Diego.

—— (1980*c*). Leech swimming: effects of interrupting the intersegmental path of swim-initiating vs pattern-generating neurons, and the ability of single ganglia to produce swim bursts. *Neurosci. Abstr.* **6**, 26.

—— (1981*a*). Neuronal basis of leech swimming: separation of swim initiation, pattern generation, and intersegmental coordination by selective lesions. *J. Neurophysiol.* **45**(4): 698–723.

—— (1981*b*). Synaptic connections between leech swim-initiating neurons and the swim central pattern generating circuit. *Neurosci. Abstr.* **7**, 137.

—— (1982*a*). Synaptic basis of swim initiation in the leech (*Hirudo medicinalis*): 1. Connections of a swim-initiating neuron (cell 204) with motor neurons and pattern-generating oscillator neurons. *J. Comp. Physiol. A. Sens. Neural. Behav. Physiol.* **148**(2), 253–64. [1983BA:66519.]

—— (1982*b*). Synaptic basis of swim initiation in the leech (*Hirudo medicinalis*): 2. A pattern-generating neuron (cell 208) which mediates motor effects of swim-initiating neurons. *J. Comp. Physiol. A. Sens. Neural. Behav. Physiol.* **148**(2), 265–80. [1983BA:66520.]

—— (1982*c*). Segmental specialization of a leech swim-initiating interneuron (cell 205). *J. Neurosci.* **2**, 972–85.

—— and Kristan Jr, W.B. (1978). Initiation, maintenance and modulation of swimming in medicinal leech by the activity of a single neurone. *J. exp. Biol.* **77**, 71–88. [1979BA:67432.]

Wegelin, R. (1966). Beitrag zur Kenntnis der Grundwasser-fauna des Salle-Elbe-Einzugsgebietes. *Zool. Jr.* (*Syst.*) **93**, 1–117. [*Haemopis sanguisuga* from drainage area, Saalbe-Elbe.]

Weil, E. (1910). The leeches in medical treatment. *Med. Press Circ.* 16 Feb. 1910. **89**, 166.

Weil, P.E., and Boye, G. (1909*a*). Recherches physiologiques sur les applications les sangsues en clinique humaine. *Semaine Med.* **29**, 421.

—— (1909*b*). Note sur les extraits disseches de têtes de sangsues. *C.r. hebd. Sáanc. Soc. Biol. Paris* **66**, 345, 516–17.

—— and Mouriquand, G. (1911). L'hemorrhagie secondaire tardive et grave consecutive à l'application de sangsues. *Bull. Soc. med. Hôp. Lyon.* **10**, 499.

—— —— and Chalier, J. (1912). Pouvoir hemolytique de l'hirudine de Sachse. *Bull. Soc. med. Hôp. Lyon.* **10**, 30.

Weiler, P. (1949). Untersuchungen über antibiotische Wirkungen an Blutegeln, Blutegelbakterien und deren Keimfreien Filtrat. *Experientia* **5**(2), 446–7.

Weisblat, D.A. (1981). Development of the nervous system, In: *The neurobiology of the leech* (ed. K.J. Muller, J.G. Nicholls, and G.S. Stent) Ch. 9, pp. 173–95. Cold Spring Harbor.

—— (1983). Cell lineage in the development of the leech nervous system. *Prog. Brain Res.* **58**, 277–82.

—— and Blair, S.S. (1984). Developmental indeterminacy in embryos of the leech *Helobdella triserialis. Dev. Biol.* **101**, 326–85.

—— Kristan, W.B. Jr. (1985). The development of serotonin-containing neurons in the leech. *Model neural networks and behaviour* (ed. A.I. Selverston), pp. 175–80.

—— —— and Kramer, A.P. (1981). Embryonic origins of identified cells in the leech CNS by tracer injection and by ablation of identified blastomeres. *Neurosci. Abstr.* **7**, 2.

—— —— and Stent, G.S. (1979). Cell lineage and regulation during leech neurogenesis as revealed by enzyme injection of identified embryonic cells. *Neurosci. Abstr.* **5**, 184.

—— —— —— (1980). Cell lineage analysis in leech neuro-genesis by tracer injection and ablation of identified embryonic cells. *Neurosci. Abstr.* **6**, 495.

—— and Shankland, M. (In press). Cell lineage and segmentation in the leech. *Proc. Roy. Soc. B.*

—— Kim, S., and Stent, G.S. (1984). Embryonic origin of cells in the leech *Helobdella triserialis. Devl. Biol.* **104**, 65–85.

—— Sawyer, R.T., and Stent, G.S. (1978). Cell lineage analysis by intracellular injection of a tracer enzyme. *Science* 202, 1295–8.

—— Harper, G., Stent, G.S., and Sawyer, R.T. (1980). Embryonic cell lineages in the nervous system of the glossiphoniid leech *Helobdella triserialis. Devl Biol.* **76**(1), 58–78. [1980BA:9769.]

—— Zackson, S.L., Blair, S.S., and Young, J.D. (1980). Cell lineage analysis by intracellular injection of fluorescent tracers. *Science, NY* **209**, 1538–41. [Cover picture, 26 Sept. 1980.]

—— Blaire, S.S., Kramer, A.P., Stuart, D.K. and Gunther, S.S. (1984). Cell lineage and cell interaction in the developing leech nervous system. *BioSci.* **34**(5), 313–17.

Weissman, B. (1955). [Preparation of leech hyaluronidase.] *J. biol. Chem.* **216**, 783.

Welch, N.J. (1975). *Marvinmeyeria lucida* — new record and addition to the leech fauna of Nebraska, USA. *Proc. Neb. Acad. Sci. Affil. Soc.* **85**, 23. (Abstr.) [Identification unconfirmed.]

Wells, G.P. (1932). Colour response in a leech. *Nature, Lond.* **129**, 686–7.

Wells, S.M. Elliott, J.M., and Tullet, P.A. (1984). Status of the medicinal leech *Hirudo medicinalis*. *Biological Conservation* **30**, 379–80.

—— Pyle, R.M., and Collins, N.M. (1983). *Hirudo medicinalis.* In: *The IUCN invertebrate red data book* pp. 205–11. International Union for Conservation of Nature and Natural Resources. Monitoring Centre, Cambridge, England.

Welsh, J.H. (1957). Serotonin as a possible neurohumoral agent: evidence obtained in lower animals. *Ann. NY Acad. Sci.* **66**, 618–30.

—— and Moorhead, M. (1960). The quantitative distribution of 5-hydroxytryptamine in the invertebrates, especially in their nervous systems. *J. Neurochem.* **6**, 146–69.

Weltner, W. (1887). *Clepsine tesselata* O. Fr. Müll. aus dem Tegelsee bei Berlin lebend vorzeigen. *Sitzungsber. Ges. Naturforsch. Freunde Berlin* **17**, 85. [*Theromyzon tessulatum* on oesophagus.]

Wendelstadt, H. (1901). See Autrum (1939*c*).

Wendrowsky, V. (1928). Über die Chromosomenkomplexe der Hirudi-neen. *Z. Zellforsch.* **8**, 153–75.

—— (1934). Croissance des gonades chez la Sangsue *Glossiphonia complanata* L. (Quelques phénomènes de phénogénétique du sexe). *Arch. Anat. microsc. Paris* **30**, 249–74.

—— (1935). Das Wachstum der Gonaden bein Rüsselegel *Glossiphonia complanata*. *J. Biol., Moscow* **4**, 325–41. [In Russian; German summary: 342.][Essentially identical to 1934 paper.]

Wenning, A. (1983). A sensory neuron associated with the nephridia in the leech. *J. comp. Physiol.* **152**, 455–458.

—— Zerbst-Boroffka, I., and Bazin, B. (1980). Water and salt excretion in the leech (*Hirudo medicinalis* L.) *J. comp. Physiol.* **B139**, 97–102.

Wenrich, D.H. (1965). John Percy Moore (1869–1965). *Yearb. Am. phil. Soc. biogr. Mem.* 191–7.

Wense, T. (1938). See Autrum (1938*c*).

Wenyon, C.M. (1926). *Protozoology*. Vol. 1. Wm. Wood and Co., New York. 778 pp.

Wesenberg-Lund, C. (1934). See Autrum (1939*c*).

Wesenberg-Lund, C. (1937). *Ferskvandsfaunaen biologisk belyst.* Invertebrat. Kobenhavn. [*Glossiphonia heteroclita* in a dissected operculate, *Bythnia tentaculata.*]

—— (1939). *Biologie des Süsswassers.* Hirudinea (Egel), pp. 339–68. Vienna.

West, C.H.K. (1972). Effects of temperature, ouabain and DNP on leech Retzius cells electrophysiology and electrotonic synapse. M.S. thesis, Ohio University.

—— and Lent, C.N. (1974). Effects of temperature on the electrophysiology of Retzius cells in the leech. *Comp. biochem. Physiol.* **47A**, 27–38.

Westheide, W. (1978). *Piscicola geometra* (Hirudinea)-Befall von Wirtstieren. Publikationen zu wissenschaftlichen Filmen. Sektion Biologie, Serie 11, Nummer 36, Film E 2484 (16mm, colour, 37m, 3½ min). Institut für den wissenschaftlichen Film, Göttingen.

—— (1980). *Erpobdella octoculata* (Hirudinea) Spermatophoren-übertragung, Kokonablage, Schlüpfen der Jungtiere. Publikationen zu wissenschaftlichen Filmen. Sektion Biologie, Series 13, Nummer 27, Film E 2562 (16 mm, colour, 122m, 11.5 min). Institut für den wissenschaftlichen Film, Göttingen.

—— (1981*a*). Fortpflanzung bei Egeln (Hirudinea). Publikatione zu wissenschaftlichen Filmen, Sektion Biologie, Serie 14, Nummer 6, Film C 1394 (16 mm, colour, 170m, 15.5 min, in German). Institut für den wissenschaftlichen Film, Göttingen.

—— (1981*b*). 'Nahrungsaufnahme bei Egeln (Hirudinea)'. Publikationen zu wissenschaftlichen Filmen, Sektion Biologie, Serie 14, Nummer 20, Film C 1416 (16 mm, colour, 160m, 15 min, German or English). Institut für den wissenschaftlichen Film, Göttingen.

Weyenbergh, H. (1878). Informe sobre una excursión zoológica a Santa Fe, practicada en 1876. *Bol. Acad. Nac. Cienc. Cba.* **II**, 217–43. [Reproduced in *Periód. Zool.* **III**, 39–64. (1879).]

—— (1879). [Sometimes erroneously written 1877, see Cordero (1937).] Alguns nuevas sanguijuelas o chancacas de la familia Gnathobdellida y Revista de esta familia. *Bol. Acad. Nac. Cienc. Rep. Argentina, Cordoba* **III**, 231–44. [= 1879 *Period. Zool. Argent.* **3**, 112–25.]

—— (1883). Tres nuevas especies de sanguijuelas. *An. Ateneo Uruguay* **5**, 427–30.

Weygoldt, P. (1966). Die Ausbildung transitorischer Pharynxapparate bei Embryonen. *Zool. Anz.* **176**(3), 147–60.

White, G.E. (1974). Parasites of the common white sucker (*Catostomus commersoni*) from Kentucky River drainage. *Trans. Am. microsc. Soc.* **93**(2), 280–2. [1975BA:62498.]

—— (1977). New distribution records of fish leeches in the Ohio River. *J. Parasitol.* **63**(6), 1138. [*Placobdella nuchalis*, nec. *P. parasitica.*] [1978BA:8877.]

—— and Crisp, N.H. (1973). The occurrence of four leeches (Hirudinea: Rhynchobdellida: Piscicolidae) on Kentucky River drainage fishes. *Trans. Ky Acad. Sci.* **34**(3), 47–8. [1974BA:43854.]

White, M., and Kolb, J.A. (1974). A preliminary study of *Thamnophis* near Sagehen Creek, California. *Copeia* 1974(1), 126–36.

White, M.J.D. (1940). Evidence for polyploidy in the hermaphrodite groups of animals. *Nature, Lond.* **146**, 132–3. [Leeches.]

White, R.H., and Walther, J.B. (1969). The leech photoreceptor cell: ultrastructure of clefts connecting the phaosome with extracellular space demonstrated by lanthanum deposition. *Z. Zellforsch. Mikrosk. Anat.* **95**, 102–8.

White, R.P., Cunningham, M.P., and Robertson, J.T. (1981). Inhibitory effects of hirudin and nimodipine on arterial contractions induced by thrombin. *Fedn Proc. Fedn Ass. Socs exp. Biol.* 40 (3 Part 1): 410. [1980BA:7785.]

Whitehead, H. (1943). Freshwater leeches of Yorkshire. *Naturalist, Hull* 1943, 107–8. (See also 1948, 39, 163–4; 1949, 20).

—— (1952). The medicinal leech (*Hirudo medicinalis* L.) in Yorkshire — another record. *Naturalist, Hull* **843**, 158.

Whitlock, M.R., O'Hare, P.M., Sanders, R. and Morrow, N.C. (1983). The medicinal leech and its use in plastic surgery: a possible cause for infection. *Br. J. Plastic Surg.* **36**, 240–4.

Whitman, C.O. (1878). The embryology of *Clepsine. Q. Jl microsc. Sci.* **18**, 215–315.

—— (1884*a*). The external morphology of the leech. *Proc. Am. Acad. Sci., Boston* **20**, 76–87.

—— (1884*b*). The segmental sense organs of the leeches. *Am. Natur.* **18**, 1104–9.

—— (1886*a*). The leeches of Japan. *Q. Jl microsc. Sci.* (new series) **26**, 317–416.

—— (1886*b*). The germ-layers of *Clepsine. Zool. Anz.* **9**, 171–6.

—— (1887). A contribution to the history of the germ layers in *Clepsine. J. Morphol.* **1**, 105–82.

—— (1889). Some new facts about the *Hirudinea. J. Morphol.* **2**, 586–99.

—— (1891*a*). Description of *Clepsine plana. J. Morphol.* **4**, 407–18.

—— (1891*b*). Spermatophores as a means of hypodermic impregnation. *J. Morphol.* **4**, 361–406.

—— (1891*c*). The naturalist's occupation: 2) a special problem. *Lect. mar. Biol. Lab. Wood's Hole* **1**, 36–52.

—— (1892). The metamerism of *Clepsine. Festschr. Z. 70. Geburtstage R. Leuckarts.* 385–95.

—— (1893). A sketch of the structure and development of the eye of *Clepsine. Zool. Jb. Anat.* **6**, 616–25.

—— (1898). Animal behavior. *Biol. Lect. Wood's Hole* 1899, 285–338.

Whittington, H.B. (1978). The lobopod animal *Aysheaia pedunculata*, middle Cambrian, Burgess Shale, British Columbia. *Phil. Trans. R. Soc.* **B284**, 165–97.

—— (1979). Early arthropods, their appendages and relationships. In: *The origin of major invertebrate groups* (ed. M.R. House) Systematics Association Special Volume No. 12, pp. 253–68. Academic Press, London.

Wiederholm, T. (1980). Effects of dilution on the benthos of an alkaline lake. *Hydrobiologia* **68**(3), 199–207.

Wielgosz, S. (1979). The effect of wastes from the town of Olsztyn on invertebrate communities in the bottom of the river Lyna. *Acta hydrobiol.* **21**(2), 149–65.

Wielgus, E. (1971). [Preliminary note on the leech fauna in fishponds in Zator near Oswiecim.] *Prz. Zool.* **15**(3), 282–4. [In Polish; English summary.]

Wierzbicka, J. (1978). Cestoda, Nematoda, Acanthocephala, Hirudinea and Crustacea from *Abramis brama*, *A. ballerus* and *Blicca bjoercna* of the Dabie Lake, Poland. *Acta parasitol. pol.* **25**(36–46), 293–306. [Polish summary.] [1979BA:49514.]

Wierzbicki, K. (1971). The effect of ecological conditions on the parasite fauna of perch *Perca fluviatilis* L. in Lake Dargin. *Ekol. Pol.* **19**(5), 73–86.

Wigglesworth, V.B. (1965). *The principles of insect physiology* (6th edn). Methuen, London.

Wilde, V. (1975). Untersuchungen zum Symbioseverhältnis zwischen *Hirudo officinalis* und Bakteria. *Zool. Anz.* **195**(5/6), 289–306. [English summary.]

Wildfowl Trust (1950). Pathology: parasites. *Wildfowl Trust Ann. Rep.* **3**, 52.

Wilkialis, J. (1962). [The leeches (Hirudinea) in the mid-stream of the Supraśl River.] *Roczn. Akad. Med., Bialystok* **8**, 267–92. [In Polish; English and Russian summaries.]

—— (1963). [Leech fauna (Hirudinea) of the Supraśl reservation.] *Prz. Zool.* **7**, 237–40. [In Polish; English summary.]

—— (1964*a*). [*Haementeria costata* Fr. Müll. on the northeastern borders of the Bialystok Voivodeship.] *Prz. Zool.* **8**(2), 152–4. [In Polish; English summary.]

—— (1964*b*). On the ecology and biology of the leech *Glossiphonia heteroclita* f. *hyalina*. (O.F. Müll.). *Ekol. Pol., Ser. A* **12**(17), 315–23. [Polish summary.]

—— (1968). Distribution of leeches along the course of the rivers Supraśl and Cazarna Hańcza in the light of habitat relations. *Ekol. Pol., Ser. A* **16**(39), 765–71.

—— (1970*a*). Investigations on the biology of leeches of the Glossi-phoniidae family. *Zool. Pol.* **20**(1), 29–54. [Polish and Russian summaries.]

—— (1970*b*). Some regularities in the occurrence of leeches (Hirudinea) in the waters of the Bialystok region. *Ekol. Pol., Ser. A* **18**(33), 647–80.

—— (1971*a*). Influence of regulation of the Laczanka River on the occurrence of Hirudinea. *Pol. Arch. Hydrobiol.* **18**(4), 359–65. [Polish summary.]

—— (1971*b*). [New stands and more particular ecological data of *Trocheta bykowskii* Gedroyć (Hirudinea).] *Prz. Zool.* **15**(2), 153–6. [In Polish; English summary.]

—— (1973*a*). The biology of nutrition in *Haementeria costata* (Fr. Müller, 1846). *Zool. Pol.* **23**(3–4), 213–25. [Polish and Russian summaries.]

—— (1973*b*). [Comments on *Trocheta bykowskii* Gedroyć culture.] *Prz. Zool.* **17**(1), 130–3. [In Polish; English summary.]

—— (1974). [*Trocheta bykowskii* Gedroyć (Hirudinea) in Beskid Slaski mountains.] *Prz. Zool.* **18**(3), 359–61. [In Polish; English summary.] [1975BA:49836.]

—— (1975). Morphology, biology, and ecology of the *Theromyzon maculosum* (Rathke, 1862). *Zool. Pol.* **25**(2–3), 163–96. [Polish and Russian summaries.]

—— and Davies, R.W. (1980*a*). The population ecology of the leech (Hirudinoidea: Glossiphoniidae) *Theromyzon tessulatum*. *Can. J. Zool.* **58**(5), 906–12.

—— —— (1980*b*). The reproductive biology of *Theromyzon rude*. *Can. J. Zool.* **58**(5), 913–16.

—— —— (1980*c*). The reproductive biology of *Theromyzon tessulatum* (Glossiphoniidae: Hirudinoidea), with comments on *Theromyzon rude*. *J. Zool., Lond.* **192**, 421–9.

—— (1984).The life-history of *Haementeria costata* (Fr. Müller) (Glossiphoniidae: Hirudinea). *Hydrobiologia* **109**, 219–27.

Wilkinson, J.M. (1976). The connective tissue coverings of the peripheral nerves in the leech. *Anat. Rec.* **184**(3), 563. (Abstr.)

—— (1976). The connective tissue coverings of leech peripheral nerves: anatomical evidences for the absence of cerebrospinal fluid in the leech. *J. comp. Neurol.* **170**(3), 381–9. [1977BA:38790.]

—— and Coggeshall, R.E. (1975). Axonal numbers and sizes in the connectives and peripheral nerves of the leech. *J. comp. Neurol.* **162**(3), 387–96. [1975BA:49432.]

Willard, A.L. (1980). Serotonin increases the probability that the leech nervous system will produce the swim motor program. *Neurosci. Abstr.* **6**, 27.

—— (1981). Effects of serotonin on the generation of the motor program for swimming by the medicinal leech. *J. Neurosci.* **1**(9), 936–44.

Willem, V., and Minne, A. (1899). Recherche sur l'excrétion chez quelques Annélides. Observations sur l'excrétion chez quelques Hirudinées. *Mém Acad. Sci. Belg.* **58**, 51–87.

Williams, E.H., Jr (1979). Leeches of some fishes of the Mobile Bay region, Alabama, U.S.A. *Northeast Gulf Sci.* **3**(1), 47–9. [1980BA:6715.]

—— (1982). Leeches of some marine fishes from Puerto Rico and adjacent regions. *Proc. helminth. Soc. Wash.* **49**(2), 323–5.

Williams, R.C.F. (1961). The distribution of freshwater leeches in the Glasgow region, with notes on their ecology. *Glasgow Nat.* **18**(3), 136–46.

Williams, T. (1982). Leeches in my breeches. *Audobon* **84**(2), 62–7.

Willomitzer, J., Lucký, Z., and Kolář, Z. (1972). Molluscicidal effectiveness of fluorinated derivatives of salicyclic acid. *Acta vet., Brno* **41**, 31–7. [Czechoslovak and Russian summaries.] [At 0.5 per cent an efficiency of 98–100 per cent against *Hemiclepsis marginata*.]

Wilson Jr, A.H. (1972). Large paired neurons in the suboesophageal ganglion of the leech, *Haemopis marmorata* (Say). Neuronal geometry and electrical coupling. M.S. thesis, Ohio University.

—— and Lent, C.M. (1973). Electrophysiology and anatomy of the large paired neurons in the sub-oesophageal ganglion of the leech. *Comp. Biochem. Physiol.* **46A**, 301–9.

Wilson, C.B., and Clark, H.W. (1912). The mussel fauna of the Maumee River. *Rep. US Comm. Fish. 1911*, and *Spec. Paps.* pp. 1–72. [Separately issued as *Bur. Fish. Doc. 757.*] [Found leeches in mantle cavity.]

Wilson, C.S. (1944). Repellents for leeches. *Naval med. Res. Inst. Nat. Naval med. Center, Bethesda* 27 April.

Wilson, D.M. (1960). Nervous control of movements in annelids. *J. exp. Biol.* **37**, 46–56.

Windsor, D.A. (1970). Faeces of the medicinal leech, *Hirudo medicinalis*, are haem. *Nature, Lond.* **227**, 1153–4.

—— (1972). *Literature of freshwater leeches of North American.* Science-aesthetics, Norwich, New York.

Wingstrand, K.G. (1951). Hirudinea. In *The mountain fauna of the Virihaure area in Swedish Lapland. Special account* (ed. P. Brinck and K.G. Wingstrand) K. Fys. Sällsk. Handl., NF *Acta Un. Lund.* (new series) **61**(2), 160–2.

Winkler, O. (1963). Beitrag zur Kenntnis der Bodenfauna der oberen Moldau vor der Errichtung der Talsperre in Lipno. *Acta Un. Carol. Biol.* **1**, 85–101. [In German; Czechoslovak summary.] [Leeches in the fauna of the Upper Moldau.]

Winnell, M.H., and Jude, D.J. (1979). Spatial and temporal distribution of benthic macroinvertebrates and sediments collected in the vicinity of the J.H. Campbell Plant, eastern Lake Michigan, 1978. Special Report number 75. Great Lakes Research Division, University of Michigan, Ann Arbor.

Winterbourn, M.J., and Brown, T.J. (1967). Observations on the faunas of two warm streams in the Taupo Thermal region. *N.Zealand J. mar. freshwat. Res.* **1**(1), 38–50. [Observations on *Alboglossiphonia* sp. in fauna with a warm stream.]

Winterstein, H., and Özer, F. (1949). Osmotischer Druck und Ionengleich-gewicht beim Blutegelmuskel. *Ztschr. vergl. Physiol.* **31**(3), 308–21.

Wisniewski, W.L. (1958). Characterization of the parasitofauna of an eutrophic lake (Parasitofauna of the biocoenosis of Druzno Laka — Part II). *Acta parasitol. Polon* **6**(1), 1–64.

Wissocq, J.C., and Malecha, J. (1974). Ultrastructure du spermatozoïde de *Piscicola geometra* (Huridinée, Rhynchobdelle). *C. r. hebd. Séanc. Acad. Sci., Paris D* **278**(4), 487–90.

—— —— (1975). Étude des spermatozoïdes d'Hirudinées à l'aide de la technique de coloration negative. *J. ultrastr. Res.* **52**, 340–61.

Witenberg, G., and Izak, G. (1969). *Investigations on the pathogenesis of lesions produced by the local leech Limnatis nilotica*. Hadassah Med. School., Hebrew Univ., Jerusalem. US Department of Agriculture, Project A10-ADP-S. Grant FG-Is-132.

Witte, F., and Witte-Maas, E.L.M. (1981). Haplochromine cleaner fishes: a taxonomic and ecomorphological description of two new species. Revision of the Haplochromine species (Teleostei: Cichlidae) from Lake Victoria, East Africa. *Neth. J. Zool.* **31**(1): 203–31.

Wohlgemuth, E., and Trnkova, J. (1979). Dispersal and effects of a preparation containing endrin in an artificial water reservoir. *Folia Zool.* **28**(1), 65–72. [French and Russian summaries.] [1979BA:70593.] [*Erpobdella octoculata* resistant to endrin.]

Wojtas, F. (1957). [Les sangsues (Hirudinea) dans la région des Monts Lyse (Lysogóry).] *Zesz. Nauk. Un. Lodz. Nauk. Mat.-Przyr., Ser. II* **3**, 51–69. [In Polish; French summary.]

—— (1958). [Sur la faune des Hirudinées de Ojcow (Resumé).] *Zesz. Nauk. Un. Lodz. Nauk. Mat.-Przyr., Ser. II* **4**, 149–58. [In Polish; French summary.]

—— (1959a). [Les sangsues (Hirudinea) de la riviere Grabia.] *Zesz. Nauk. Un. Lodz. Sect. III* **58**, 1–64. [In Polish; French summary.]

—— (1959b). [Les sangsues des Tatry, des Pieniny et de Podhale.] *Zesz. Nauk. Un. Lodz. Nauk. Mat.-Przyr., Ser II* **5**, 133–46. [In Polish; French summary.]

—— (1960). [Remarques concernant la reproduction des sangsues de l'espèce *Cystobranchus respirans* (Troschel).] *Zesz. Nauk. Un. Lodz.*

Nauk. Mat.-Przyr., Ser II **7**, 153–9. [In Polish; detailed French summary.]

—— (1961). [The leech *Cystobranchus respirans* (Troschel), a little known fish parasite (Hirudinea).] *Prz. Zool. V* **4**, 361–2. [In Polish.]

—— (1962). [Sur la structure des organes génitaux de quelques sangsues du genre *Erpobdella* de Blainville.] *Prace Wydz. III Nauk Mat.-Przyr. Lódz. TN, Lodz* **78**, 1–17.

—— (1968). *Trocheta bykowskii* Gedroyć (Hirudinea) w Bieszczadach Zachodnich. *Prz. Zool.* **12**(3), 290–1.

—— (1970). [Beschreibung des Eikokons von *Trocheta bykowskii* Gedroyć, 1913 (Hirudinea).] *Zesz. Nauk. Un. Lodz. Ser. II*, **40**, 43–6. [In Polish; German summary.]

—— (1972). [Fauna of the Niebieskie Źródla. Leeches (Hirudinea).] *Zesz. Nauk. Un. Lodz. Ser. II, Nauk. Human.-Przyrodnicze* No. 46, 13–14. [In Polish; German summary.]

—— (1974). Pijawki (Hirudinea) Bieszadow Zachodnich. *Zesz. Nauk. UL. Ser. 2* **56**, 13–16.

Wolfe, D.E., and Nicholls, J.G. (1967). Uptake of radioactive glucose and its conversion to glycogen by nervous and glial cells in the leech central nervous system. *J. Neurophysiol.* **30**(6), 1593–609.

Wolner, E. (1966). Über die Wirking von Bretylium auf die Azetylcholin-kontraktur des Blutegels. *Arch. Int. Pharmacodynam. Ther.* **162**(2), 487–92.

Wolnomiejski, N., and Wolnomiejska, K. (1968). Leeches (Hirudinea) in the southern part of the Lake Jeziorak. *Zesz. Nauk. Un. Mikolaja Kopernika Torun.* **18**, 29–44. [Polish summary.]

Wolterstorff, W. (1900). See Autrum (1939c).

Woo, P.T.K. (1969a). Trypanosomes in amphibians and reptiles in southern Ontario. *Can. J. Zool.* **47**(5), 981–8. [Metacyclic forms occur in leeches.]

—— (1969b). The life cycle of *Trypanosoma chrysemydis*. *Can. J. Zool.* **47**(6), 1139–51.

—— (1969c). The development of *Trypanosoma canadensis* of *Rana pipiens* in *Placobdella* sp. *Can. J. Zool.* **47**(6), 1257–9.

—— (1970). Origin of mammalian trypanosomes which develop in the anterior-station of blood-sucking arthropods. *Nature, Lond.* **228**, 1059–62.

—— (1978). The division process of *Cryptobia salmositica* in experimentally infected rainbow trout (*Salmo gairdneri*). *Can. J. Zool.* **56**(7), 1514–18. [1978BA:72560.]

—— (1979). *Trypanoplasma salmositica:* experimental infections in rainbow trout, *Salmo gairdneri. Exp. Parasitol.* **47**, 36–48.

Wood, E.J., Mosby, L.J., and Robinson, M.S. (1976). Characterization of the extracellular hemoglobin of *Haemopis sanguisuga* (L.). *Biochem. J.* **153**(3), 589–96. [1976BA:66961.]

Wood, J.G. (1965). Electron microscopic localization of 5-hydroxy-tryptamine (5-HT). *Texas Rept. Biol. Med.* **23**, 828–37.

Wood, R.D., and Longest, W.D. (1975). Fish parasites of McKellea Lake, Tennessee. *J. Miss. Acad. Sci.* **20**, 60.

Woodruff, G.N., Walker, R.J., and Newton, L.C. (1971). The actions of some muscarinic and nicotinic agonists on the Retzius cells of the leech. *Comp. gen. Pharmacol.* **2**(5), 106–17.

Woods, C.S. (1959). Appendix, pp. 377–8 to K.H. Mann (1959). [Irish leeches.]

—— (1976). First records of *Trocheta subviridis* Dutrochet and the status of other arhynchobdellid leeches (Hirudinea) in Ireland. *Proc. R. Ir. Acad. (B)* **76**, 359–67. [1977BA:39050.]

Wordeman, L. (1982). Kinetics of primary blast cell production in the embryo of the leech *Helobdella triserialis.* Honors thesis, Department of Molecular Biology, University of California, Berkeley.

Worth, C.B. (1951). Description and discussion of the biting of an Indian land leech (Annelida: Hirudinea). *J. Bombay nat. Hist. Soc.* **50**, 423–6.

Woskresenskii, A.E. (1858). O meditsinskikh piyavkakh v srednikh guberniyakh Rossii i Malorossii. *Voen.-Med. Zhurn.* **21**, otd. 2. [In Russian.]

—— (1859). *Monographiya vrachebn'kh piyavok* . . . SPB I–XVI + 1–500. [In Russian.]

Wrede, W. (1927). See Autrum (1939*c*).

Wright, P. (1981). The leech in peril of extinction. *The Times* (London) p. 4., 21 October 1981.

Wróblewski, A. (1960). O wystepowaniu pijawki lekarskiej (*Hirudo medicinalis* L.) w Wielkopolsce i na Pomorzu. *Bad. Fisjogr. Pol. Zach., Poznan* **4**(3–4), 109–15.

Wrona, F.J., Davies, R.W., and Linton, L. (1979). Analysis of the food niche of *Glossiphonia complanata* (Hirudinoidea: Glossiphoniidae). *Can. J. Zool.* **57**: 2136–42. (French summary).

—— Davies, R.W., Linton, L., and Wilkialis, J. (1981). Competition and co-existence between *Glossiphonia complanata* and *Helobdella stagnalis* (Glossiphoniidae: Hirudinoidea). *Oecologia* **48**(1), 133–7.

Wu, C.F. (1930). A revised list of the Chinese leeches. *Bull. Dep. Biol. Yenching Un.* **1**, 45–8. [= *Peking Soc. nat. Hist. Bull.* **4**, Pt. 3, 45–8.]

Wu, Shi-Kuei (1979). The leeches (Annelida: Hirudinea) of Taiwan. Part 1.

Introduction and descriptions of two hirudinid species. *Q. J. Taiwan Mus.* **32**, 193–207.

—— (1981). The leeches (Annelida: Hirudinea) of Taiwan. Part 2. *Hirudinaria manillensis* (Lesson). *Q. J. Taiwan Museum* **34**, 207–11.

Würgler, E. (1918). Beiträge zur Kenntnis der Reparationsprozesse bei Hirudineen. *Viertel Jahrsschr. Naturforsch. Ges. Zurich.* 3, **4**, 552–65.

—— (1920). Beiträge zur Kenntnis der Reparationsprozesse bei Hirudineen. *Jena. Z. Naturw.* (NF) **49**(= 56), 252–360.

Yagminene, I.B. (1979). [Anthropogenic action on the quantitative and qualitative development of the macrozoobenthos in the lower part of the Neman River, USSR.] *Liet. Tsr. Mokslu. Acad. Darb. Ser. C Biol. Mokslai* **1**, 65–74. [In Russian.]

Yaki, K., Bern, H.A., and Hagadorn, I.R. (1963). Action potentials of neurosecretory neurons in the leech *Theromyzon rude. Gen. comp. Endocr.* **3**, 490–5.

Yakimoff, W.L. (1916). *Trichomonas* v Kishechnom Kanale turkestanskoi piyavki (*Limnatis turkestanica*). *Predvar. Soobsch. Russk. Zool. Zh.* **1**(9), 305–6.

—— (1917). *Trichomonas* de l'intestin de la sangsue du Turkestan (*Limnatis turkestanica*). *Bull. Soc. Path. exot.* **10**, 293–4.

Yaksta, T., and Coggeshall, R.A. (1973). Neuromuscular transmitters in a simple nervous system. *Texas Rep. Biol. Med.* **31**, 607. (Abstr.)

Yaksta-Sauerland, B.A., and Coggeshall, R.A. (1973). Neuromuscular junctions in the leech. *J. comp. Neurol.* **151**(1), 85–100.

Yamagishi, H. (1975). [The segmental nerve discharge during the swimming movement in the leech]. *Zool. Mag.* (Tokyo) **84**(4), 372. [1977BA:28309.]

Yanagisawa, H., and Yokoi, E. (1938). The purification of hirudin, an active principle of *Hirudo medicinalis. Proc. Imp. Acad. Jap.* **14**, 69–70.

Yanese, T., Fujimoto, K., and Nishimura, T. (date undetermined). The fine structure of the dorsal ocellus of the leech, *Hirudo medicinalis. Mem. Osaka Gakugei Un. lib. Arts Educ. B* **13**, 117–19.

Yang, J. *et al.* (1983). Frequency domain analysis of electrotonic coupling between leech Retzius cells. *Biophys. J.* **44**(1), 91–9.

—— and Kleinhaus, A.L. (1984). Effects of tetraethylammonium chloride and divalent cations on the afterhyperpolarisation following repetitie firing in leech neurons. *Brain Res.* **322**, 380–4.

—— and Lent, C.M. (1983). Calcium depletion produces sodium-dependent, sustained depolarizations of Retzius cell membranes in the leech central nervous system. *J. Comp. Physiol. A Sens. Neural. Behav. Physiol.* **150**(4), 499–508. [1983BA:65297.]

—— Johansen, J., and Kleinhaus, A.L. (1984). Procaine actions on tetrodotoxin sensitive and insensitive leech neurons. *Brain Res.* **302**, 297–304.

Yang Tong (1980*a*). [On the coelom and vascular system of Hirudinea.] *Acta zool. sin.* **26**(3), 213–19. [In Chinese; comprehensive English abstract.]

—— (1980*b*). [On a new species of *Poecilobdella* (Hirudinea: Hirudinidae) from Hubei Sheng, China.] *Zool. Res.* 1(4), 541–6. [In Chinese; comprehensive English abstract.][1983BA:3652.]

—— (1981). [Two new species of parasitic leeches from freshwater fishes in China.] *Acta zootax. sin.* **6**(1), 27–30. [In Chinese; comprehensive Chinese summary.]

—— (1983). [On the genus *Dina R.* Blanchard 1892 and a new species from Xiangjiang River, China]. *Acta zootax. sin.* **8**(2), 129–34. In Chinese; English summary.

—— (1984). [On the genus *Piscicola* Blainville, 1818 and a new species from Yellow Sea (Hirudinoidea: Piscicolidae)] *Acta Zootax. sin.* **9**(1), 30–33. [In Chinese; English abstract.]

—— (1986*a*). [On *Batracobdella canricola* Oka, 1928.] *Acta hydrobiol. sin.* **10**(2). [In Chinese; English abstract.]

—— (1986*b*). [Studies on two marine leeches from an antarctic fish of the genus *Notothenia.*] *Acta zool. sin.* **32**(2). [In Chinese; English abstract.]

—— and Davies, R.W. (1985). Parasitism by *Placobdella multilineata* (Hirudinoidea: Glossiphoniidae) and its first record from Asia. *J. Parasitol.* **71**(1), 86–8.

—— —— (1985). The morphology of *Placobdella multilineata* Hirudinoidea: site of Crocodilia (French summary). *Can. J Zool.* **63**(3), 550–1.

Yankovskaya, A.I. (1965). [Fauna of the warm springs in the eastern Pamir.] *Tr. Zool. Inst. Leningr.* **35**, 43–56. [In Russian. Helobdella stagnalis.]

Yaroshenko, M.F., Naberezhnyy, A.I., and Vladimirov, M.Z. (1972). Hydrobiological research in the Moldavian SSR. *Hydrobiol. J.* [Engl. transl. of *Girdrobiol. Zh.*] 8(6), 30–5.

Yau, King-Wai (1975). Receptive fields, geometry and conduction block of sensory cells in the leech central nervous system. Ph.D. thesis, Harvard University, Cambridge, Massachusetts.

—— (1976*a*). Physiological properties and receptive fields of mechanosensory neurones in the head ganglion of the leech: comparison with homologous cells in segmental ganglia. *J. Physiol.* **263**(3), 489–512.

—— (1976*b*). Receptive fields, geometry and conduction block of sensory neurones in the central nervous system of the leech. *J. Physiol.* **263**(3), 513–38.

Yonezawa, S. (1942). [Leech muscle assay for acetylcholine.] *Okayama Igakkai Zasshi* **54**, 691 and 861. [In Japanese.]

Yoshida, T., and Yoshimura, M. (1955). [Leech muscle in assay for acetylcholine.] *Folia pharmacol. japon.* **51**, 715.

Young, A.M. (1984). Bleeding antiques. Part 3. Leeching antiques. *Antique Collecting* **19**, 27–30.

Young, J.O. (1972). The Turbellaria of some Friesland lakes with incidental records of Gastropoda and Hirudinea. *Zool. Bijdr.* **13**, 59–70. [1974BA:43933.]

—— (1974). Life cycle of some invertebrate taxa in a small pond together with changes in their numbers over a period of three years. *Hydrobiologia* **45**(1), 63–90. [*Helobdella stagnalis*.]

—— (1980). A serological investigation of the diet of *Helobdella stagnalis* (Hirudinea: Glossiphoniidae) in British lakes. *J. Zool., Lond.* **192**, 467–88.

—— (1981*a*). A serological study of the diet of British, lake-dwelling *Glossiphonia complanata* (L.) (Hirudinea: Glossiphoniidae). *J. nat. Hist.* **15**, 475–89.

—— (1981*b*). A comparative study of the food niches of lake-dwelling triclads and leeches. *Hydrobiologia* **84**, 91–102.

—— (1983). A stimulus for egg production in *Glossiphonia complanata* (Hirudinoidea: Glossiphoniidae). *Freshwat. Invertebr. Biol.* **2**(2), 112–15. [1984BA:11676.]

—— and Harris, J.H. (1974). The occurrence of some invertebrate animals in the littoral zone of some lowland lakes in Cheshire and Shropshire. *Naturalist* **928**, 25–32.

—— and Ironmonger, J.W. (1979). The natural diet of *Erpobdella octoculata* (Hirudinea: Erpobdellidae) in British (UK) lakes. *Arch. Hydrobiol.* **87**(4), 483–503. [1980BA:35956.]

—— —— (1980). A laboratory study of the food of 3 species of leeches occurring in British (UK) lakes. *Hydrobiologia* **68**(3), 209–16. [1980BA:8448.]

—— —— (1981). A quantitative study of the comparative distribution of non-parasitic leeches and triclads in the stony littoral of British lakes. *Int. Revue Ges. Hydrobiol.* **66**(6), 847–62. [1983BA:17127.]

—— —— (1982*a*). A comparative study of the life histories of three species of leeches in two British lakes of different trophic status. *Arch. Hydrobiol.* **94**(2), 218–50.

—— —— (1982*b*). The influence of temperature on the life cycle and occurrence of three species of lake-dwelling leeches (Annelida: Hirudinea). *J. Zool., Lond.* **196**, 519–43.

Young, S.R., Dedwylder, R. II, and Friesen, W.O. (1981). Responses of the medicinal leech to water waves. *J. comp. Physiol.* **144**, 111–16.

Yuki, H., and Fishman, W.H. (1963). Purification and characterization of leech hyaluronic acid-endo-ß-glucuronidase. *J. biol. Chem.* **238**, 1877.

Yurkowski, M., and Tabachek, J.L. (1979). Proximate and amino acid composition of some natural fish foods. In *Finfish nutrition and fishfeed technology* (ed. J.E. Halver and K. Tiews) Vol. 1, pp. 435–48. Proc. World Symposium, Hamburg, 20–23 June 1978.

—— —— and Boese, H.R. (1974). Lipids of leeches *Nephelopsis obscura*. *J. Am. Oil Chem. Soc.* **51**(7), 516A.

Zach, O. (1952). Anatomische und histologische Besonderkeiten der Würmer. *Mikrokosmos* **41**, 173–8.

Zackson, S.L. (1982). Cell clones and segmentation in leech (*Helobdella triserialis*) development. *Cell* **31**, 761–70. [1983BA:10949.]

—— (1984). Cell lineage, cell-cell interaction, and segment formation in the ectoderm of a glossiphoniid leech embryo. *Develop. Biol.* **104**, 143–60.

Zacwilichowska, K. (1965). Benthos obrzeza zbiornika Goczal-Kowickiego w latach 1958–1959. *Acta hydrobiol., Krakow* **7**, 83–97. [Only record to date of *Cystobranchus mammillatus* in Poland, at Slask Gorny.]

Zajiček, D. (1971). [On the epizootiology of flukes of the genus *Apatemon* Szidat, 1928 (Trematoda: Strigeidae) in ducks in the South Bohemian fish-pond systems.] *Vet. Med.* (Prague) **44**, 53–60. [In Czechoslovak; English summary.]

—— and Valenta, D. (date undetermined). *Erpobdella octoculata* L. (Hirudinea), the reservoir host of *Microsomacanthus paririla* (Kowalewski, 1904) in Czechoslovakia. *Věst čsl. zool. Spol.* **33**, 272–7.

Zaloznyi, N.A. (1973). [Study of oligochaetes and leeches in West Siberia.] In *Vodoemy Sibiri i Perspektivy Ih. Rybokhozyaistvennogo Ispol'-zovaniya*, Tomsk, USSR, pp. 182–3; Izdatel'stvo Tomskogo Universi-teta. [In Russian.]

—— (1974). Maloschetinkob'e chervi i piyavki vodoemov Zapadnoi Sibiri. *Avtoreph. Diss., Tomsk* 1–20.

—— (1976). [Aquatic oligochaete and leech fauna in West Siberia.] In *Problemy Ekologii* (ed. B.G. Ioganzena) Tom 4. Tomsk, USSR, pp. 97–112. Izdatel'stvo Tomskogo Universiteta. [In Russian; English summary.]

Zamaraev, V.N. (1958). [Data on the leech fauna of the environs of Kalinin City.] *Bull. Mosk. Obsch. Ispytat. Prirody. Kalininsk Otd.* **1**, 51–4. [In Russian.] [12 species, inc. *Hirudo medicinalis* which was numerous in L. Kuzminkovskoe.]

—— (1962). Pitanie bol'shoi lozhnokonskoi piyavki (*Haemopis sangui-suga*). *Pervoe Nauk. Sovesch. Zool. Ped. Inst. R.S.Ph.S.R. Tez. Dokl., Moscow* 20–1.

Zangheri, P. (1966). Repertorio sistematico e topografico della flora e fauna vivente e fossile della Romagna. *Memorie Mus. Civ. Stor. Nat. Verona, Mem. f.s.* **1**, 483–854. [*Hirudo* in Italy.]

Zapkuvene, D.V. (1970). [Medical leech occurrence and reserves in the Lithuanian SSR.] *Liet. TSR Mokslu Akad. Darb. Ser. C* **53**(3), 91–7. [In Russian; English summary.]

—— (1972*a*). [Breeding and growth of medical leeches under laboratory conditions. I. Breeding of *Hirudo medicinalis* f. *serpentina* and *H. medicinalis* f. *officinalis*.] *Lietuvos TSR Mokslu. Akad. Darb. Ser. C* **3**(59), 71–6. [In Russian; English and Lithuanian summaries.]

—— (1972*b*). [Breeding and growth of medical leeches under laboratory conditions. II. Growth of *Hirudo medicinalis* f. *serpentina*.] *Lietuvos TSR Mokslu Akad. Darb. Ser. C* **3**(59) 77–84. [In Russian; English and Lithuanian summaries.]

—— (1972*c*). Rasprostranenie Meditsinskoi Piyavki (*Hirudo medicinalis* f. *serpentina* Moquin-Tandon, 1846) v Latviiskoi SSR i Op't ee iskusstvennogo Razvedeniya. *Avtoreph. Diss., Vil'nius* 1–22. [In Russian.]

—— (1980). [Effect of UV irradiation on the body of *Hirudo medicinalis* 1. Survival and growth]. *Lietuvos TSR M kslu Akad. Darb. Ser C* **1**, 47–54. [In Russian.]

—— (1982). [Effect of UV irradiation on the medicinal leech: 2. Salivary glands]. *Lietuvos TSR. Mokslu Akad. Darb. Ser. C* **4**, 86–92. [In Russian; Lithuanian summary.] [1984BA:27885.]

—— and Sinyavichene, D.P. (1976). [Development of the salivary glands of medicinal leeches during starvation.] *Lietuvos TSR Mokslu Akad. Darb. Ser. C* **2**, 53–66. [In Russian; Lithuanian and English summaries.] [1977BA:57172.]

Zenkevich, L.A. (1944). [Notes on the evolution of the locomotory apparatus of animals.] *Zh. Obsch. Biol.* **3**, 129–70. [In Russian.]

—— (1962). [Several peculiarities of the formation of the longitudinal muscle of leeches.] *Volros' Obschei Zool. Med. Parazitol., Moscow* 101–15. [In Russian.]

—— and Naumov, S.P. (1955). *Kratkii Kurs Zool., Moscow.* [In Russian.]

Zerbst-Boroffka, I. (1970). Organische Säurereste als wichtigste Anionen im Blut von *Hirudo medicinalis*. *Z. vergl. Physiol.* **70**, 313–21.

—— (1973). Osmo- and Volumenregulation bei *Hirudo medicinalis* nach Nahrungsausfnahme. *J. comp. Physiol.* **84**(2), 185–204.

—— (1975). Function and ultrastructure of the nephridium in *Hirudo medicinalis* L.: III. Mechanisms of the formation of primary and final urine. *J. comp. Physiol. B* **100**(4), 307–16. [1976BA:2005.]

—— (1978). Blood volume as a controlling factor for body water homeostasis in *Hirudo medicinalis*. *J. comp. Physiol. B* **127**(4), 343–8. [1979BA:60931.]

—— and Haupt, J. (1975) Morphology and function of the metanephridia in annelids. *Fortschr. Zool.* **23**, 33–47. [Comparison with *Lumbricus*.]

—— Bazin, B., and Wenning, A. (1982). Nervenversorgung des Exkretionssystems und des Lateralgefasse von *Hirudo medicinalis* L. *Verh. Deuts. Zool. Ges.* **75**. (Incomplete reference)

—— Wenning, A., and Bazin, B. (1982). Primary urine formation during diuresis in the leech, *Hirudo medicinalis* L. *J. comp. Physiol.* **146**, 75–9.

—— —— Kollman, R., and Hildebrandt, J.P. (1983). Antidiuretischer Faktor beim Blutegel, *Hirudo medicinalis. Verh. Dtsch. Zool. Ges.* 1, **243**.

[See also Boroffka.]

Zhacin, V.I., and Gerd, S.V. (1961/63). [*Fauna and flora of the rivers, lakes and reservoirs of the USSR.*] Moscow. [In Russian.]

Zhang, J. *et al.* (1982). [Studies on land leech repellent.] *Chung Hua Yu Fang I Hsueh Tsa Chih* **16**(6), 331–4. [In Chinese; English summary.]

Zhatkanbaeva, D. (1980). [Leeches of Kurgaldzhin Lake — a second intermediate host of trematodes of the family Strigeidae]. *Lietuvos TSR Mokslu Akad. Vilnius* 1980, 40–1 [Abstract; in Russian.]

Zhuravlev, V.L. (1974). [Selective reactions of the neurons of the leech on rectangular steps of depolarizing current of determined duration.] *Vestn. Leningrad Un. Ser. Biol.* **3**, 85–9. [In Russian; English summary.] [1975BA:14433.]

—— and Safonova, T.A. (1974). [Characteristics of electrical reactions of different types of leech sensory neurons to direct stimulation of the cell body.] *Fiziol. Zh. SSSR IM. I. M. Sechenova* **60**(7), 1030–6. [In Russian; English summary.] [1975BA:67052.]

—— —— (1975*a*). [Electrical characteristics of the leech sensory neurons.] *Neurofiziol.* **7**(3), 295–301. [In Russian.] [1976BA:20035.]

—— —— (1975*b*). [Trace effects in different neurons of the ganglia of the horse leech after rhythmic stimulation.] *Dokl. Akad. Nauk SSSR Ser. Biol.* **222**(3), 750–2. [In Russian.] [1976BA:48917.]

Zia, F.C. (1927). Anatomy of the leech (*Whitmania laevis* Baird). *Contr. Biol. Labor. Sci. Soc. Nanking (Zool.)* **4**(2), 1–20.

Zick, K. (1931). Zur Frage der Verbreitung des medizinischen Blutegels (*Hirudo medicinalis* L.) in Deutschland. *Zool. Anz.* **96**, 328–30.

—— (1932). See Zick (1936) in Autrum (1939*c*).

—— (1933). Weiteres über Zucht und Fortpflanzung des medizinischen Blutegels. *Zool. Anz.* **103**, 49–55.

Zimmermann, P. (1967). Fluoreszenzmikroskopische Studien über die Verteilung und Regeneration der Faserglia bei *Lumbricus terrestris* L. *Zeit. Zellforsch. Mikroskop. Anat.* **81**, 190–220.

Zinn, D.J., and Kneeland, I.R. (1964). Narcotization and fixation of leeches (Hirudinea). *Trans. Am. microsc. Soc.* **83**, 275–6.

Zipser, B. (1976). Neurons innervating sex organs in the leech. *Neurosci. Abstr.* **II**, 362.

—— (1979*a*). Identifiable neurons controlling penile eversion in the leech.

J. Neurophysiol. **42**(2), 455–64. [1979BA:15811.]

—— (1979*b*). Voltage-modulated membrane resistance in coupled leech neurons. *J. Neurophysiol.* **42**(2), 465–75. [1979BA:15812.]

—— (1980*a*). Horseradish peroxidase nerve backfilling in leech, *Haemopis marmorata. Brain Res.* **182**(2), 441–5.

—— (1980*b*). Identification of specific leech neurones immunoreactive to enkephalin. *Nature, Lond.* **283**(5750), 857–8.

—— (1982). Complete distribution patterns of neurons with characteristic antigens in the leech central nervous system. *J. Neurosci.* **2**, 1453–64.

—— (In press). Preparation of monoclonal antibodies and their advantages in identifying specific neurons, In: *Advances in Cellular Neurobiology.*

—— and Schley, C. (In press). Two differently distributed central nervous system antigens to which a single monoclonal antibody binds can be differentiated through different fixations. *N.Y. Acad. Sci.* 1984. (Incomplete reference)

—— Stewart, R., Flanagan, T., Flaster, M., and Macagno, E. (In press). Do monoclonal antibodies stain sets of functionally related leech neurons? *Cold Spring Harbor Symp. quant. Biol.* **48**. (Incomplete)

—— Hockfield, S., and McKay, R. (1981). Immunological identification of specific neurons. In: *The neurobiology of the leech* (ed. K.J. Muller, J.G. Nicholls, and G.S. Stent) pp. 235–247. Cold Spring Harbor.

—— and McKay, R. (1981). Monoclonal antibodies distinguish identifiable neurones in the leech. *Nature, Lond.* **289**(5798), 549–54.

—— —— and Farrar, J. (1980). Individual nerve cells are immunologically distinguishable in the leech. *Neurosci. Abstr.* **6**, 338.

Zirpolo, G. (1923). See Autrum (1939*c*).

Ziser, S.W. (1978). Seasonal variations in water chemistry and diversity of the phytophilic macroinvertebrates of three swamp communities in southeastern Louisiana. *Southwest. Nat.* **23**(4), 545–62.

Žitňan, R. (1967). [Nematodes, Acanthocephala and Hirudinea of fish of flowing waters in the lowland of Potiska Nizina.] *Biol. Bratisl.* **22**(5), 381–5. [In Czechoslovak.]

—— (1968). Nematoda, Acanthocephala und Hirudinea bei Fischen im Flusse Hron (CSSR). *Stud. Helmintol.* **2**, 21–32.

Zoltowski, M. (1965*a*). [The leeches (Hirudinea) of the Elk River.] *Zesz. Nauk. WSR Olsztyn* **20**, 135–43. [In Polish; English and Russian summaries.]

—— (1965*b*). [The distribution of the leech, *Haementeria costata* (Fr. Müll.) in the Elk Lakeland.] *Zesz. Nauk. Wyzsz. Szkoly Rolniczej Olsztynie* **20**(431), 145–50. [In Polish; English and Russian summaries.]

Zubchenko, A.V. (1979). Parasitic fauna of Anarrhichadidae and Pleuronectidae families of fish in the northwest Atlantic. *ICNAF Res. Doc.* 79/VI/98, 41–6. [*Platybdella anarrhichae* north to Baffin Island.]

Zubchenko, I.A., and Petrova, T.A. (1973). [Absorption of amino acids by leeches, *Hirudo medicinalis.*] *Biol Nauk.* **16**(2), 7–11.

Zwilling, R. (1968). Zur Evolution der Endopeptidasen, VI. Specifische Hemmung von Proteasen aus Invertebraten durch Hirudin. *Hoppe Seyler's Z. Physiol. Chem.* **349**, 1787–8. [Probably referring to the proteolytic inhibitors, bdellins, rather than hirudin.]